Spatial Analysis

A Guide for Ecologists

Third Edition

Designed for researchers in ecology at all levels and career stages, from students and postdoctoral fellows to seasoned professionals, this third edition reflects the significant advances in quantitative analysis of the past decade. It provides updated examples and methods, with reduced emphasis on older techniques that have seen limited use in recent ecological literature. The authors cover new and emerging approaches, including Hierarchical Bayesian analysis and spatio-temporal methods. A key feature is the integration of ecological and statistical concepts, highlighting the critical role that this type of analysis plays in ecological understanding. The book provides up-to-date summaries of methodological advancements in spatial and spatio-temporal analysis, along with insights into future developments in areas such as spatial graphs, multi-level networks and machine learning applications. It also offers practical examples and guidance to help researchers select, apply and interpret the appropriate methods.

Mark R. T. Dale is Professor of Ecosystem Science and Management at the University of Northern British Columbia, Canada, where he has served in several administrative roles. His research interests are based on the spatial structure of plant communities and include the development of quantitative methods to answer ecological questions. His work on plant community structure and the dynamics of spatial pattern development has led to significant contributions in the application of graph theory and the study of ecological networks. He is the author of *Applying Graph Theory in Ecological Research* (Cambridge University Press, 2017) and co-author of *Quantitative Analysis of Ecological Networks* (Cambridge University Press, 2021).

Marie-Josée Fortin is University Professor in the Department of Ecology and Evolutionary Biology at the University of Toronto. She is a Fellow of the Royal Society of Canada and holds a Tier 1 Canada Research Chair in Spatial Ecology. Her research is at the interface of spatial ecology and conservation biology, including modelling connectivity in urban environments, ecosystem services and forest resilience. She is co-author of *Quantitative Analysis of Ecological Networks* (Cambridge University Press, 2021).

Spatial Analysis

A Guide for Ecologists

Third Edition

MARK R. T. DALE
University of Northern British Columbia

MARIE-JOSÉE FORTIN
University of Toronto

Shaftesbury Road, Cambridge CB2 8EA, United Kingdom

One Liberty Plaza, 20th Floor, New York, NY 10006, USA

477 Williamstown Road, Port Melbourne, VIC 3207, Australia

314–321, 3rd Floor, Plot 3, Splendor Forum, Jasola District Centre, New Delhi – 110025, India

103 Penang Road, #05-06/07, Visioncrest Commercial, Singapore 238467

Cambridge University Press is part of Cambridge University Press & Assessment,
a department of the University of Cambridge.

We share the University's mission to contribute to society through the pursuit of
education, learning and research at the highest international levels of excellence.

www.cambridge.org
Information on this title: www.cambridge.org/9781009158671

DOI: 10.1017/9781009158688

First edition © Marie-Josée Fortin and Mark R. T. Dale 2005
Second edition © Mark R. T. Dale and Marie-Josée Fortin 2014
Third edition © Mark R. T. Dale and Marie-Josée Fortin 2025

This publication is in copyright. Subject to statutory exception and to the provisions
of relevant collective licensing agreements, no reproduction of any part may take
place without the written permission of Cambridge University Press & Assessment.

When citing this work, please include a reference to the DOI 10.1017/9781009158688

First published 2005
9th printing 2013
Second edition 2014
Third edition 2025

A catalogue record for this publication is available from the British Library

A Cataloging-in-Publication data record for this book is available from the Library of Congress

ISBN 978-1-009-15867-1 Paperback

Cambridge University Press & Assessment has no responsibility for the persistence
or accuracy of URLs for external or third-party internet websites referred to in this
publication and does not guarantee that any content on such websites is, or will
remain, accurate or appropriate.

For EU product safety concerns, contact us at Calle de José Abascal, 56, 1°,
28003 Madrid, Spain, or email eugpsr@cambridge.org

Contents

Preface		*page* xi
1	**Ecological Processes**	1
	Introduction	1
	1.1 Spatial Processes	1
	1.2 Ecological Processes	4
	1.2.1 Spatial Patterns along Gradients	5
	1.2.2 Spatial Associations among Species	6
	1.3 Plant Community Spatial Structure	9
	1.4 Spatial Processes by the Level of Organization	13
	1.5 How to Use This Book	16
2	**Spatial Concepts and Notions**	20
	Introduction	20
	2.1 The Spatial Context	23
	2.2 Ecological Data	23
	2.3 Spatial Structure: Spatial Dependence and Spatial Autocorrelation	26
	2.4 Spatial Scales	28
	2.5 Sampling Design	31
	2.5.1 The Sample Size (the Number of Observations 'n')	31
	2.5.2 Spatial Resolution: Grain and Extent	31
	2.5.3 The Size and Shape of the Sampling Units	32
	2.5.4 Sampling Design	33
	2.5.5 The Location in the Landscape	33
	2.5.6 Spatial Lag	33
	2.5.7 Multiscalar Analysis	35
	2.5.8 Effects of Edges	36
	2.6 Stationarity	37
	2.7 Scaling	38
	2.8 Spatial Neighbours	40
	2.8.1 Lattice-based Neighbours	40
	2.8.2 Topological Neighbours	40
	2.8.3 Distance-based Spatial Neighbours	44
	2.8.4 Directional Angle-based Spatial Graphs	48

	2.9	Spatial Statistics		49
		2.9.1 First-order Statistics		49
		2.9.2 Second-order Statistics		50
	2.10	Ecological Hypotheses and Spatial Analysis		52
	2.11	Randomization Tests for Spatially Structured Ecological Data		54
		2.11.1 Restricted Randomizations		55
		2.11.2 Markov Chain Monte Carlo		58
	2.12	Concluding Remarks		58
3	**Spatial Analysis of Point and Quadrat Data**			**60**
	Introduction			60
	3.1	Mapped Point Data		60
		3.1.1 Introduction: Three Basic Patterns		61
		3.1.2 Distances to Neighbours		62
		3.1.3 Refined Nearest Neighbour Analysis		62
		3.1.4 Second-order Point Pattern Analysis		63
		3.1.5 Bivariate Data		66
		3.1.6 Thinning or Combining		68
		3.1.7 Multivariate Point Pattern Analysis Data		69
	3.2	K-function Analysis for Inhomogeneous Point Patterns		73
		3.2.1 Regional Expected Values around Events		73
		3.2.2 Regional Values for Subareas of the Study Plot		74
	3.3	Mark Correlation Function		75
	3.4	Point Patterns in One and Three Dimensions		76
		3.4.1 One Dimension		76
		3.4.2 Three Dimensions		79
	3.5	Circumcircle Methods		80
	3.6	Areal Unit Analysis		81
		3.6.1 Quadrat Variance Methods		81
		3.6.2 Two or More Species		84
		3.6.3 Two or More Dimensions		86
	3.7	Spectral Analysis and Related Techniques		88
	3.8	Wavelets		89
	3.9	Concluding Remarks		90
4	**Spatial Analysis of Sample Data**			**97**
	Introduction			97
	4.1	Join Count Statistics		98
		4.1.1 Join Count Statistics for k-categories		100
	4.2	Global Spatial Statistics		101
		4.2.1 Spatial Covariance		101
		4.2.2 Spatial Autocorrelation Coefficients for One Variable		102
		4.2.3 Variography		109

	4.3	Sampling Design Effects on the Estimation of Spatial Pattern	117
	4.4	Spatial Relationship between Two Variables	118
	4.5	Local Spatial Statistics	119
	4.6	Spatial Scan Statistics	122
	4.7	Spatial Interpolation	124
		4.7.1 Proximity Polygons	125
		4.7.2 Trend Surface Analysis	125
		4.7.3 Inverse Distance Weighting	126
		4.7.4 Kriging	128
	4.8	Concluding Remarks	134

5 Spatial Partitioning: Spatial Clusters and Boundary Detection 136

Introduction		136
5.1	Patch Identification	136
	5.1.1 Patch Properties	136
	5.1.2 Spatial Clustering	138
	5.1.3 Fuzzy Classification	140
5.2	Boundary Delineation	142
	5.2.1 Ecological Boundaries	142
	5.2.2 Boundary Properties	143
	5.2.3 Boundary Detection and Analysis for One-Dimensional Transect Data	145
	5.2.4 Boundary Detection based on Two-Dimensional Data	148
5.3	Boundary Overlap Statistics	156
5.4	Hierarchical Spatial Partitioning	158
	5.4.1 Edge Enhancement with Kernel Filters	159
5.5	Concluding Remarks	161

6 Spatial Autocorrelation and Inferential Tests 163

Introduction		163
6.1	Models of Autocorrelation in One Dimension	164
6.2	Dealing with Spatial Autocorrelation in Inferential Models	169
	6.2.1 Simple Adjustments	169
	6.2.2 Adjusting the Effective Sample Size	170
	6.2.3 More on Induced Autocorrelation and the Relationships between Variables	174
	6.2.4 Correlation and Related Measures	176
6.3	Randomization Procedures	180
	6.3.1 Restricted Randomization and Bootstrap	180
	6.3.2 Monte Carlo Markov Chain	182
6.4	Considerations for Sampling and Experimental Design	183
	6.4.1 Sampling	184
	6.4.2 Experimental Design	186
6.5	Concluding Remarks	187

7		**Spatial Regression and Multiscale Analysis**	189
		Introduction	189
	7.1	Spatial Causal Inference	189
	7.2	Correlation between Spatially Autocorrelated Variables	192
	7.3	Mantel Test	193
		7.3.1 Partial Mantel Tests and Multiple Regression on Distance Matrices	197
	7.4	Spatial Regressions	200
		7.4.1 Spatial Filtering Using Autoregressive Models	204
		7.4.2 Other Spatial Filtering Models	205
		7.4.3 Spatial Error Regression	206
		7.4.4 Geographically Weighted Regression	206
		7.4.5 How to Remove Spatial Autocorrelation from the Residuals	208
	7.5	Canonical (Constrained) Ordination	210
	7.6	Multiscale Analysis	211
		7.6.1 Generalized Moran's Eigenvector Maps	214
		7.6.2 Multiresolution Spectral Decomposition Analysis with Wavelets	218
	7.7	Concluding Remarks	225
8		**Spatio-temporal Analysis**	227
		Introduction	227
	8.1	Spatial Statistics Reassessment	229
	8.2	Spatio-temporal Join Count	229
	8.3	Spatio-temporal Analysis of Clusters and Contagion	230
	8.4	Spatio-temporal Scan Statistics	237
	8.5	Polygon Change Analysis	237
	8.6	Analysis of Movement	241
	8.7	Spatio-temporal Networks	249
		8.7.1 Phenology	251
	8.8	Spatial Aspects of Synchrony	253
	8.9	Concluding Remarks	256
9		**Spatial Diversity Analysis**	258
		Introduction	258
	9.1	Space and Diversity	258
		9.1.1 Spatial Scale	259
		9.1.2 Spatial Location and Environmental Gradients	260
		9.1.3 Spatial Heterogeneity	260
		9.1.4 Spatial Dependence	261
	9.2	Application: Why Spatial Diversity	261
		9.2.1 α-Diversity	262
		9.2.2 β-Diversity	265
		9.2.3 γ-Diversity	270

9.3	Combinations and Composition: Agreement and Complementarity	270
	9.3.1 Species Combinations	271
	9.3.2 Nested Subsets, Constrained Compositional Diversity	275
9.4	Spatial Diversity: Putting It All Together	278
9.5	Concluding Remarks	282
	9.5.1 Temporal Aspects	282
	9.5.2 Complexity	282
	9.5.3 Space and Time	283

10 Points and Lines, Graphs and Networks 284

Introduction	284
10.1 Lines Alone: Fibre Pattern Analysis	286
10.1.1 Aggregation and Overdispersion of Fibres	287
10.1.2 Fibres with Properties	288
10.1.3 Curving Fibres	291
10.2 Points and Lines Together	292
10.3 Points and Lines: Spatial Graphs and Spatial Networks	295
10.3.1 Spatial Nodes	295
10.3.2 Neighbour Networks	296
10.3.3 Signed and Directed Graphs and Networks	299
10.3.4 Creating Subgraphs	301
10.4 Network Analysis of Areal Units	301
10.5 Spatial Analysis of Flow	309
10.6 Testing Hypotheses with Spatial Graphs	311
10.7 Concluding Remarks	312

11 Spatial and Temporal Analysis with Multilayer Networks 313

Introduction	313
11.1 Multilayer and Multiplex Networks	315
11.2 Multilayer Metrics for Emergent Properties	316
11.2.1 Node Degree and Related Measures	316
11.2.2 Walks and Paths	317
11.2.3 Centrality and Node Ranking	318
11.2.4 Clustering	319
11.2.5 Spectral Properties	321
11.2.6 Resilience, Robustness and Fragility	322
11.3 Null Randomization Procedures	322
11.3.1 Replica Node Randomization	322
11.3.2 Independent Layer Randomization	323
11.3.3 Randomization Preserving Multidegree Sequences	323
11.4 Getting the Most from Multilayer Networks	323
11.4.1 Behavioural and Disease Ecology	323
11.4.2 Metawebs	325
11.4.3 Multispecies Connectivity	327

		11.5 Multilayer Networks and Spatio-temporal Analysis	329
		11.6 Concluding Remarks	330
12		**Closing Comments and Future Directions**	331
		12.1 Reminders and Challenges	331
		12.1.1 Reminders	332
		12.1.2 Challenges	334
		12.2 Back to Basics	336
		12.3 Numerical Solutions: Software Programs and Programming	337
		12.4 Statistical and Ecological Tests	338
		12.5 Complementarity of Methods	338
		12.6 Looking Ahead	342
		12.6.1 Ongoing Development	342
		12.6.2 The Bayesian Approach	344
		12.6.3 Spatial Causal Inference	354
		12.6.4 Artificial Intelligence: From Machine Learning to Deep Learning to AI	355
		12.6.5 Geometric Algebra	358
		12.7 Other Future Directions	359
	References		361
	Index		394

Preface

Spatial analysis, and its extension into spatio-temporal analysis, has been a rapidly growing field for at least two decades (consider that our first edition was 20 years ago: Fortin & Dale 2005!). The growth can be attributed to (1) ecologists' increasing awareness of the essential role of spatial and temporal structure for understanding ecological systems, (2) concern about the effects of the all-to-obvious alteration of landscapes around us and (3) increasing sophistication and technical scope of resources to make such analyses possible. The large range of choices for analysis brings its own problems in the form of questions about which methods to use and what the conditions are for their correct application. It is easy to use them incorrectly: (1) by not realizing the difficulties related to different spatial and temporal scales; (2) inadequately accounting for autocorrelation which creates puzzles and uncertainty for parametric analysis; and (3) lack of clarity about spatio-temporal structure's effects on ecological processes and how it can best be incorporated into experiment and analysis. This book is designed to be helpful for all those difficulties by going beyond specific spatial statistics to treat spatio-temporal analysis more generally and by showing which methods are most appropriate for given circumstances and how those chosen are best interpreted.

This purpose makes the book broad in scope, especially with an increased emphasis on spatio-temporal data, including dynamic spatial graphs and spatio-temporal networks. The intent is to help both those who are new to the topic and those that are familiar with some (or even many) aspects of such analysis but are unsure of how to start. Our goal is, therefore, to provide a broad overview of well-established methods, thus easing in through the more familiar, but then moving on to those that are unfamiliar and recent or currently under development. The less familiar includes approaches that have arisen in fields other than ecology, such as geography, geology and epidemiology, and we also look ahead to future developments from more distance disciplines such as applied mathematics and machine learning (Chapter 12).

In order to answer our ecological questions, we begin by detecting patterns in the data, and then investigate further to understand the processes that give rise to the patterns we observe. Pattern recognition is an important part of the endeavour of discovery that includes observation, experimentation, analysis with interpretation and various forms of modelling. In trying to understand ecological processes, we must acknowledge that most of what we study occurs within complex systems, with complexity taking many forms: great biological diversity, different kinds of hierarchical organization, many levels and layers of multispecies interactions and stochastic components in multi-level dynamics. Consider the complexities of the processes that give rise to the spatial structure and temporal dynamics of 20 species of trees in a temperate forest ... and then those for a tropical forest with hundreds of tree species ... and then for all the insects and fungi in that tropical forest ... Challenges indeed for spatial analysis and spatio-temporal interpretation!

This book stems from years of teaching at our respective universities, as well as research into many of the methods represented here and their application to diverse data by ourselves and colleagues including students and post-doctoral researchers. We acknowledge and thank all those who have helped with our teaching, methodological research and data analysis, but there are really too many to thank in lists of names (thank you all!). The one exception is to thank Cheryl Smyth for her meticulous attention to detail in preparing the text, figures and references. Lastly, we will acknowledge with great appreciation all the sources of funding that supported our research efforts and this project, including NSERC, U of T, UNBC, ...

1   Ecological Processes

Introduction

Every ecological process takes place in a spatial context where the pre-existing spatial structure of ecological systems and environmental conditions affect or determine the resulting spatial outcomes. Hence, studying ecological processes by averaging over locations (and times) can be misleading because it ignores the effects of heterogeneity and other spatial aspects that are crucial for understanding species' responses to a dynamic world. Indeed, organisms do not live in uniform environments; they encounter environmental gradients and patchiness of abiotic and biotic origin, as well as perturbations and disturbance, creating two- or three-dimensional mosaics containing patches with boundaries and ecotones between them. All these result in spatial structures that affect the processes and their networks of interactions according to the organization level (individual, population, community, etc.). In ecological studies, explicit considerations of spatial structures have come to be increasingly important as components for understanding ecological processes. Spatially explicit studies must go beyond mere comparisons of regional attributes and include locations, distances and other spatial relationships. Here, we provide an introductory discussion of the relationships between ecological processes and those spatial characteristics.

1.1   Spatial Processes

In mathematics, a *stochastic process* is a collection or 'family' of random variables governed by at least one parameter, such that each outcome has a distribution associated with it (the values are random, not determined). A spatial process is, therefore, a mathematical system with stochastic rules that generate events or values of variables in a spatially explicit framework where there is a location for each event or value. Consider a simple spatial point process that creates a set of 16 randomly located point events, each having an attribute labelled 1 to 5, in a square of 100 units by 100 units. The rules could be:

Repeat the following steps (1 to 4) 16 times, starting with $i = 1$ and going to $i = 16$:

(1) randomly select a real number between 0 and 100 for the ith x-coordinate;
(2) randomly select a real number between 0 and 100 for the ith y-coordinate;
(3) randomly select an integer from 1 to 5 for the label of the ith event; and
(4) record (and plot on a diagram) the location and value of the ith event.

Provided that the numbers are generated correctly, the events are independent of each other, as are the events' labels. With some random number algorithms, we get a different set of numbers every time the algorithm is implemented with different starting conditions; and any one of these is an individual realization of the same process. If a uniform distribution is used for the random coordinates, any point in the plane has an equal probability of having an event. If both spatial coordinate variables follow bell-shaped distributions, the events will have higher density in the centre. In other cases, the rules can be structured so that spatial location affects the label (e.g. higher label values for central locations) or the labels may depend on relative positions (e.g. neighbouring labels tend to be similar). The resulting set of events has statistical properties determined by the rules of the process that generates them, although their observed values will vary from one realization to another.

Spatial processes are often treated similarly to stochastic processes that occur in time. Whereas time has asymmetric, possibly causal, relations of 'before' and 'after', space usually has no inherent directionality and is treated in two or three dimensions, rather than just one.

Homogeneous describes a process that is invariant under translation and *isotropic* describes one that is invariant under rotation (Ripley 1988). Again, the terms refer to the underlying process, and the characteristics may not be manifested in its realization.

Like a temporal process, a spatial process can be:

- **Stationary**, where the statistical characteristics of the process that generated the variable of interest do not change with location (for *weak stationarity*, the mean and covariance structure are invariant; for *strong stationarity*, the distribution itself or all its moments are invariant); or
- **Non-stationary**, where the statistical characteristics of the process that generated the variable of interest do change with location (the mean, variance or covariance structure).

The realization of a process can exhibit non-stationarity, however, in many ways: a trend in any direction or patchiness at one or more scales, with or without directionality. Spatial inference requires an assumption of stationarity of some kind (Ripley 1988) because that is what allows prediction from one location to another.

To illustrate a spatial process, consider a rectangular area A of a plane into which we place n events, each a dimensionless point. The magnitude of the process is the mean number of events per unit area, here $\lambda = n/A$. If the events are random and independent, every sub-unit of the area will contain an event with a probability proportional to its area. The number of events per areal unit will follow a Poisson distribution with parameter λ. In a Poisson distribution, both the expected value (the overall mean) and

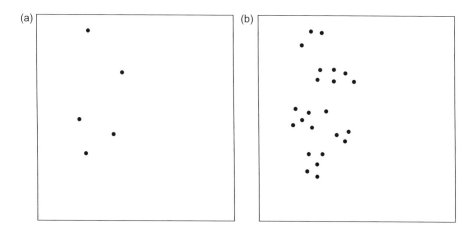

Figure 1.1 (*a*) Five random points in one-half of the area. (*b*) Five clusters from those individual points.

the theoretical variance equal λ. This is the null model of complete spatial randomness (hereafter CSR). It is important to realize that a completely random stationary process can give rise to a spatial arrangement of events that does not look random, especially if the number of events is small. Figure 1.1 shows five events placed at random in a square; if random, one in eight realizations (about 13%) will have all five events located in only one-half of the square. This may seem a bit contradictory or at least puzzling because it does not look like the result of anything stationary.

This apparent non-uniformity resulting from a homogeneous random process is different from the situation in which patchiness is created by the inhomogeneous structure of the process itself. In a homogeneous process, the density of points is constant, whereas, in an inhomogeneous process, the density depends on location. Still, a homogeneous process can give rise to an inhomogeneous outcome in many ways; several are based on the CSR process which gives rise to the Poisson pattern just described. Each event created by CSR can produce, in turn, a cluster of events, with its number and location governed by a second stochastic process (Figure 1.1*b*). Another mechanism is to have a process like CSR with intensity parameter λ but allowing the process intensity to vary with location 's', so that $\lambda(\mathbf{s})$ is itself the result of a second stochastic process. If the processes at both levels are Poisson, we end up with Cox's 'doubly stochastic Poisson process' (Kingman 1993). Biologically, this situation can arise when a second generation is derived from a population of parents which were distributed following CSR, provided the offspring disperse from their own parents independently, but with their average location being the location of their parent. The positions of the offspring are then conditional upon the positions of the parental generation (Kingman 1993). Processes like this one result in a pattern that has a distribution of event counts per areal unit for which the variance is greater than the mean, indicating patchiness of some kind. In that case, each realization of the process (the pattern observed) is not invariant under translation, and so the **pattern** is apparently inhomogeneous, even though the **process** (and the pattern over many realizations) is **homogeneous** and **stationary**: over all possible realizations, every small subarea has an equal probability of being occupied and the expected value (mean) and variance are constant. (In a non-stationary process, the parameters of mean and variance can be different at different locations.)

Inhomogeneous and non-stationary processes can be classified according to whether direction affects the probabilities:

Isotropic (no directional effect); or
Anisotropic (with differences according to direction, e.g. stripes).

For biological systems, as for mathematical ones, a spatial process gives rise to events or values of variables with definite locations, but what generates the patterns are biological factors, like dispersal or mortality. Consider a seed tree in the middle of a clear-cut: its seed rain is a non-stationary spatial process as seed abundance varies with distance and by direction due to wind direction, resulting in a pattern of seedlings that is expected to be inhomogeneous and anisotropic.

In mathematics, the generation of values and the values generated are both referred to as a spatial process, but we will distinguish between the generating process and the set of values it produces, the latter being the spatial pattern or the realization of the process. Although we introduced spatial stochastic processes with labelled point events in the plane, more structural models should be considered as ecological processes of interest, including finite line segments or curving fibres, random walks on a network, and random relabelling of a spatial network or lattice. The evolution of a spatial network (nodes with locations joined in pairs by edges; Chapter 10) can itself be a random process emulating biological systems (Barthelemy 2018), and, with branching, it can emulate ecological phenomena like clonal growth. The concept of a random walk on a network can be related to the ecological processes of diffusion or dispersal; consider an aquatic organism spreading through a riverine system or a novel pathogen spreading through a spatially structured population. Correlated random walks, in which successive steps are not fully independent are used extensively to model animal movement such as foraging (Lewis *et al.* 2021; see Chapter 8).

The concept of relabelling the nodes of a spatial network, for example from 'closed' to 'open' for sites in a landscape, can be related to percolation once a critical proportion of 'open' nodes allows flow across the network (see Barrat *et al.* 2008). Similarly, the random relabelling of nodes in a spatial network or from 'living' to 'dead' can be a null model for the self-thinning of a population. Of course, a particular non-random relabelling of nodes can give rise to cellular automata, including Conway's 'Game of Life' (Gardner 1970).

1.2 Ecological Processes

The processes that are most relevant for spatial analysis in ecology are both biological and environmental. These include natural and anthropogenic disturbances that trigger a

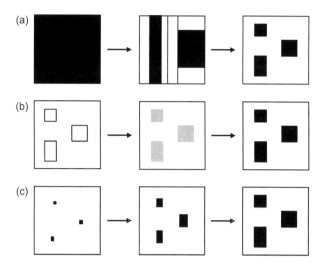

Figure 1.2 The same pattern from different processes: (*a*) fragmentation, (*b*) density increase and (*c*) nucleation followed by patch growth.

cascade of biological processes affecting organisms, such as dispersal, establishment, growth and species interactions (intra- and interspecific ones). These biological processes, and their interactions, also interact with abiotic processes to produce the spatial structure and spatial dynamics of subsequent processes. We might hope to deduce the past processes from the current spatial patterns (Watt 1947), but this is not always possible: the same process may result in different patterns, and different processes may give rise to indistinguishable patterns (Figure 1.2). The patchiness can result from the growth structure of the organisms, such as corals or aspen groves, or from topographic structure, such as knob-and-kettle landscapes of alternating hills and hollows. In addition, several mechanisms may contribute to a single process, such as the biotic and abiotic factors that produce paludification in *Sphagnum*-dominated systems (Rietkerk *et al.* 2004).

Spatial analysis, by definition, focuses on the pattern observed at a single time, but we know that the processes are dynamic and that communities change, suggesting the need for spatio-temporal analysis (Chapters 8 and 11). Ecologists also study stochastic disturbances that recur (Shoemaker *et al.* 2020), like fire or infestation-induced mortality, because they can create obvious patchiness and have significant impacts on other processes. In some instances, mere senescence can have effects like those of abiotic disturbance, pathogens, or herbivory. Often the ecological process and spatial pattern that arises from it interact, producing patches of different species composition and age structure, affecting the dynamics of future processes (Sturtevant & Fortin 2021).

1.2.1 Spatial Patterns along Gradients

We can start with a familiar phenomenon: species occurrences on a one-dimensional environmental gradient. The gradient is a monotonic change of one environmental factor over physical distance and it may create obvious zonation in the community.

This spatial structure is non-randomness in space with some predictability and therefore falls into the category of 'spatial pattern', although there are no repeating units (Dale 1999). A gradient creates predictable and directional variation, which allows the appearance of species and then their disappearance farther along the gradient. The pattern lies in where the species enter and leave the gradient and in their ranges and densities, where present.

On gradients, the observed spatial pattern may allow *some* inferences about the processes that created it and the current pattern affects future processes. The potential location of any species is determined by its physiological responses to the gradient but the observed locations result from the interaction of physiology with other processes including competition, facilitation and predation. The current arrangement of species determines which species may interact with one another: nearby organisms may compete most strongly but they also have the greatest potential for positive interactions (Bertness & Calloway 1994).

The usual model of a species' response to an environmental gradient is a symmetric unimodal curve as a function of the factor's intensity or of the physical distance along the gradient. The symmetric unimodal response may be rare in nature; asymmetry is more common and some responses are bimodal (Austin 1987). The skewness of a unimodal response depends on the scaling of the environmental factor's axis. Furthermore, many landscapes are fragmented, such that species abundance fluctuates within the geographical range due to the availability of habitat (Fortin *et al.* 2005).

The rate of change in a controlling factor may itself vary along a gradient, or organisms may respond unequally to the same amount of change in the factor, depending on the position. Consequently, identifiable levels may exist on a gradient where species replacement occurs rapidly over small distances. How species are arranged on environmental gradients may reveal characteristics of the community organization. For example, interspecific competition affects the spatial pattern on a gradient and the inability of two competitors to coexist can result in the beginning of one species' range following immediately after the ending of another's range (Figure 1.3*a*). On the other hand, if species replacement allows a zone of competitor coexistence, the density of one species will decrease as the others increase in that coexistence zone (Figure 1.3*b*). We can also look at models for multispecies replacement on such gradients. One model suggests that *groups* of species replace each other along the gradient, producing clusters of upper and lower boundaries (Figure 1.4*a*). An alternative is that the species occur independently so that the boundaries are not clustered (Figure 1.4*b*). Spatial analysis can distinguish among possible arrangements in systems that are well-structured by gradients, as will be described in Chapter 5.

1.2.2 Spatial Associations among Species

Organisms respond to the proximity of other species. In most situations, the individuals of different species are neither randomly nor independently arranged. For plant communities, the term 'association' can refer to the tendency of the plants of different species to occur together more often than expected, 'positive association', or less

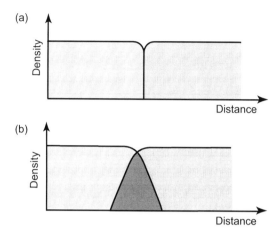

Figure 1.3 Two species' abundance replacing each other along a gradient: (*a*) without spatial overlap (sharp boundary) when species cannot coexist and (*b*) with spatial overlap (gradual boundary, ecotone) where species can coexist.

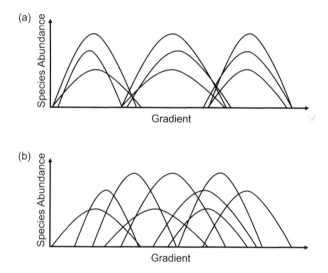

Figure 1.4 Two examples of arrangements of many species' densities on an environmental gradient with competition as one factor. (*a*) With clusters of boundaries and (*b*) no boundary clusters.

often, 'negative association'. The association between species can be based on shared or divergent ecological requirements and capabilities or on the ability of one species to modify the environment to make it more (positive) or less (negative) suitable for the other. Examples of positive influence include facilitation by 'nurse plants' that enhance regeneration in stressful environments (Bertness & Calloway 1994). Examples of negative influence include situations in which plants are affected by allelopathy where chemicals from one plant reduce the growth of another. Negative influence also includes competition for resources, but that interaction depends on the relative sizes of the individual plants and may be less asymmetric than allelopathy, which tends to be strongly directional (Zhang et al. 2021).

The association of species is usually considered in pairs, and the network of pairwise associations forms a graph (species as nodes, associations as edges, see Chapter 10), sometimes called the phytosociological structure (Dale 1985). Rajala et al. (2019) have reviewed the detection of interspecific interactions using bivariate point pattern analysis (Chapter 3) and found that detectable interactions are generally rare but most common in species-poor communities and that the most abundant species tend to have the most detectable interactions (Rajala et al. 2019). Keil et al. (2021) reviewed the measurement and analysis of spatial species association and its potential relationship with biodiversity and concluded that spatially explicit approaches were more useful than spatially implicit methods.

Pairwise treatment of species association needs to accommodate the fact that the relationship of any species pair may be influenced by the presence or absence of a third (or fourth or fifth ...) species, and we should consider multiple species associations, where the frequencies of combinations of species are examined (Chapter 9). Such associations may be closely related to indirect interactions among species, although associations are deduced from spatial frequencies and not observed as active interactions. Indirect interactions occur when the direct interaction between species nodes is modified by a third species; in Figure 1.5, the edge A↔B is modified by species node C, indicated by the indirect edge C→ (A↔B). The edge A↔B can also be modified by the interaction edge of two other species, D↔E, creating the indirect edge (D↔E) → (A↔B) (Figure 1.5). In essence, a network edge can act as a node in the next level of interactions. Familiar indirect interaction structures include the 'trophic cascade' or 'apparent competition', and hypo- or hyper-predation effects (Dale & Fortin 2021, figures 4.18 and 4.19) and a complex ecological example with several types of interactions at several levels is given in Dale and Fortin (2021, figure 4.13). The topic of multispecies combinations and multispecies interactions will be pursued further in Chapter 7 (spatial relationships) and Chapter 9 (spatial diversity). While multispecies methods clearly involve greater complexities, the insights they can produce can often justify the effort (Clark et al. 2014; Warton et al. 2015; Ovaskainen et al. 2016).

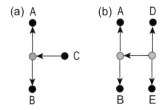

Figure 1.5 Indirect interactions (dashed edges): (*a*) The interaction between two species (solid nodes) acts as a node (grey) affected by the third species. (*b*) The interactions (grey nodes) between two pairs of species (solid nodes) act as nodes for a second-level interaction.

1.3 Plant Community Spatial Structure

An orthodox view of plant community development by ecological succession begins with an intense and extensive disturbance (fire, insect outbreak, harvesting; Sturtevant & Fortin 2021). When the plant community is completely removed, leaving only bare (mineral) substrate, the subsequent development is called primary succession. When some residual material (soil) remains, what follows is called secondary succession. This dichotomy sounds like a clear distinction, but many disturbances create patches of different severity, with different sizes and shapes. Glaciation–deglaciation cycles tend to leave mainly linear features, like scrapes, moraines and eskers, with some isodiametric features like drumlins and knob-and-kettle topography. Forest fires are notoriously uneven, leaving some areas more or less untouched (remnants) and others burned to the mineral layer. Fire-created patches are usually elongated in the direction of the wind prevailing at the time, but large fires can burn for many days with winds from several directions, leaving a complicated spatial footprint. Insect outbreaks can be heterogeneous, even in monocultures, and act selectively in mixed communities. Some outbreaks may be incomplete, including both mortality and recovery in multi-year infestations, leaving complex patchy structures. All this variability in disturbance affects the spatial pattern in developing communities, usually in uneven and complicated ways (Sturtevant & Fortin 2021).

Whatever the disturbance, several processes can contribute to the ecological succession that follows, and different views of how these successional processes interact are tied in with concepts of the nature of the 'community'. One view is that the plant community is like an organism, developing through a series of predictable phases towards a 'climax' self-replacing community (Clements 1916). The contrasting view (Gleason 1927) is that the plants and propagules of different species act more-or-less independently in response to the availability of establishment sites and the environment. This allows for different combinations of species to be the end-points of succession in similar regions or for different successional pathways to converge to similar communities (Glenn-Lewin and van der Maarel 1992). It is possible for similar starting conditions to develop into different compositional end-points. For spatial analysis, the question arises whether a clear difference in the spatial structure of mature plant communities will be found, based on which view is the best description. We suggest that the Gleasonian model might produce much more variability in species combinations and physical structure within any small area than the Clementsian one. Further consideration of this question can wait for our discussion of spatial aspects of species diversity (Chapter 9).

The processes that may be invoked to explain some of these temporal patterns observed in successional sequences include the following:

- **Facilitation**, when the plants early in succession modify the environment in a way that enhances the recruitment of later species;
- **Inhibition**, when the early plants' influence on the environment decreases its suitability of later species' recruitment so that later species establish only when the first group dies off or is reduced by disturbance events;

- **Tolerance**, where the environmental modifications have little effect on subsequent recruitment; and
- **Self-inhibition**, where early-stage plants make the environment less suitable for their own recruitment, as with shade-intolerant trees typical of some early successional sequences.

All four of these may have a spatial effect, although 'tolerance' is a kind of null model and, while the spatial effects may be primarily local, the local effects may influence over large areas (Solé 2007). The spatial version of facilitation is nucleation (Yarranton & Morrison 1974), where the plants of one species act as the nuclei for the establishment of others, like 'nurse plants' reducing the heat load on seedlings or bird perches enhancing seed deposition. Inhibition and self-inhibition may have very localized effects that could be detected by spatial analysis, just as nucleation may produce a clear spatial signature.

Regeneration is a key process in the development of plant communities, and it depends on the availability of propagules (a seed source close enough to be effective) and of substrate suitable for germination and development. These both have spatial aspects that affect success: the number and distances of seed sources and the number, locations and sizes of patches of suitable seed bed substrate. Competition is a second key process for community development, particularly arising from previously or simultaneously established plants. It has a strong effect both numerically, by affecting growth rates and survivorship, and spatially, affecting the distances between plants and the relative sizes of neighbours. These ecological processes all contribute to the spatial structure of the community and have implications for the application of spatial analysis and the interpretation of the results. The critical factor is the relationship between the ecological processes and the spatial patterns we investigate.

We have alluded to Watt's (1947) comments on the relationship between pattern and process in plant communities, laying a foundation for plant ecology over the following decades. One major theme is cyclic change at small scales within a well-defined plant community. Watt described several communities with aggregations of species that can be considered as repeating phases of a mosaic which have recognizable periods of regeneration alternating with periods of degradation. While these phases may develop at different rates and for different durations, the whole community remains essentially the same with consistent processes giving rise to a repeating sequence.

The phases appear to form a temporal cycle and so an understanding of the community 'as a working mechanism' should be based on the relationships among the phases. Watt's ideas on the relationship between pattern and process have influenced the development of plant ecology and have affected our conceptual models of community dynamics.

A general model is the 'patch-gap' model which is applied most frequently to forests. In simple form, it suggests that significant tree recruitment takes place only below canopy gaps, so that the regeneration depends on gap formation, whether autogenic like senescence or allogenic like wind-throw. This model depends explicitly on disturbance to drive the cycle, and the rate of regeneration depends on the size of the gap and the conditions in the gap thus created.

Potential non-linear effects of patch or gap size also affect this model: doubling the patch size may more than double the number of seeds and other propagules that are intercepted. Therefore, the distribution of gap sizes can affect the outcome of the processes of dispersal and regeneration. The same is often true of density effects, although density and neighbour distances may be confounded, creating other non-linear responses.

The relationship between patches may be affected by feedback loops in the processes. Negative feedback loops are a common mechanism for maintaining homoeostasis (e.g. a thermostat) and positive feedback loops are found in self-reinforcing systems (e.g. lighting a fire). *Feedback switches* in ecology include the phenomenon by which one plant association modifies the environment to be more suitable for itself and less suitable for others.

Wilson and Agnew (1992) have provided a classification of feedback switches, but most of the examples are 'one-sided': the plants of association X change environmental factor Z where they are present, enhancing their ability to invade adjacent areas (Figure 1.6a; for the square brackets: [X] means abundance and [Z] means strength). A familiar example is invasion by *Sphagnum*, where the moss acidifies the environment and raises the water table, allowing the sphagnum-dominated area to expand by a process known as paludification. In a two-sided 'reaction' switch, there is a spatial division between an area with X and an area without, '~X'; association X increases variable Z in its own area, which causes a decrease in Z elsewhere (Figure 1.6b), and the decrease in [Z] (down-arrow) reinforces the absence of X. If the vegetation stripes

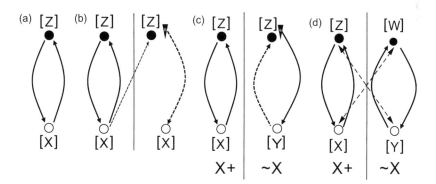

Figure 1.6 Four types of feedback switches: (*a*) 'One sided': the plants of association X change environmental factor Z, enhancing their own success. (*b*) Association X and variable Z are mutually reinforcing in their own area (left panel), decreasing Z elsewhere (right panel), and the decrease in [Z] (down-arrow) reinforces the absence of X. (*c*) X increases factor Z in its own patches (left) and association Y decreases Z in its own patches (right); the decrease in [Z] (down-arrow) reinforces the abundance of Y. (*d*) X changes factor Z in its own patches (left panel) and association Y changes factor W in its own patches, (right panel). The pairs of X and Z and Y and W are mutually reinforcing, but the interactions of X and W and Y and Z are negative.

in an arid region capture moisture at their upslope edges, less moisture is available for the areas downslope. This process exaggerates the differences because the stripes capture moisture to support plants and the bare regions do not.

Two-sided switches are also possible. In the 'symmetric' version with spatial division, X increases factor Z in its own patches and association Y decreases Z in its own patches (Figure 1.6c); the decrease in [Z] (down-arrow) reinforces the abundance of Y, [Y]). In the 'two factor' version, X changes factor Z in its own patches and association Y changes factor W in its own patches (Figure 1.6d). The pairs of X and Z and Y and W are mutually reinforcing, but the interactions of X and W and Y and Z are negative.

The potential spatial effects of these switches can help interpret the results of spatial analysis because positive feedback can reinforce the mosaic system in which it occurs, increase the sharpness of a boundary between two plant community types as tiles of the mosaic, or change the rate of community type replacement (Wilson & Agnew 1992; Kéfi et al. 2016). Evolution within a metapopulation can have a similar effect in amplifying or even creating differences which show up as patterns, although evolution may smooth out underlying heterogeneity, reducing the intensity of the spatial pattern (Urban et al. 2020).

The amplification or reinforcement of spatial differences is one mechanism by which landscapes have memory. *Landscape memory*, or legacy, describes how strongly current ecological processes are influenced by the landscape's past (Hendry & McGlade 1995; Khalighi et al. 2022), and it can be seen as spatially explicit temporal autocorrelation, positive or negative. On the one hand, if recently burned areas of a forest do not reburn until fuel has accumulated for decades, fire recurs only at long intervals, giving negative autocorrelation. On the other hand, if some areas are immune from fire because of topographical characteristics, whereas others burn frequently, the result is positive autocorrelation. Hendry and McGlade (1995) concluded that the emergent spatial structure depends on the amplification of local interactions through the mechanisms of ecological memory (Peterson 2002; Johnstone et al. 2016). Where landscape memory is an important factor, the effects of neighbouring tiles may be a critical factor in the system dynamics and community interactions (Fortin et al. 2012b; Khalighi et al. 2022).

Self-organization, or autogenic pattern formation, is the last in this set of related concepts. The phenomenon shows that simple rules, acting very locally, can produce clear patterns over large extents (see Solé & Bascompte 2006; Kéfi et al. 2016). Simple spatial models produce hexagonal arrays of spots, rings or stripes like brousse tigrée (Thiéry et al. 1995). Colonial insects are often used as examples in this context as producing elaborate and robust structures from simple behavioural modules. Plant-dominated systems have many examples like those described by Watt (1947), with a strong spatial component in their processes. We have mentioned that wetlands often have elevated strings of established plants separated by long narrow pools oriented across the water flow. While the development and orientation of all these structures may be driven by abiotic forces, the form produced is affected by the characteristics of the plants themselves. In any case, self-organization involves a balance between positive feedback at some spatial scales and negative feedback at other scales and locations.

We have described some of the processes in communities that affect the spatial structure and spatial dynamics because these concepts provide the background for an informed interpretation of the results of spatial analysis. These apply most directly to studies of plants but they also provide a conceptual framework for the organisms that depend on or inhabit that community, whether ground beetles, pollinators, songbirds or large mammals.

1.4 Spatial Processes by the Level of Organization

Spatial structure can vary with the level of organization, from individuals and populations to communities and landscapes (Levin 1992). We described phenomena that amplify existing differences, thus enhancing boundaries in a spatial mosaic, which has consequences for spatial structure. The first is that small local differences may be amplified through time into differences of ecological significance. The second is that boundaries between 'phases' of a spatial mosaic, its tiles, can be very important for ecological processes, even if they are not immediately obvious.

All these processes are dependent on scale and level of organization (Dungan *et al.* 2002). The spatial scales should be evaluated because the dynamics that follow a disturbance or environmental change depend on the duration and the area affected (Levin 1992, 2000) as well as the landscapes 'complexity' (Newman *et al.* 2019). Anything we consider to be 'cyclic succession' is noticeable because of the spatial grain and temporal pace at which it occurs. The scale of the observations and the scale of the observer both have effects. Cyclic species replacement in a grassland may be invisible because the spatial scale is too fine; cyclic replacement among saxicolous lichens may be invisible because the temporal scale is too long. If our observational scales are too narrow, we may miss or misrepresent the processes we are trying to understand (Estes *et al.* 2018).

In addition to processes at the population level (competition, predation and so on) and the phenomena of switches and self-organization, spatial structure affects other levels of organization, such as metapopulations and metacommunities (Hanski 2009; Leibold & Chase 2018). A *metapopulation* is a network of local populations, each occupying a patch in a heterogeneous environment, with a dynamic balance of local patch extinctions and re-establishment through re-colonization from other subpopulations. The balance is affected by risks and rates of extinction, dispersal speed and distance range and the factors determining the probability of re-establishment. Metapopulation studies are based on a spatial network of sites and connections, whether explicit or merely implicit, based on simple assumptions about connectivity in that network (see Bode *et al.* 2008).

Similarly, a *metacommunity* consists of component metapopulations forming local communities linked by dispersal between them. The form of spatial patchiness complicates that simple description, however, and several archetypes (previously 'paradigms') have been proposed for metacommunities (Leibold & Chase 2018) – neutral dynamics, species sorting, mass effects and patch dynamics. These consider factors such as local extinction, competitive ability, dispersal mechanisms and environmental heterogeneity within and among patches (for details see Leibold & Chase 2018). In most versions of the archetypes of metacommunities, community composition depends on spatial structure (their table 2.1), and there is good evidence of Clementsian distributions of species along spatial gradients (their table 4.1). These archetypes provide further motivation for spatial analysis as a basis for understanding these systems.

How the various effects play out in real community systems is not fully resolved, but the concepts are important for the development of appropriate approaches to spatial analysis of communities and for the interpretation of the results. They are especially important for the spatial analysis of diversity in real systems (see Chapter 9), and how the differences between local communities may depend on physical distance and the processes of dispersal and ecological interactions (Figure 1.7; Zelnik et al. 2024). On an environmental gradient, compositional difference should increase linearly with geographic distance (Figure 1.8a). This resembles 'isolation by distance' in population genetics, unless the mass effect is sufficiently rapid and strong to counteract environmental differences, thus selecting out different species on different sections of the gradient. In that circumstance, or in an environment that is either uniform or just randomly heterogeneous, compositional difference should not exhibit a trend with distance, although actual values will depend on factors such as the degree of heterogeneity, the strength of the mass effect and the speed of

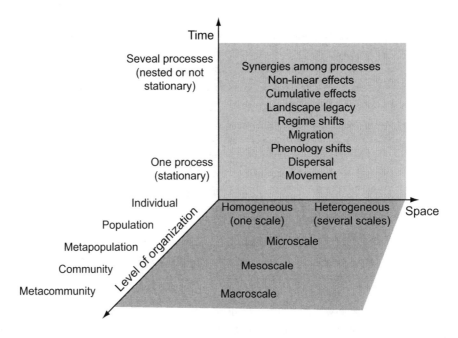

Figure 1.7 Three main dimensions of ecological data: level of organization, spatial scale and temporal scale.

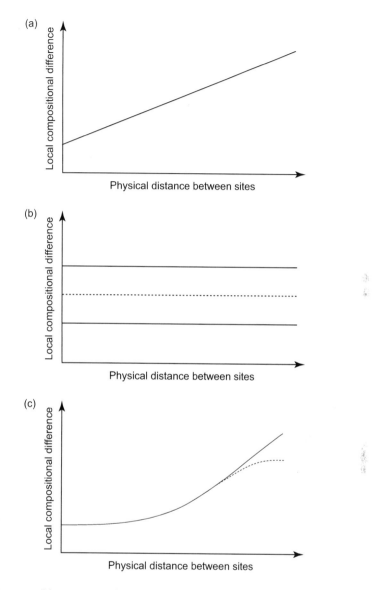

Figure 1.8 Three possible responses of local compositional difference (y-axis) to geographic distance (x-axis): (a) differences increase linearly with distance; (b) no trend with distance, dashed line is the mean value with variability occurring between the two solid lines; and (c) dispersal overrides compositional differences over short distances, but not at larger distances where it may continue to increase or level off (dashed line; see text).

species dispersal (Figure 1.8b). A third possibility is that dispersal, which is distance-limited in its effect, can override the accumulation of compositional differences over shorter spatial ranges but not over larger ranges, producing a curve that is flat initially but then begins to rise. The long-distance end of the curve may or may not reach a maximum set by extrinsic factors (Figure 1.8c).

Having looked at organizational levels from populations and communities to metapopulations and metacommunities, we should extend from species interactions and food webs to the spatially structured 'meta-' equivalents. These are all networks of some kind – interaction networks with species (or equivalent) as nodes and nodes joined in pairs according to their interaction (predation, pollination, etc.) or spatial networks with sites or samples as nodes joined in pairs according to proximity or similarity. Network types for this context can be:

- *Metanetwork* (broadly speaking), representing the ecological interactions in a metacommunity over all sites and combinations of species;
- *Metanetwork* (specifically speaking), as a 'network-of-networks' consisting of a spatial network of sites each with its own ecological interaction network. These are commonly shown for bipartite non-trophic interaction networks such as plants and pollinators (Dale & Fortin 2021, figure 4.22), and;
- *Metaweb* (also *meta-foodweb*; Barter & Gross 2017), representing trophic networks, sometimes referring to all potential trophic interactions in a delimited area (Dunne 2006) but elsewhere referring to observed predator–prey relations (Ceron *et al.* 2021).

All these structures will change with the scale at which they are studied, and not only because of changing species (see Galiana *et al.* 2018, 2022). We will reserve the topic of metanetworks for Chapter 11 because the best approach to these structures is through multilayer networks.

We close this section with a summary table of some of the types of spatial patterns and spatial analysis associated with different levels of ecological organization. Many of the categories are the same for several levels but the details of the methods used may have to change with the level.

1.5 How to Use This Book

The design of this book is to provide both specific and in-depth explication of data types or ecological characteristics of interest and general guidance on the reasoning behind the development of spatial analysis in ecology. The basic concepts and notions for spatial analysis described in Chapter 2 will inform our thinking on how to approach the phenomena and questions to be studied (Table 1.1). The rest of the book is organized around particular approaches, such as point pattern analysis, and more specialized topics, such as autocorrelation or spatial diversity. The combination of levels of coverage should provide sufficient background and detailed understanding to allow well-informed choices for analysis and interpretation.

Clearly, our knowledge about the ecological processes related to the questions we ask will influence the kind of data we collect and our choice of analysis (Table 1.2). Consider the establishment of plants of a single population – initial regeneration may be spatially clumped but with growth and competition, size differences may produce asymmetric competition between neighbours creating non-random mortality which

1.5 How to Use This Book

Table 1.1 Summary of the key spatial processes and corresponding methods to analyse them

Level of Organization	Spatial Process	Spatial Analysis
Organisms	patchiness	spatial autocorrelation
	movement	space occupancy, home range size
	dispersal	range size, dispersal record
Population	demography	spatial autocorrelation
	abundance	spatial autocorrelation
Metapopulation	dispersal	spatial network dynamics
	patches	spatial overlap analysis
	occupancy	join count analysis
Species interaction	patchiness	spatial network dynamics
	seg-/aggregation	bivariate spatial analysis
Community	patchiness	multivariate spatial analysis
	seg-/aggregation	spatial associations
Metacommunity	dispersal	spatial network dynamics
	mosaic	spatial beta-diversity
	neighbour patch id	state & transition of memory
Metanetwork	patchiness	network-of-networks analysis
	gradients	trends with environment and distance
	neighbour effects	diversity and autocorrelation
Metaweb	patchiness	network-of-networks analysis
(trophic meta-net)	gradients	trends with environment and distance
	neighbour effects	diversity and autocorrelation

increases the distance between survivors, thus reducing clumping. The best procedure may be to map, categorize and measure the plants from the beginning and then use several versions of point pattern analysis, such as Ripley's K for locations, marked point pattern analysis for living versus dead and mark correlation analysis to investigate the sizes of neighbours as a function of distance (Chapter 3). To decide, consider the options described in that chapter appropriate to the kind of data available, remembering the assumptions on which valid inference needs to be based.

For this non-random mortality example, as for many, the key question is whether we need to collect data through time. Can we manage without and study just the spatial structure? Methods for assessing spatial structure based on spatial census data, such as the point pattern analysis just mentioned, are covered in Chapter 3; and for samples in Chapter 4. For spatial structure in the form of clusters or boundaries by partitioning space, we can consider the methods described in Chapter 5. Studying spatial relationships through spatial autocorrelation and spatial regression is covered in Chapter 6, and multiscale analysis in Chapter 7. The treatment of spatial relationships and interactions in the form of graphs or networks is provided in Chapter 10.

When time is included as a dimension, Chapter 8 covers many of the basic approaches for spatio-temporal analysis, with Chapter 11 looking specifically at applications of multilayer networks that include multi-time 'stacks' of spatial networks. Spatial aspects of ecological diversity may proceed without a temporal component, but many studies of spatial diversity include time explicitly (Chapter 9).

Table 1.2 Ecological processes and related spatial analysis according to the types of data being static (one snapshot: x–y coordinates) or dynamics (multiple snapshots; t for time events)

	Static		Dynamic	
	1 Species	2 or more Species	1 Species	2 or more Species
Microscale to Mesoscale				
Movement Dispersion or Dispersal	x–y • Point pattern (Ripley's K) • Resource function • Home-range (area vs. cloud or network of points)	x–y • Point pattern (Bivariate-, Multivariate- Ripley's K)	x–y + t • Point pattern (Bivariate Ripley's K) • Resource function • Home-range change through time or movement within	x–y + t • Point pattern (Bivariate-, Multivariate- Ripley's K) • Spatial joint dynamics
Connectivity Spatial abundance	Nodes/patches: • Spatial connectivity • Spatial autocorrelation • Metapopulation	• Homophilous vs Heterophilous connectivity • Species co-occurrence • Ordination • Spatial clustering • Spatial covariance • Metacommunity	Nodes/patches: • Spatio-temporal connectivity • Spatio-temporal autocorrelation • Metapopulation	• Changes in homophilous vs Heterophilous connectivity • Community trajectory • Metacommunity
Species assemblages Species interactions Food webs Disease	• Diversity of ages or life-history stages • Relations among ages or life-history stages	• Community • Metacommunity • Predator–prey • Motifs • Directed graphlets	• Diversity of ages or life-history stages • Relations among ages or life-history stages	• Community trajectory • Metacommunity • Predator–prey cycles • Motifs • Directed graphlets
Mesoscale to Macroscale				
Geographical range	• Polygon convex hull	• Spatial overlap	Range shift: • Polygon change analysis • Moran's Eigenvector Maps • Wavelet decomposition	• Spatial overlap change
Multiscale	• Moran's Eigenvector Maps • Wavelet decomposition	• Moran's Eigenvector Maps		• Moran's Eigenvector Maps

1.5 How to Use This Book

Interspecific analysis to investigate localized facilitation or inhibition during succession and the effects on diversity's trajectory, could begin with marked point pattern analysis of mapped stem data, or quadrat covariance methods for species density in quadrats, and then proceed to examine their temporal sequence.

Consider testing the mosaic cycle hypothesis described in Section 1.3: methods for the detection of boundaries (Chapter 5) and for the analysis of neighbour networks (Chapter 10) or spatial transitions (Chapter 8) may be the most appropriate choices, with the mosaic tiles as the spatial units (see figure 10.11 of Dale 2017). One goal of this book is to provide advice on the range of methods and how they can be matched to the ecological questions and hypotheses of interest. Those hypotheses will involve several potentially complex processes and phenomena, as have been introduced here.

The concepts of spatial processes and their ecological counterparts are required for spatial analysis and its interpretation and we use them throughout this book, but without the technical details to be found in more mathematical texts on stochastic processes. Most important is to understand the ways in which spatial processes and ecological processes go together in the spatial context.

For spatial analysis, knowing the ecological processes that may be active allows us to realize that the results can provide insights into the spatial processes, even if the relationship between pattern and process is not simple. Given the variety of phenomena such as feedback switches, chaos, and spatial self-organization, in addition to the more familiar processes like competition and facilitation, it is possible that several processes can produce indistinguishable patterns, despite the strong relationship between pattern and process. Therefore, although spatial analysis can eliminate some hypotheses, it can seldom be used to confirm mechanisms definitively. On the other hand, knowledge of the kinds of processes that may be acting should affect the choice of data to be collected, the methods of analysis used and the interpretation of the results of that spatial analysis. The goal of the analysis, after all, is to develop a better understanding of these ecological processes.

2 Spatial Concepts and Notions

Introduction

The processes in natural systems and their resulting patterns occur in ecological space and time. To gain a better understanding of the ecological functions of these systems, we first need to identify the spatial and temporal scales at which they occur. While the spatial and temporal dimensions of ecological phenomena are inherent in the conceptual framework of ecology, it is only recently that these spatio-temporal dimensions have been incorporated explicitly into ecological theory, sampling design, experimental design, and formal models (Levin 1992, 2000). Furthermore, all phenomena of ecological interest have both spatial locations (i.e., geographic coordinates) and aspatial characteristics (not requiring locations to be meaningful). Therefore, we have choices for the analysis of ecological data:

- the spatial locations can be included explicitly to evaluate spatial structure and pattern;
- the spatial locations can be incorporated directly into the evaluation of aspatial characteristics; or
- the aspatial characteristics can be analysed separately by ignoring, or controlling for, their relative positions that can be defined either by neighbours or locations in some coordinate system.

Spatial analysis has a historical basis that combines methods from a wide range of disciplines including geography, statistics, ecology, economics, epidemiology and environmental sciences. This history is reflected in the broad range of both inputs and outputs accommodated in spatial analysis (Table 2.1). To be truly spatial, studies must do more than merely compare attributes at different locations by including spatial relationships among the locations like distances or other relational spatial characteristics, such as individual dispersal, migration among locations; disturbances, pests and disease spread among locations, in the analysis. The range of spatial methods can be classified in several ways, but it is difficult to find a classification that is 'simultaneously exclusive, exhaustive, imaginative, and satisfying' (Upton & Fingleton 1985).

We propose a classification of six categories (see Box 2.1):

(1) Spatial structure: description and evaluation;
(2) Spatial interpolation;

Table 2.1 Spatial analysis: Inputs and outputs

Inputs (Data types):
  Mapped points with locations but nothing else
  Mapped points with labels or quantities
  Mapped straight lines
  Mapped curving lines
  Points and lines: spatial graphs
  Points and directed edges: spatial networks
  Patches or polygons within a matrix: location and shape
  Patches or polygons with labels or quantities
  Patches forming tessellation
  Polygons forming tessellation
  Contiguous samples in grid or transect: presence or counts
  Contiguous samples in grid or transect: categories or values
  Spaced samples in grid or transect: values
  Fine grid of values: raster data
  Fine grid of two or more variables: raster data

Outputs (examples of results from analyses that one interprets)
  Single number
  Single number with assessment (p-value or index with confidence intervals)
  Series of numbers as a function of distance (K-function)
  Series of numbers as a function of distance with assessments (K-function with confidence intervals)
  Series of numbers as a function of area (K-function)
  Series of numbers as a function of area with assessments (K-function with confidence intervals)
  Map of numbers for locations (for samples, grids, raster, Getis' map)
  Map of numbers with assessments (Kriging)
  Map with new 'things': edges, patches, clusters (edge detection)
  Functional relationship equation with assessment
  Structural equation (causal) with parameter estimates and assessments

(3) Spatial partitioning and spatial clustering;
(4) Spatial regression and spatial simulation;
(5) Spatial interaction; and
(6) Spatio-temporal analysis and modelling.

Many spatial statistics are already available in all these categories and new methods are constantly being developed. This book cannot cover all, hence we selected the most common and important for ecological research. We acknowledge that we are omitting several important fields of research such as spatial issues related to information theory and spatio-temporal modelling.

Further, for some fields of research, we will cover only key aspects that are related to spatial analysis. For example, macroecology is a very active subdiscipline studying the relationships between organisms and the environment at large scales, focusing on abundance, distribution and diversity (Brown 1995). Although some aspects of macroecology, such as the spatial turnover of species, are clearly linked to topics in this book, others are less closely related, and so we will concentrate on the spatial aspects.

> **Box 2.1** Classification of the Spatial Analyses
>
> **Spatial Structure:** (a) This can refer to the degree of dependence in the values of a variable between neighbouring locations, usually as a function of distance, Euclidean or otherwise. Most such analyses are 'global' with values of a single statistic which summarizes the entire study area, they can be 'local' when subsets of the locations are used to calculate a value for each sampled location. (b) Spatial structure can also refer to the topology of the system under study, whether due to the physical relationships of the subunits that constitute the system or the connections that join them one to another.
>
> **Spatial Interpolation:** Using the known values from locations that have observations and the degree of spatial autocorrelation based on the distance between locations, the values of the variable can then be estimated for locations that do not have observations. Extrapolation refers to the situation where the unsampled locations of interest are beyond the range of the sampled locations with known values, whether outside the whole sample area or outside a convex hull of the samples; interpolation is where the unsampled locations are within the area covered by the locations with known values.
>
> **Spatial Partitioning and Spatial Clustering:** Creates spatial clusters using either clustering methods that group sampling locations together based on the degree of similarity of the variable(s) measured or boundary detection methods that separate sampling locations by identifying the high rates of change, which may represent possible boundaries between sampling locations.
>
> **Spatial Regression and Spatial Simulation:** Modelling that includes spatial dependency and spatial location in such a way that closest values have the greatest effect on the result for a specific location. Autoregressive models, spatial error models, geographically weighted regression, and so on, are used to evaluate the relationship of one set of variables to another.
>
> **Spatial Interaction:** Examines the flow of material or energy or information among locations and the factors that affect the flow such as distance, density, and resistance. This kind of analysis requires an underlying topology of the connections between locations and therefore leads to the requirements and possibilities of graph theory, the branch of mathematics that deals with structure in the abstract and in a spatial context.
>
> **Spatio-temporal Analysis and Modelling:** Include the spatial estimation of parameters and their temporal changes using spatio-temporal statistics, spatio-temporal networks and polygon change analysis.

Many texts describe spatial statistics, but no single book can provide everything that might be needed. Several advanced books cover the mathematics of the methods (Cressie & Wikle 2011; Dutilleul 2011) but the material may not be accessible to many ecologists and may not relate to specific ecological concerns. These techniques always have some potential for misapplication, leading to incorrect inferences.

Because software changes rapidly, we will not comment in detail on software packages for the techniques we describe. Instead, we suggest seeking technical guidance on how to use R as a well-established environment for analysis in ecological research (Borcard et al. 2018; Plant 2019). We will provide the code for a few selected methods, and more is available elsewhere, covering a range of related applications found in textbooks (e.g. Kolaczyk & Csárdi 2014; Fletcher & Fortin 2018; McElreath 2020; Oyana 2021).

We intend to present the concepts needed to perform valid spatial analyses and interpretations. We include various real and simulated data sets to illustrate the behaviour of the methods and the relationships among them. We concentrate on the spatial aspects of ecological data analysis to provide helpful examples and some guidance. The intended audience are graduate students and other practicing researchers, who have some familiarity with basic statistics and related approaches to ecological analysis but who are not themselves experts in spatial statistics. We proceed here with a series of key concepts and notions that are the foundations for the spatial statistics presented in later chapters.

2.1 The Spatial Context

Explicit considerations of spatial structure have come to play an important role in our efforts to understand and manage ecological processes and species persistence. Therefore, assessing ecological patterns, both spatial and temporal, is an important first step in our quest to comprehend the complexity of nature. Description of spatial patterns is not the final goal but rather the beginning of a process that leads to insight and which generates new hypotheses to be tested. Indeed, ecological research is an iterative process that provides, at each stage, insights into underlying ecological processes through the quantification of ecological patterns.

The match between pattern and process is far from perfect because changes in process intensity can create different patterns and several different processes can result in the same pattern signature (Figure 2.1). Hence, many processes may create a mosaic of intermingled and confounded patterns. Furthermore, processes shape the resulting spatial heterogeneity that then affects the intensity and types of ecological processes that act on them subsequently. This is known as the spatial legacy or memory of the landscape (Fortin et al. 2012b). The feedback effects between processes and patterns are, therefore, difficult to distinguish. Prior knowledge of the spatial and temporal domains of these processes can help to guide the choice of scale for investigating spatial patterns.

2.2 Ecological Data

Ecological data include many kinds of observations from qualitative records (e.g. species) and semi-quantitative values (e.g. pH), to quantitative measures, whether

Spatial Concepts and Notions

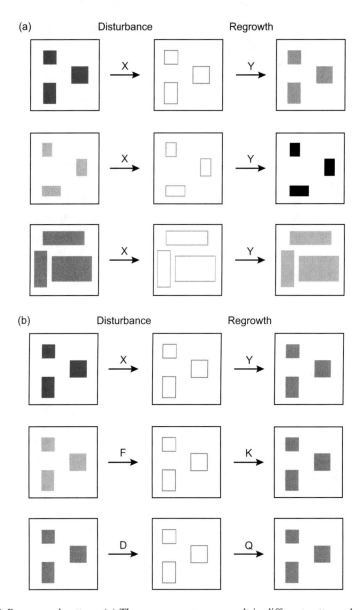

Figure 2.1 Process and pattern: (*a*) The same process can result in different patterns, based on legacy. (*b*) Different processes can give the same patterns, based on legacy (cf. Figure 1.2). (*c*) Different processes can give different patterns with no legacy effect.

discrete (e.g. counts) or continuous (e.g. weight). These observations can be for individuals (as point data for discrete organisms, Figure 2.2), along a line (transect data), over an area (surface data as within a sampling unit) or in a volume (as for phytoplankton productivity). When sampling units are used, these can either be spatially adjacent and contiguous or separated by constant or variable distances (Figure 2.2). In either case, the measurements raise precision and accuracy concerns.

2.2 Ecological Data 25

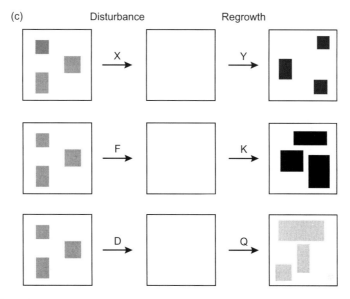

Figure 2.1 (cont.)

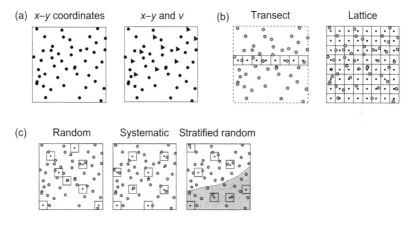

Figure 2.2 Spatial sampling strategies to collect ecological data: (*a*) point data methods: exhaustive survey of the geographic *x–y* coordinates of all the individuals of a species (left panel) or more species (right panel; here two species (dot and triangle), where *v* indicates the attribute of each individual – e.g. the species' name); (*b*) contiguous sampling units: transect and lattice; and (*c*) sparse sampling units: random, systematic and stratified random. See text for more details.

The quality of the observations is a function of (1) for quantitative measurements, the precision and accuracy of the instrument or of an observer to count species abundance or to estimate percent cover with the same accuracy over time; (2) for qualitative data, the ability of the observer to identify species correctly; (3) for positional data of either the individuals or sampling units, the precision and accuracy of the instrument used (GPS, telemetry, laser, tape measure, etc.); (4) the precision in data gathering and transfer to digital form (accuracy of transcription); and (5) the appropriate match between the sampling unit size and the variable measured. All these accuracy levels and potential types of errors will affect the identification and quantification of spatial patterns. All these accuracy concerns cannot be eliminated but they can be minimized or acknowledged in analysing and interpreting spatial structure.

2.3 Spatial Structure: Spatial Dependence and Spatial Autocorrelation

Most ecological data have some degree of spatial structure and some of that structure may follow what is known as the 'first law of geography': 'Everything is related to everything else, but near things are more related than distant things' (Tobler 1970). In ecological data, the basic structure of similarity that typically declines with distance may be complicated by patchiness in the system. The spatial structure of ecological phenomena can take several different forms: a directional *trend* or gradient (Figure 2.3a); a non-random arrangement of values that is *aggregated*, either irregularly patchy or cyclic (Figure 2.3b); and an arrangement of values that is *random*, implying that adjacent values are independent (Figure 2.3c). Either exogenous or endogenous processes can generate any of these patterns. Several factors can act together, either additively or multiplicatively, or otherwise when the factors are non-linear, such as a threshold response to habitat fragmentation. Hence numerous spatial patterns can be identified when the variables of interest (say, species abundances) respond to an exogenous process (such as disturbance) or underlying environmental conditions (spatial heterogeneity). For example, soil patchiness can produce regions of high plant density within which individuals are either random or over-dispersed; here, any local similarity is due to the species responding to external processes, which have their own spatial structure. When endogenous processes, like dispersal, are dominant, the observed pattern is an inherent property of the variable of interest.

Spatial dependence is a lack of independence among the values of a variable at particular locations. It includes both a species' response to exogenous processes and the spatial autocorrelation due to its own endogenous processes (Wagner & Fortin 2005; see Cressie 1993; Haining 2003). Where exogenous processes predominate, we say that spatial dependence is 'induced' by the underlying variable that is itself spatially autocorrelated. This phenomenon is 'induced spatial dependence'. Spatial patterns usually result from a mixture of both exogenous (induced) and endogenous (inherent) processes. While autocorrelation is a correlation among values of a single variable; 'spatial autocorrelation' means that the correlation is a function of locations in space or the distances between them. For endogenous processes, it can be referred to as 'inherent' spatial autocorrelation. In some examples, high similarity at small scales declines with distance, and the phenomenon is described as 'distance decay' because of the kind of function that describes it.

Spatial dependence is usually assessed by comparing values separated by a set of given distances (spatial lag), say at 1, 2 or 3 metres. Consider wind-borne seed dispersal from a single tree: spatial autocorrelation is due to dispersal, as seed

2.3 Spatial Structure: Dependence, Autocorrelation

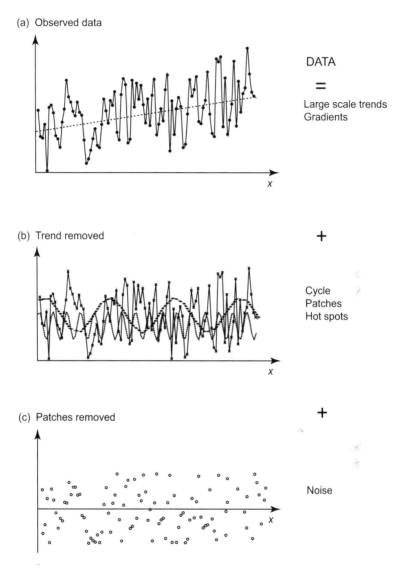

Figure 2.3 Nested spatial patterns (signals) imbedded in ecological data: (*a*) if the data are gathered along a temperature gradient, tree height can increase linearly with scale; (*b*) both topography and spatial dispersal processes can generate patchy patterns at intermediate, landscape, scale; and (*c*) there is only random noise at the micro, local, scale.

abundance decreases with distance, the degree of spatial autocorrelation also declines. At short distances from the tree, values of seed abundance should be similar at nearby locations, giving positive autocorrelation; as the distance increases, the values are less likely to be similar. They may become apparently independent, with no spatial autocorrelation, or even dissimilar, giving negative autocorrelation. Over large areas, this may produce a repeating patchy pattern with two spatial scales: within-patch and between patches in the landscape.

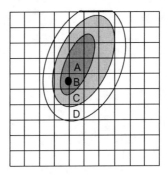

Figure 2.4 Seed abundance from a tree source. The filled circle indicates the location of the tree source from which seeds are dispersed by wind. As the distance from the tree increases, the number of seeds decreases (as indicated by the grey-shaded gradient: dark grey for high abundance; light grey for low abundance; white for no seeds). Positive spatial autocorrelation exists between adjacent sampling units A and B; no significant spatial autocorrelation exists between A and C; and negative spatial autocorrelation exists between A and D.

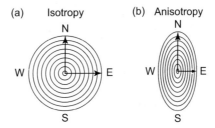

Figure 2.5 Pattern directionality. (*a*) Isotropic and (*b*) anisotropic spatial patterns. Each isoline indicates the same value of the variable decreasing from the highest value at the centre to the lowest value at the periphery.

The degree and shape of spatial autocorrelation can also vary with direction. In the seed dispersal example, prevailing winds may produce an elongated patch of seeds downwind (Figure 2.4). This pattern is *anisotropic* (magnitude and range vary with direction); the opposite is *isotropic* (similar in all directions) (Figure 2.5). Various internal and external processes create anisotropic patterns: topography, gradients, streams and so on. A favourite example is the 'brousse tigrée' (tiger bush) of striped scrubland that develops on gentle slopes in some arid regions (Wu & Qi 2000) but string bogs (Rietkerk *et al.* 2004), and wave-regenerated forests (Sprugel 1976) are equally well-known examples.

2.4 Spatial Scales

The presence of spatial patterns in ecological systems reflects the degree of predictability for characteristics, based on the spatial location. Ecological processes create the spatial pattern we observe, but spatial pattern affects processes including those that produce the pattern. When the scale at which the processes are realized is unknown, analysing the spatial pattern can help our understanding of ecological complexity.

These patterns range from obvious windfall gaps in a forest to the diffuse heterogeneity of species patchiness in a prairie. Spatial homogeneity is a convenient null model but is very rare in reality. Spatial heterogeneity depends on the spatial scale used to analyse the data: an area may appear to be homogeneous at a large extent and coarse resolution, but at finer resolution, heterogeneity emerges.

The term 'scale' covers several concepts, including the physical extent of the processes (the range) and the resolution of the data (the grain) (Figure 2.6). Our perception of spatial structure is directly related to both the study area's extent and sampling unit size (Wiens 1989). Ecological data often include patterns at several spatial scales which are confounded (Figure 2.3): (1) trends at larger scales, (2) patchiness at intermediate and local scales and (3) small-scale random fluctuations or noise. We try to disentangle the spatial scales of these processes using various analytical techniques. Our ability to identify spatial patterns and their underlying processes is affected by three factors (Figure 2.7; Dungan et al. 2002): (1) the extent of expression of the processes; (2) the design for data collection (sample versus population data); and (3) the statistical tools applied to the entire sampling area (global statistics) or each sampling location (local statistics). The spatial statistics presented in this book characterise the spatial pattern of the quantitative data at the plot or landscape levels.

Several processes can combine to produce the spatial pattern observed (Figures 2.1 and 2.3); if the combination is additive, the pattern is the sum of those from the individual processes acting independently. In Figure 2.3, a trend is augmented by

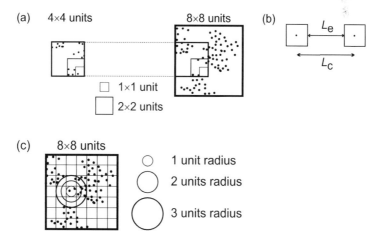

Figure 2.6 Sampling design: (*a*) changes in spatial extent and sampling unit size; (*b*) spatial lag (L_e indicates the distance 'edge-to-edge' between the sampling units, L_c indicates the distance 'centroid-to-centroid' between the sampling units); and (*c*) a multiscalar analysis using three radii sizes where both the extent and sampling unit are kept at the same scale. See text for more details.

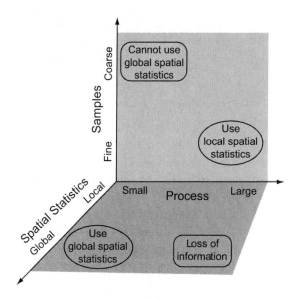

Figure 2.7 Three main components that interact and affect our ability to identify and characterize spatial patterns accurately: the scale of expression of processes, the sampling design being used at the plot level or landscape level and the spatial statistics characterizing either the spatial structure of each sampling location (local spatial statistics) or the entire study area (global spatial statistics).

patchiness caused by limited species dispersal and by a small-scale random component. Additive spatial patterns can be analysed by removing each contribution identified and then re-examining the residual. A linear trend can be removed by linear detrending and the residuals can be subjected to a second detrending with another type of structure (quadratic, cubic, etc.). This sounds good but, in practice, it is uncertain that only the targeted trend is removed: in the absence of prior knowledge or a detailed hypothesis, one should not detrend the data.

Nowadays, studies focusing on climate change and global change are carried out either at regional or global scales. Over such large regions, the resulting spatial patterns may not be non-additive spatial patterns as the processes that generated them are non-stationary processes. Hence, patterns can combine multiplicatively, most easily in binary data like presence–absence records (see Dale 1999, figure 3.5), when several factors, each of which can cause a species to be absent, act independently and at different scales. With regional differences, the various processes may result in multiscale patterns that also differ by region. To analyse these patterns properly, we first partition the data into homogeneous areas as best we can (see Chapter 5) and then proceed to analyse each subarea.

Lastly, in studies carried out over such large extents, it is rare to collect quantitative data as the required sampling effort would be too intense. Instead, the data used in large study extents are usually categorical (e.g. landcover-type data). To characterise the spatial pattern of categorical data, landscape metrics can be used (Frazier &

Kedron 2017). We will not present these metrics and refer the readers to papers and textbooks for details (e.g. Kupfer 2012; Turner & Gardner 2015; With 2019; Urban 2023).

2.5 Sampling Design

Any sampling design imposes a structure by the essence of its temporal and spatial units. Hence, to be efficient, a sampling design needs to be crafted carefully: (1) define explicitly the spatial and temporal domains of expression of the process(es) under study; (2) determine that the spatial and temporal resolution of the sampling design can capture the process; and (3) ensure that the spatial and statistical analyses are appropriate for the data (Dungan *et al.* 2002). Most ecological studies are based on a design-based sampling strategy (e.g. random, systematic, stratified; Williams & Brown 2019). A good sampling scheme balances the tension between samples that are too close together, giving less information because of autocorrelation, and samples that are too sparse, allowing processes at other scales to introduce variability. To understand how sampling design affects different statistics, we will review aspects common to all: the sample size, the sizes of the study area and the sampling unit, the shape and spacing of sampling units, and the overall sampling design.

2.5.1 The Sample Size (the Number of Observations 'n')

The sample size, n, is one of the most important decisions in any study. In spatial analysis, the choice is based on the minimum requirement for subsequent analysis (Fortin *et al.* 1989; Fortin 1999*a*; Table 2.2). To detect significant autocorrelation, at least 30 locations are required (Legendre & Fortin 1989) and the reliable estimation of spatial structure and its parameters in variography (Chapter 4) may require 50–100 sampling locations.

2.5.2 Spatial Resolution: Grain and Extent

The detection of spatial patterns is affected by the observations' spatial scale (Figure 2.3), which has at least two aspects: the size of the study area, or 'extent', and the size of the sampling unit or 'sampling grain' (Figure 2.6*a*). The extent is the entire area and should match the domain of the ecological process under study and the grain is the smallest spatial resolution of a measurement. Spatial statistics are sensitive to both (Fortin 1999*a*; Dungan *et al.* 2002). As a guideline, O'Neill *et al.* (1996) suggested a study area two to five times larger than the extent of the largest process being studied. If the study area is smaller, not enough of the pattern is included to characterize it. If the extent is too large, different processes or different intensities may affect subregions, creating greater variability. This is especially true for remotely sensed data because the study area is unlikely to correspond with a naturally defined

Table 2.2 Requirements, assumptions, and rules of thumb for spatial statistics

Methods	Objectives	Assumption (Requirements)	Rule of thumb or limits
Network	Relationships of locations	Exhaustive mapping of all events	Planar graphs limit the number of edges in a network
Aggregation indices	Testing for spatial structure	Stationarity	Cannot differentiate among spatial structures
Block variance methods	Computing the intensity and range of spatial structure	Contiguous sampling units	Computed up to $1/10$ of the length of the transect
Ripley's K	Computing and testing the intensity and range of spatial structure	Exhaustive mapping of all events	Extents having rectangular shape are favoured for computation; Edge effect needs correction; Statistic computed up to $1/3$ to $1/2$ of the shortest extent
Moran's I, Geary's c	Description of spatial structure; testing for the presence of spatial autocorrelation	Stationarity	Edge effect needs correction; Statistic computed up to $1/3$ to $1/2$ of the shortest extent; Minimum of 20–30 sampling locations
Semi-variance	Description of the spatial structure; estimation of spatial parameters	Pseudo-stationarity [assume stationarity in the search region]	Edge effect needs correction; Statistic computed up to $1/3$ to $1/2$ of the shortest extent; Minimum of 50 sampling locations; Search neighbourhood of either the range distance or 12–25 neighbouring sampling locations
Mantel and partial Mantel tests	Correlation between spatially autocorrelated variables	Stationarity	Overall, synthetic, value of the linear relationship between distance matrices, not the raw data
Wavelets	Description of the spatial structure at multiple scales; Detection of boundaries	Contiguous sampling units and no missing data; Stationarity not required	Multiscale analysis uses powers of 2 (2^k); the number of scales k depends on the number of observations

homogeneous area. The smaller the extent, the more likely it is that patches will be truncated by the study area limits.

2.5.3 The Size and Shape of the Sampling Units

The size and shape of the sampling units affect the accuracy of the detection of the spatial pattern; their combined effects or the inappropriate size and shape may hamper pattern detection. The size of the sampling unit (grain) sets the smallest spatial resolution (Figure 2.6a). When studying a landscape using remotely sensed data, the grain is the pixel resolution, which most likely will not match the ecological process of interest. O'Neill et al. (1996) recommended sampling units, the grain resolution, to be two to five times smaller than the features of interest. Hence, the sampling unit should be large enough to contain more than one individual, but not so big as to increase within-unit variability or to render the smallest spatial scale undetectable (Figure 2.6a). In general, a smaller grain should be favoured because small sampling units can be aggregated into larger ones without losing information, while the reverse is not true.

Choosing a sampling unit that is more-or-less isotropic like a square or hexagon has the implicit assumption that the pattern is also isotropic. Ecologists sometimes use rectangular sampling units to reduce within-sampling unit variability along a gradient, but such anisotropic units can artificially generate the appearance of an anisotropic spatial pattern (Fortin 1999a). When it is not known whether the pattern is isotropic or not, we recommend small isotropic sampling units so that the spatial pattern can be characterized more accurately.

2.5.4 Sampling Design

Once the extent and grain are determined, some choices remain (see Figure 2.6): should the units be contiguous or spaced and if spaced at which distance? Contiguous units in a transect or lattice allow a finer description of the spatial pattern because no information is missing due to gaps in the sample. Furthermore, contiguous units sampled over the entire extent are exhaustively sampled where the data represent the entire population of units. This does not guarantee that the data are representative of the entire ecological process. Then, spaced units (e.g. design-based sampling strategies: random, systematic, or stratified; Figure 2.2) leave gaps in the array, meaning information is missing. Such spaced sampled data could be spatially interpolated (Chapter 4) to estimate the values at unsampled locations.

2.5.5 The Location in the Landscape

The location of the study area in the landscape can affect or bias the results of spatial analysis (Plante et al. 2004). If the process generates repetitive patterns (such as alternating patches and gaps), the study area's positioning relative to peaks and troughs will affect the assessment of the spatial structure.

2.5.6 Spatial Lag

The spatial lag between the sampling units is directly related to the previous decisions about the sample size (n), the extent and the grain, and the shape of the sampling unit. As n increases, the spatial lag decreases; as the extent increases, the spatial lag increases. The considerations and choices for the spatial lag among sampling units include the following:

(1) Contiguous sampling units can have no edge-to-edge lag but only centroid-to-centroid lag equivalent to the unit's size.

Spatial Concepts and Notions

(2) For spaced (non-contiguous) sampling units the spatial lag can be measured either from edge to edge (L_e) or centroid to centroid (L_c) of the units (Figure 2.6b).

The choice of the spatial lag should also be guided by the goals of the study:

(1) To detect, characterize, quantify and test the significance of a spatial pattern for gaining insight into ecological processes, the spatial lag among sampling units should be smaller than the size of the patch (Figure 2.8a), with several samples within each patch.
(2) To establish the relationship between two kinds of ecological data, the presence of autocorrelation will impair any inferential tests that require data independence (Legendre & Legendre 2012).

The design-based sampling strategy will also affect the ability to detect spatial patterns or not:

(1) Random sampling ensures that each sampling unit is drawn independently and will be representative of the population (Fortin et al. 1989) but it cannot guarantee that the data are not spatially autocorrelated (Fortin & Dale 2009). In the presence of spatial autocorrelation, it is almost impossible to obtain truly independent data but spacing the units at distances larger than the patch (Figure 2.8b) could minimize the spatial dependency of the data.
(2) While systematic sampling is easy to implement in the field, the choice of spatial lag is crucial in the detection of spatial patterns (Fortin et al. 1989). For example, the spatial lag could match the pattern in the data, as depicted in Figure 2.9 where, with a lag of 5 m (locations 'a'), the pattern of the data will appear to be uniform;

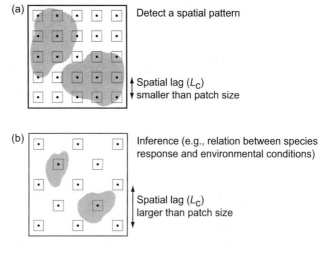

Figure 2.8 Spatial sampling design. (a) Spatial lag must be set according to the objective of the study: to detect a spatial pattern, the spatial lag should be smaller than the patch sizes (grey polygons); (b) to perform inference statistics, the spatial lag should exceed the patch size (grey polygons).

2.5 Sampling Design

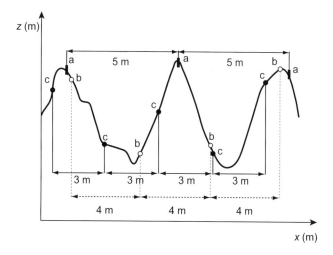

Figure 2.9 Importance of spatial lag while using a systematic sampling design: when the spatial lag is 5 m ('a' sampling locations) the spatial structure identified is uniform, flat; when the spatial lag is 4 m ('b' sampling locations) high and low periodicity is detected; and when the spatial lag is 3 m ('c' sampling locations), patchiness can be characterized.

with a lag of 4 m (locations 'b'), the periodicity of high and low values is detected; and with a lag is 3 m (locations 'c'), a patchiness of the data can be detected.

(3) When the scale of the process is unknown, sampling with several lag distances is recommended, increasing the chances of identifying the pattern (Fortin *et al.* 1989; Webster & Oliver 2007). For spaced samples, we recommend a pilot round of sampling to inform a second round, sometimes called 'adaptive' or 'progressive' sampling. One version of this procedure is to locate additional samples where local estimates have high variance, as in Kriging methods for interpolation (Chapter 4).

2.5.7 Multiscalar Analysis

Species' responses to environmental variables vary according to several ecological processes and spatial heterogeneity. For songbirds, the most important processes vary according to distance to the nest: at the microscale they are resource use and movement, at the mesoscale they are species interactions and dispersal, and at macroscale it is migration (Desrochers *et al.* 2010). Therefore, the key question that ecologists are asking is: what is the optimal scale of study species? To address this question, researchers have used a multiscalar approach to optimize the relationship between species' relative abundance and the number of land cover types estimated at various search windows of increasing radii while keeping the extent and the grain at the same scale (Figure 2.6c; Jackson & Fahrig 2015; Holland & Yang 2016; McGarigal *et al.* 2016).

2.5.8 Effects of Edges

The interaction between the extent and the sampling unit size generates 'edge effects' which can bias the estimation of spatial patterns at small and large distances (Figure 2.10; Cressie 1993). In particular, a sampling unit or a sampling point at the corner of a sample plot can have neighbours in only two directions, not four. Several of the spatial statistics presented in Chapters 3 and 4 that evaluate patterns at increasing distances between sampling locations are affected by edge effects. In Figure 2.8, sampling units along the border of the study area (in grey) have fewer neighbours at short distances than at intermediate ones: we have 200 pairs at a 1.5 unit separation, almost 600 pairs at 4.5 units and very few at large distances (2 at 10.5 units). The appropriate edge effect correction procedure should be selected according to the data type and the spatial statistics used (Cressie & Wikle 2011). When the resources are not available for a buffer sample, or when the surroundings of the study area are not homogeneous, the calculation of spatial statistics should be limited to the centre of the extent, using the samples around the border only as neighbours. Another technique is the computation of 'torus distances', created by 'joining' the opposite

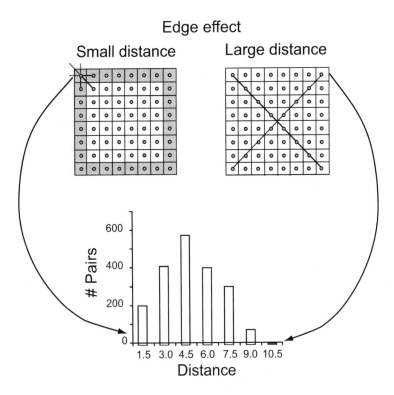

Figure 2.10 Edge effect affecting the sampling units along the border of the study area (the grey sampling units): at small distance (left panel; 1.5 units apart) and large distance (right panel; say 10.5 units apart), where the number of pairs of sampling locations used to estimate the spatial pattern is lower than for intermediate distances (say 4.5 units apart).

borders of a rectangular plot, north to south and west to east, creating a doughnut shape, so that the samples at the northern and southern borders are near neighbours as are those at the east and west borders. The distribution of pairs of sampling locations is then more uniform, minimizing the edge effects. The torus transformation should only be used when the process can be assumed to be stationary, a concept described in Chapter 1, but which we discuss here in greater detail.

2.6 Stationarity

Stationary describes a process in which the properties are independent of the location and direction in space, so that the parameters of the process, such as mean and variance, are constant throughout the study area. Stationarity is a property of the process or a model, not the data, and spatial statistics should be calculated over areas for which stationarity can be assumed. In heterogeneous processes with changing mean and variance, the assumption of stationarity is violated (Figure 2.11) and the spatial pattern detected by most spatial statistics may be inaccurate, making spatial inferences invalid (Legendre 1993; Dale & Fortin 2009). Because stationarity is a property of the process (not the data), it cannot be tested directly, but we can assess homogeneity by computing the mean and the variance for windows of varying sizes.

As mentioned in Chapter 1, depending on the scale and the characteristics of a model's realization, a stationary process can give rise to an apparently non-stationary pattern. Figure 1.1 shows a randomly generated pattern, which appears to exhibit non-stationarity. The five clumps of events are all in one half of the plane, although the probability of a clump occurring is uniform throughout. The probability of this kind of pattern is on the order of 1/10; so, it is rare but not very surprising.

A set of data from field sampling is much like a single realization of an underlying model and apparent inhomogeneity may result from processes that are, in fact, stationary. To understand the power for rejecting the null hypothesis of stationarity, we can determine in advance the strength of the spatial pattern that would be required to reject the null hypothesis, given the sample size or effort expended. Combined with a pilot study to determine some of the characteristics of the spatial structure, this prior

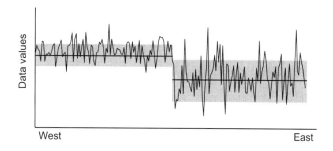

Figure 2.11 The concept of stationarity: a process level issue. From the west to the east, the mean (solid lines) and variance (grey bands) change.

knowledge of the sample size needed will provide guidance for the study design and analysis.

Spatial statistics that summarize the spatial pattern for the entire study area with a single number, 'global' spatial statistics, should not be applied to a study area with non-stationary processes. While it is impossible to test directly whether an underlying process is stationary, it is possible to check whether model residuals appear independent and identically distributed throughout. A heterogeneous covariance structure of the residuals may indicate that the stationarity assumption is violated. In that case, we may subdivide the data into spatially homogeneous subregions so that each subregion can be analysed separately (see Chapter 5). Another solution is to use 'local' spatial statistics to estimate the characteristics of the pattern for each sampling location (Anselin 1995; see also Chapter 4). In this book, global spatial statistics will be referred to loosely as 'spatial statistics'; local statistics will be designated explicitly.

To sum up: stationarity is the condition that the defining characteristics of the processes underlying spatial patterns are constant across the region of interest, including that the species–environment relationship does not change with location or scale. Most modelling methods to account for residual spatial structure assume stationarity as the basis for subsequent inference. Of course, the situation and our analysis and interpretation become more 'interesting' when we cannot assume stationarity or are forced to acknowledge that it may not be obtained.

2.7 Scaling

It is a common phenomenon that pattern structure changes with scale (consider Figure 1.1*b*: in the whole study area, the points appear to be in clusters, but in a smaller study plot that included only one cluster, the points appear to be over-dispersed). *Scaling* describes how the properties of a process change with the scale at which it is studied, for both scale as grain and scale as extent. The same process studied at different scales can produce different results, so studies of scaling in ecological systems attempt to evaluate quantitatively how these properties change (Wiens 1989). Consider questions like these:

- How does scaling affect species interaction networks? (Galiana *et al.* 2018);
- How do spatial, temporal and phylogenetic scales (grain and extent) affect the mechanisms of community assembly of microbial communities? (Ladau & Eloe-Fadrosh 2019);
- How does the scale of ecological processes affect biodiversity patterns, ecosystem dynamics and their interaction? (Zelnik *et al.* 2024); or,
- How does scaling affect aspects of stability (resilience, resistance and invariance) in ecological systems? (Clark 2020*a*).

The structure of spatial non-stationarity, as just discussed, will obviously affect how changes in spatial scale change the assessment of a process; Newman *et al.* (2019) listed non-stationarity as one of the three intrinsic limitations on progress in landscape ecology. If the grain is held constant, but the extent increases, spatial heterogeneity tends to increase as more habitats and, thus, more (different) species are included in the sample.

Another effect is observed when the grain is increased, or small sample units are aggregated into larger units. Newman *et al.* (2019) referred to the problem of 'coarse graining' as first in their list of limitations for spatial ecology: how to aggregate fine-scale information without introducing bias in the statistics. A similar phenomenon is the 'modifiable areal unit problem', where aggregated data have different properties than the data they are based on (Openshaw 1984). (This is sometimes referred to as MAUP and, yes, the temporal version is the modifiable temporal unit problem, MTUP.) A related problem has the unfortunate name of 'the ecological fallacy': the aggregated characteristics of a group are assumed to be true at the level of the individual unit. The word 'ecological' in the name comes from epidemiology referring to studies at the population and environmental level, rather than by individual.

The third limitation in Newman's list is the 'middle number problem' with systems that have elements that are too few and varied to be averaged, but too many and varied to all be treated individually. This is a problem first, or most obviously, encountered in physics; but, in ecology, it is typified in circumstances that have many variables that are non-uniform in space and time and with relative importance that change with scale (see Newman *et al.* 2019 for details and examples).

Several other challenges related to scaling in ecology have been identified. Viana and Chase (2019) examined the scaling of metacommunity core assembly processes of dispersal, drift and niche selection. These were found to be dependent on scale, with niche selection increasing in importance with increasing extent, but decreasing with increasing grain. Looking at ecological interactions and spatial scale, Galiana *et al.* (2022) found that network complexity scales with an area, and not just due to increasing diversity (species richness) but also due to the number of links and the structure of the network. The implication is that habitat loss will lead to the simplification of the communities and their metanetworks. Blackman *et al.* (2022) noted that biodiversity patterns and food web dynamics are not directly scalable to each other, even using the same spatial and temporal scales, so they must be jointly studied to be properly compared. Most of the challenges mentioned so far are evident in natural systems and natural landscapes, considering the scales of the process itself, the sampling design, and the statistics applied (Dungan *et al.* 2002), but scaling in fragmented landscapes has the additional challenge of multiple scales in a single landscape. Fletcher *et al.* (2023) have considered this problem and recommend sampling at the finest grain that can capture the underlying mechanisms and thus the phenomena that operate at multiple spatial extents, in order to understand the interrelationships among scales in a fragmented landscape. Related to the scaling challenges of landscape fragmentation is the fact that conservation efforts, conservation strategies, and conservation success all vary according to the scaling of the various components and processes of the system of focus (see Fletcher & Fortin 2018, section 2.2.2).

Many more studies could be cited, and in much more detail, but this overview should suffice to highlight the importance of understanding the effects of scaling on studies of spatial processes, on choosing methods for spatial analysis and on interpreting the results. To explicitly consider the challenges related to scale will limit the potential of spatial misalignment (Fletcher *et al.* 2023) between data and patterns, especially when studying processes that act at different scales throughout a landscape. Indeed, nowadays, we sometimes tend to think of landscapes as the scale of greatest interest, but small-scale local processes and interactions at short distances can be crucial, and so it is often an important first step to appropriately define the determination of spatial neighbours (that is next!).

2.8 Spatial Neighbours

2.8.1 Lattice-based Neighbours

There may be cases in which the sampling units are contiguous squares or form a lattice with regular spacing. In those circumstances, the tessellation links between sampling units can be described as chess piece moves: the rook makes links in four cardinal directions (i.e. four neighbours), the bishop makes only diagonal links (i.e. four neighbours), and the queen makes links in all eight directions (i.e. eight neighbours). By using topological space to establish the spatial relationship among sampling units, we are assuming implicitly that Euclidean distances among sampling units are less important, and that we are interested only in their relative positions. When the absolute distances among sampling units or their positions in some frame of reference (e.g. the study area limits) are important, then Euclidean space should be used.

2.8.2 Topological Neighbours

Several rules or algorithms are available to define which points are neighbours to be joined in pairs in a two-dimensional plane. Each algorithm results in a mathematical object, a *graph*, consisting of a set of points, the *nodes*, joined in pairs by lines, the *edges*; this is a spatial graph because the nodes have spatial locations. The simplest algorithm determines each event's closest neighbour and joins these pairs to give the spatial network of nearest neighbours (NN). The structure usually has disconnected components, as shown in Figure 2.12. The nearest neighbour definition can be narrowed to include only those pairs that are mutually nearest neighbours (MNN), producing a network with fewer edges (the heavy lines in Figure 2.12) in which some events have no neighbour. The network of mutually nearest neighbour pairs is a *subgraph* of the graph of all first nearest neighbours because all nodes and edges in MNN are also in NN.

The first nearest neighbour network can be elaborated by including first and second nearest neighbours, or first, second and third, and so on. With a different approach, we can describe a hierarchy of neighbour networks that begins with mutually nearest neighbours and ends with the Delaunay triangulation. The second network in that hierarchy is NN and the third is the Minimum Spanning Tree (MST; Figure 2.13). A spanning tree is a graph with no cycles (polygons formed by edges) that includes all

2.8 Spatial Neighbours

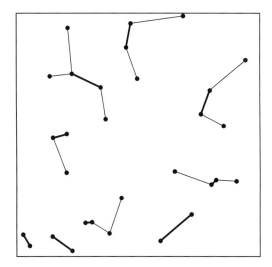

Figure 2.12 Network of first nearest neighbours for an artificial example of point pattern (NN). Mutual nearest neighbour pairs have bold lines and form a subgraph (MNN) of the example.

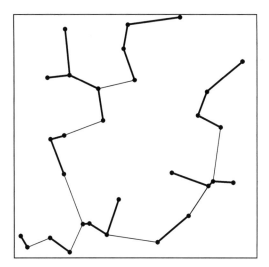

Figure 2.13 Minimum Spanning Tree for the same pattern as shown in Figure 2.12. The bold lines are those in the nearest neighbour graph.

nodes; the Minimum Spanning Tree is the one with the smallest total length of edges. The Minimum Spanning Tree contains all nodes and edges of the nearest neighbour network. To visualize the relationship between the two, consider connecting the disconnected components of the nearest neighbour network, using the shortest edges available; this produces the MST. The fourth network is the relative neighbourhood graph (RNG; Figure 2.14; Toussaint 1980), formed by joining all pairs of nodes,

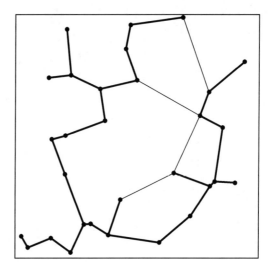

Figure 2.14 The relative neighbourhood graph for the same pattern as shown in Figure 2.12.

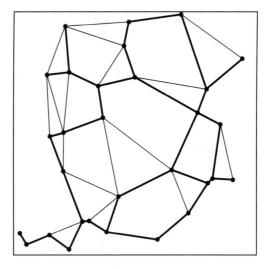

Figure 2.15 Network of a Gabriel graph for the same pattern as shown in Figure 2.12. The bold lines are those edges of the Relative Neighbourhood Graph.

A and B, for which the lens formed by the radii of the two circles AB, centred on A and B, contains no other node. The next in the hierarchy is the Gabriel graph (GG; Figure 2.15; Gabriel & Sokal 1969). It is formed by joining all pairs of nodes, A and B, for which the circle on diameter AB is empty.

The last network in this hierarchy is the Delaunay triangulation (DT; Figure 2.16). It is formed by joining, with triangles of edges, all triplets of nodes A, B and C, for which the circumcircle of the triangle ABC contains no other node. It can be shown that

2.8 Spatial Neighbours

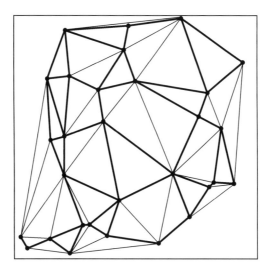

Figure 2.16 Network of a Delaunay triangulation (DT) for the same pattern as shown in Figure 2.12. The bold lines are those edges of the Gabriel graph.

the average number of neighbours in the Delaunay network approaches six, no matter the spatial arrangement of the events (Upton & Fingleton 1985). One disadvantage of the DT network for ecological applications is that it often has very long edges around the periphery; these are sometimes removed based on an upper distance threshold, producing a reduced Delaunay graph. The Delaunay triangulation is related to familiar tessellations of polygons known as Dirichlet domains, Thiessen polygons or Voronoi polygons (Figure 2.17). Any pair of events that have an edge in the triangulation also have a common boundary to their polygons in the tessellation. The ecological interpretation comes from the fact that each polygon contains all parts of the plane closer to its own event than to any other. Where the event is an organism, its polygon determines the resources it can pre-empt and survival may depend on the polygon area.

Delaunay tessellations can be computed using the function *delaunay* of the R package *spatstat* as well as the function *tri2nb* and the R package *spdep* while Voronoi polygons with the function *dirichlet* of the R package *spatstat* as well as the function *voronoi.mosaic* of the R package *tripack*.

This hierarchy of networks is unified by the fact that each is a subgraph of the next; going up, edges are only added. In ecological terms, this means that more and more events are neighbours to be considered. This hierarchy of neighbour networks can be the basis for multivariate point pattern analysis and for the analysis of marked point patterns (Chapter 3), counting the number of like–like joins (multivariate classification) or calculating correlation coefficients (quantitative marks) in the hierarchy of networks. In most applications, a 're-labelling' randomization procedure will be an effective technique to evaluate the significance of any result.

The neighbour networks can transfer analysis methods from irregular to lattice arrangements of observations. For example, multivariate cluster analysis creates

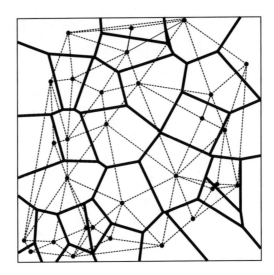

Figure 2.17 The Delaunay network (fine dashed lines) as Dirichlet domains, Thiessen polygons or Voronoi polygons (bold lines).

groupings with greater similarity within groups than among. In 'spatially constrained clustering', the grouping of similar objects is conditional on being adjacent in space, with adjacency usually defined on a lattice, but it can be defined by the neighbour networks for irregular arrangements. Spatially constrained clustering has many applications in ecology, and it is closely related to determining spatial boundaries, as in delimitating landscape habitat patches (Chapter 5).

In general, it may be helpful to apply the same analysis to the data using all of the six networks of the hierarchy because the differences among them can provide valuable insights. In some instances, however, applying only a subset of the neighbour networks may be sufficient; the MNN network may be too sparse for some analyses and the GG or DT too dense, with a lack of independence among comparisons (all those triangles!). Dale (2017) has provided an interesting example with MST and the GG assessing the relationships of diversity, species composition and nestedness for tenebrionid beetles on islands in the Aegean. The focus was on the autocorrelation of community characteristics between neighbouring islands (details in Dale 2017).

Many other neighbour networks are possible and available for applications in ecological research. Other rules for neighbour networks remain to be explored, but the general principles for application will resemble those in this section. As an alternative to the 'topological' neighbour rules, we can consider distance threshold rules; and that comes next.

2.8.3 Distance-based Spatial Neighbours

A different hierarchy of neighbour networks is created by a simple series of threshold distances, joining all pairs of nodes closer than a given threshold. Increasing the

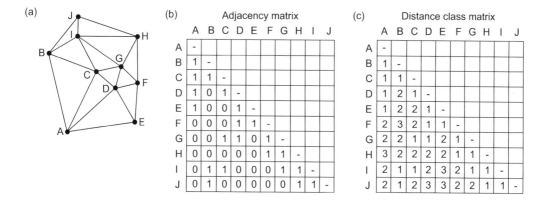

Figure 2.18 (a) Adjacency based on the topology among 10 sampling locations (A–J). The x–y coordinates of each sampled location (here represented by dots) are used to determine a Delaunay tessellation network linking all the sampling locations. From this series of links, we get a binary adjacency matrix (b) of the first nearest neighbours (1, connected; 0, not connected). Distance class matrix (c) can then assess the k nearest neighbours (1, first nearest neighbour; 2, second nearest neighbour; and so on).

distance produces a network that includes all the edges derived from any smaller threshold. Many spatial statistics are based on pairs of units 'binned' into distance classes (Chapter 4), implicitly based on such a hierarchy. Distance-based algorithms are the same for sampling units as for points, with sampling units reduced to centroids (Figure 2.18). Any of the neighbour networks have an associated *adjacency matrix*, which indicates all pairs of nodes joined by an edge: the first-order neighbours have an entry 1 and higher-order neighbours not joined by an edge have a 0 in the adjacency matrix, even if they are connected by a path. The *distance class matrix* records the minimum number of steps (path length) between connected nodes; the values $d = 1, 2, \ldots, m$ and where m is the longest of those shortest paths (the graph diameter). For Figure 2.18, (A, B) = 1 but (A, H) = 3, which is the maximum.

When the units are contiguous squares, the links can be designated as chess moves: the rook (castle) in four cardinal directions, the bishop only diagonally, and the queen in all eight directions. Using topological space suggests implicitly that Euclidean distances are less important than relative positions. If the absolute distances among sampling are important, then Euclidean space should be used.

In some cases, restricting the range of distances may be helpful: a two-threshold network has edges only for pairs of nodes that fall between two chosen values, D_{min} to D_{max}. This allows the evaluation of more narrowly defined questions about the spatial structure. We might be interested in habitat patches in a landscape that are nearby but not right next to focal patches of habitat. These patches would be nodes of a spatial network with edges of length within the defined range (Fall et al. 2007). If the full range of lengths is divided into exclusive distance classes, several two-threshold networks are created, partitioning the original into subgraphs, each with only those edges that fall into their own distance class.

When the absolute distances among sampling units or their positions in some frame of reference are important, then Euclidean space should be used. Euclidean distances among sampling units can be computed using the *x–y* coordinates of the centroid of the sampling units, resulting in a matrix of straight-line distances (Figure 2.19*a*). Based on these Euclidean distances, a distance class connectivity matrix can be created for several distance classes (Figure 2.19*a*). When other kinds of a priori knowledge of the point or surface pattern process are available, the connectivity matrix (Figure 2.19*b*) that indicates which sampling units are neighbours (1, neighbours; 0, otherwise) may be replaced by a weight matrix where real values (usually between 0 and 1) stress that nearby neighbours are more important than farther ones (Figure 2.19*b*). The most common such weight is the inverse distance (1/*d*) or the inverse of distance squared $(1/d^2)$.

Some 'rules of thumb' for the number of distance classes are based either on sample size or spatial lag. Legendre and Legendre (2012) suggested Sturge's rule, based on sample size, to determine the number of histogram classes: for *n* samples the recommended number of classes is:

$$D = 1 + 3.3 \times \log_{10} \left[\frac{n(n-1)}{2} \right]. \qquad (2.1)$$

Once the number of classes is selected, the Euclidean distances can be distributed among them in two ways. The first is to distribute them by equal distance intervals (the spatial lag, Figure 2.20*a*) and the second is to distribute them to ensure equal numbers of pairs in all distance classes (Sokal & Wartenberg 1983; Figure 2.20*b*). The equidistant approach is the most frequently used and is more intuitive, allowing comparison of results among different studies at a given distance interval. However, the number of pairs per distance class varies, which affects the reliability of the estimation of spatial structure (see Chapter 6). Equal frequency supports better estimation of autocorrelation at each distance because that minimizes the edge-effect problem presented in Section 2.5.8. The equal frequency approach is less common, either because it is less well known or because it makes it harder to compare results between study areas.

When using equidistant classes, one can select the distance interval instead of the number of distance classes. The minimum distance interval must be equal to or larger than the sampling unit length (or the longest side length in the case of a rectangular sampling unit). When the sampling units form a regular lattice, this minimum distance interval (the length of a sample unit, say 1 m) will make links equivalent to the rook's move, so that each sample unit has only four neighbours (Figure 2.20). To include all eight neighbours (equivalent to the queen's move), the minimum distance interval needs to include the diagonal, that is $\sqrt{2} = 1.412$ (Figure 2.20). Some considerations for selecting a distance interval are: (a) too small a distance interval will result in too many classes with fewer pairs of observations in each, making estimation less reliable; and (b) too large a distance interval gives too few distance classes, each covering a larger area and reducing the magnitude of spatial structure detected.

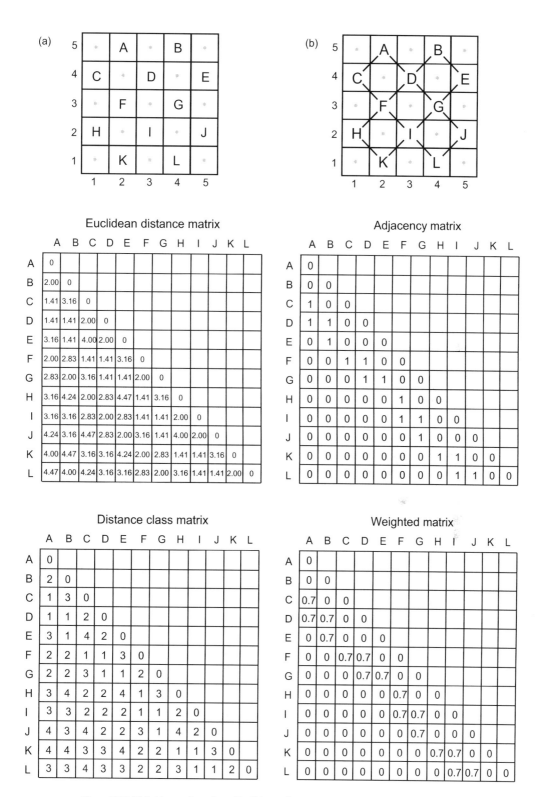

Figure 2.19 Neighbours based on Euclidean distance among 12 sampling locations (A–L). (a) The Euclidean distance matrix among the 12 sampling locations and the corresponding distance class matrix when using a distance interval class of 1.5 units. (b) The binary adjacency matrix is based on the first nearest neighbours and the corresponding weight matrix as a function of distance ($1/d = 1/1.41$ units, which gives a weight of 0.7).

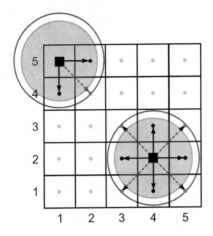

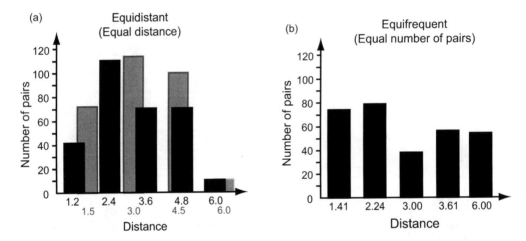

Figure 2.20 Distance class determination based on distance (*a*) or number of pairs (*b*) of sampling locations. (*a*) Defining the distance interval class at 1.2 units (in solid black) gives five equidistant classes (the diagonal locations at 1.41 units are not included in the first distance class); the distance interval class at 1.5 units (solid grey) gives four equidistant classes (including the diagonal locations in the first distance class) yielding more pairs in the first distance class. (*b*) With the equifrequent distance classes, each has about 50 pairs but the Euclidean distance between classes varies: 1.41, 0.83, 0.76, 0.61 and 2.39.

2.8.4 Directional Angle-based Spatial Graphs

The spatial arrangement of geomorphological features (topography, hydrology) and directionality of physical processes (wind, stream flow) influence spatial patterns, creating patches that are elongated (anisotropic) rather than isotropic (Figure 2.5). The definition of spatial neighbours should take direction as well as distance into account for anisotropic systems and their data. The additional criterion is the angle of the line between the locations. The spatial neighbours are then determined as follows

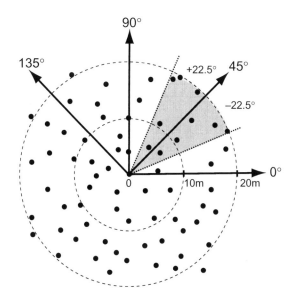

Figure 2.21 Angle-based neighbour determination including both Euclidean distance classes and angle classes for directional neighbour evaluation.

(Figure 2.21): (1) distance; and (2) angle (usually 0°, 45°, 90°, 135°) with a tolerance angle zone on either side (usually 22.5°). Because the combined criteria result in fewer pairs of points than the only distance-based approach, larger distance classes are recommended so that more points are considered.

2.9 Spatial Statistics

'Spatial statistics' includes all the statistical methods that describe spatial patterns or evaluate, estimate and predict spatial processes. Not all spatial statistics have the same goal or assume the same underlying processes, nor do they all use the same data or the same mathematical approach (Dale *et al.* 2002). Therefore, each of these statistics has its own requirements, assumptions and guidelines for application (Box 2.1).

The classification of spatial statistics into the six categories listed at the beginning of this chapter reflects both the historical context in which the methods were developed and their mathematical approach (Dale *et al.* 2002). The first element of our classification, describing and evaluating the spatial structure, is a basic step with an achievable goal using first-order or second-order statistics.

2.9.1 First-order Statistics

First-order spatial statistics are a family of indices assessing aggregation in species abundance data, using the mean of local abundance measures (hence first order) and looking for trends in intensity. These indices can determine whether a spatial pattern is present, but not its magnitude. The common variance-to-mean ratio for quadrat counts distinguishes between three types of pattern based on the quantitative relationship between the variance and the mean of unit counts: random, when the mean and the variance are about equal; patchy, when the variance is greater than the mean; and regular, when the variance is smaller than the mean (Chapter 3). As shown elsewhere (Dale 1999), the logic of this first-order method has some flaws and we will focus on the many second-order statistics which are commonly used by ecologists.

2.9.2 Second-order Statistics

Second-order statistics measure the local spatial pattern and the degree of spatial structure by computing the deviations at neighbouring locations as a spatial variance. Most of the spatial statistics that characterize and test spatial structure in this way are presented in Chapter 3. The basis of most second-order statistics is the concept of spatial weights to characterize the links between units, which can be used to determine several characteristics, including identifying the spatially adjacent sampling units, several search window types, neighbour determination rules and connectivity algorithms.

These statistics can be further classified by the ecological processes to which they apply. Some processes act on the locations of individual organisms; these are called point pattern processes (see Figure 2.3). A second group of processes affects the quantitative values of variables as a pattern in continuous space, and several spatial statistics apply to these continuous processes, such as spatial variance estimators (Chapter 3) and spatial autocorrelation (Chapter 4). These spatially continuous processes can be sampled by contiguous or separated sampling units (Figure 2.2b & c). A third set of processes involves qualitative changes within an area, and surface pattern methods are used to analyse them, such as join count statistics (Chapter 4). Such categorical processes usually require contiguous sampling units. When several processes are active, the ecological data usually show a multiscale spatial structure and require analysis with wavelets or ordination techniques for either univariate or multivariate data (Chapter 7).

After the spatial pattern is characterized, spatial parameters can be estimated to model the process for prediction or interpolation (the second category in our list, see Chapter 5), or to test the relationships among data (the fourth category; Chapter 6). For estimating parameters or testing the relations among variables, the goal of the study and prior knowledge about the data should help decide whether the spatial structure should be explicitly modelled or detrended before further steps (see Chapters 6 and 7). As an example, consider the relationship between soil moisture and plant growth. Both variables have a spatial structure shown by multiple regression on the x and y coordinates of locations (Figure 2.22). The relationship between the variables (Figure 2.22b) is significant only for the raw data (left panel) but not when detrended by multiple regression (right panel). This exercise shows that we should test for

2.9 Spatial Statistics

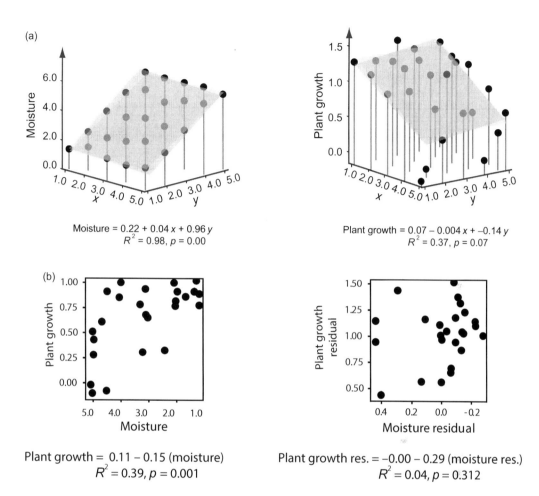

Figure 2.22 Example of the relationship between soil moisture and plant growth (artificial data). (a) Multiple regressions, using the x and y coordinates of the sampling locations as independent variables (soil moisture; plant growth). (b) Relationship (linear regression) between the two variables using the raw data (left panel) and residuals from the multiple regressions (right panel). Note that the relationship changes from significant to non-significant as the data are detrended for the spatial structure.

significant spatial structure before claiming a causal relationship. This investigation falls into the category of interactions among variables. Here, the spatial structure is due to the spatial dependence of both variables on the slope, and this may be a spurious correlation. If the residuals from the multiple regressions retain some spatial pattern, the ecological variables may both have had spatial dependence at one spatial scale and spatial autocorrelation at another. That is why it is important to define the scale of the question so that the appropriate statistics can be applied. The remaining categories of our classification are spatial partitioning to create spatially homogeneous areas (the third category; Chapter 5); and spatio-temporal analysis (the sixth category; Chapters 8 and 11), with locations in time as well as in space.

2.10 Ecological Hypotheses and Spatial Analysis

The relationship between hypotheses and analysis is not simple in most ecological studies, especially in a spatial context. The hypothesis will determine the data and the analysis; but the results of the analysis will generate further hypotheses, leading to more observations and analysis in an iterative sequence. Such sequences can be costly in time and effort, particularly if data collection is required at each stage. In some cases, the sequence can be replaced by a well-planned combination of data collection and analysis that takes advantage of a hierarchy of null hypotheses and may be better than the simple null hypothesis of complete spatial randomness (CSR). McIntire and Fajardo (2009) recommended using space as a surrogate for unmeasured processes, taking advantage of the combination of a priori hypotheses, background knowledge and precise spatial analysis to understand patterns and the underlying processes. Space in this context is a surrogate for unsampled variables and it should be used as the last resort. Further knowledge from pilot studies or the literature can help formulate the pattern's predicted shape (trends or patchiness), degree (low or high) and range (species dispersal ability or isolation-by-distance).

Hierarchical hypothesis frameworks work best when each level of testing is independent of the others. Independence is not always possible and we may end up, as in the case of a series of nested scale-specific hypotheses, with tests that are not mutually independent. Hierarchical structure requires careful consideration of the meaning of the significance levels of test results. Hierarchical analysis may parallel a hierarchy of randomization procedures. Goovaerts and Jacquez (2005) used a typology of null models for the detection of spatial clusters of disease based on (1) (a) risk uniform but spatially independent versus (b) risk uniform but spatially correlated versus (c) risk heterogeneous and spatially correlated; and (2) at-risk population not accounted for (disease incidence) versus at-risk population accounted for (disease prevalence). This scheme provides six different null hypotheses for consideration and simulation.

Translating an ecological idea or question into well-formulated testable spatial or statistical hypotheses can be a difficult challenge. As an example, consider point pattern analysis (Chapter 3) to study the spatial population dynamics based on the locations of the plants (the point events). Popular methods of point pattern analysis include several versions of Ripley's K function (Ripley 1976, 1988; Chapter 3), which detects the scales of under-dispersion (clumping) and over-dispersion. The ecological hypothesis is that intraspecific competition increases the spacing of plants through time due to density-dependent mortality. The seedlings are initially clumped due to high seed density and the clumps self-thin as local-density-dependent mortality occurs. Translating this into spatial hypotheses and spatial analysis results might be the following:

(1) An initial analysis should detect overall patchiness at a range of scales, but most evident at distances related to the intervals between the patches of high density. The null hypothesis is just CSR.

2.10 Ecological Hypotheses and Spatial Analysis

(2) Because the density that causes mortality is very local, small-scale overdispersion develops as the shortest interplant distances are eliminated. Evidence of the underdispersion at the original scale of patchiness will persist, but its magnitude declines due to fewer plants in those patches. The null hypothesis is still CSR.

The ecological hypothesis predicts results incompatible with CSR: initially the analysis will show clumping at all scales but with a maximum corresponding to the distance between patches of high density; the results of later sampling will show over-dispersion at short distances and a maximum under-dispersion at the same scale as previously, but with decreased magnitude. That seems easy! We can determine whether the results are consistent with the predictions, as statistical alternative hypotheses. This will allow us to evaluate the original hypothesis, although the chain of logic is inferential. More than one process can give rise to the same spatial pattern and, while results provide corroboration, alternative explanations are possible, even if improbable. We might strengthen our case by determining whether mortality is highest among plants with the closest neighbours or the highest local density.

The data for this population example are simple: the locations of point events in two dimensions, observed at two points in time. The relationship between ecological hypotheses and spatial analysis can be much more complicated, both in the complexity of the ecological situation and in the number of variables included. Consider a forest with a high density of fallen logs and the questions we might ask about the relationship between the logs and small mammal trails. Many questions could be converted into statistical hypotheses for spatial analysis; here are a few:

- Is there a pattern to the locations and other characteristics of the fallen logs? What are the important features of that pattern? Is the pattern isotropic? Is it stationary? Are there edge effects?
- Are the locations of the small mammal trails independent of the locations or other characteristics of the logs? Are the trails apparently random and independent of each other once the dependence on the logs is factored out?
- Is the dependence of trails on logs directionally specific? Is the dependence stationary? Are there edge effects?
- Are big logs more influential on the trails' locations than small logs? Are older logs more influential than younger logs?

The kinds of questions that can be asked become more complicated as more features are included; we could add contour and slope, the positions of the standing tree boles, the rock outcrops, streams and other biological influences, and so on.

Although hierarchical hypothesis testing has much to recommend it, it is not common because it may be difficult to translate an 'easy' ecological hypothesis into clearly defined hypotheses and outcomes for spatial analysis. Collecting the data to assess competing hypotheses may be difficult or expensive. The difficulty is compounded because different processes can give rise to similar patterns (Figure 2.1) and the alternatives must be explored. Yet, prior knowledge of physiology, ecology, dispersal ability and system topology can help to develop and refine suitable null and alternative hypotheses about the shape and key distances of the spatial structure.

2.11 Randomization Tests for Spatially Structured Ecological Data

When ecological hypotheses are translated into statistical hypotheses, it is crucial to have the appropriate statistics and procedures to assess significance, usually by comparing a statistic computed from the data to an appropriate reference distribution. When the reference distribution is known, parametric tests can be used, provided that the required conditions, like independence, are met.

When the reference distribution is not known, randomization procedures can generate a reference distribution from the data (Figure 2.23*a*; Manly 2018). Bootstrap procedures and Monte Carlo simulations can be used when inference to the population level is required (Figure 2.23*b*; Efron & Tibshirani 1993; Manly 2018). Randomization tests are an attractive alternative because significance is evaluated using distributions generated from the data, and ecologists are often faced with small data sets not eligible for parametric tests. These randomization tests are not without

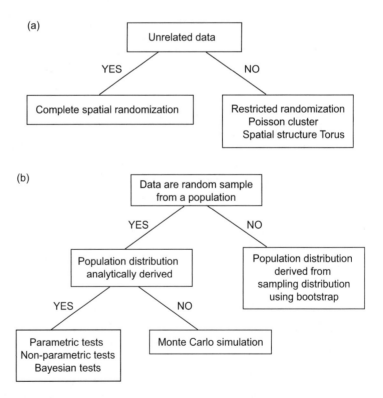

Figure 2.23 Decision trees to select appropriate significance tests at (*a*) the sample level and (*b*) the population level.

assumptions; they assume that the data are independent (equivalent to CSR) so that all re-arrangements (such as re-ordering) are equally likely.

Suppose we wish to evaluate the correlation of diameters of nearest neighbour trees in a forest plot. How do we test whether it is significantly high or low? The basic procedure to generate a reference distribution from the observed data is as follows:

(0) Calculate the test statistic γ_0 from the original data, n observed diameters at n locations.
(1) Take the original set of values, the diameters, and re-allocate them randomly and independently, one to each of the original tree locations; 'shuffling' the values among the locations (actually called 'shuffle' in R or Python).
(2) Re-compute the statistic for the randomized data and record its value.
(3) Repeat steps 1 & 2 many times (e.g. for $j = 1$ to $N = 10{,}000$), creating a distribution for the statistic, $\{\gamma_j\}$.
(4) Compare the value from the data, γ_0, with the distribution from the randomizations, $\{\gamma_j\}$; reject the null hypothesis if it is more extreme than a pre-determined proportion of the randomizations (e.g. 249 out of 10,000).

We can examine all possible permutations of the original data if n is very small (Good 1993); for example, $n = 7$ has only 5,040 permutations. When n is large, only a subsample can be computed because the number of permutations is very large ($n = 100$ has 10^{158} permutations). The precision of the probability determined by this procedure depends on the number of randomizations and so $N = 10{,}000$ iterations or more is recommended (Manly 2018).

The flexibility of randomization tests allows us to analyse complex ecological data using custom experimental designs for which classical tests have not been developed. Ecologists can also develop their own statistics, providing the possibility of reliable tests in novel situations. For example, the boundary statistics presented in Chapter 5 were developed to investigate the properties of coherent boundaries, which would not be possible within standard parametric procedures.

2.11.1 Restricted Randomizations

Both parametric and randomization tests are affected by the lack of independence in our data due to time, space, behaviour, phylogeny and so on. Parametric tests require independent error terms and, with independence, each data point brings a full degree of freedom. When the lack of independence is due to spatial relationships, the spatially autocorrelated data points do not bring a full degree of freedom each, but a fraction inversely proportional to the strength of autocorrelation (Legendre 1993; Dale & Fortin 2002, 2009). The issue is critical for ecological data analysis and will be addressed in Chapter 6. However, because it applies in randomization procedures for spatial data, CSR is not an appropriate comparator and the incorporation of some spatial structure is required (Cressie & Wikle 2011; Figure 2.24). This creates *restricted randomizations* that now include some of the spatial dependency already in the data (Fortin *et al.* 2012*a*; Manly 2018). This can be accomplished in several ways:

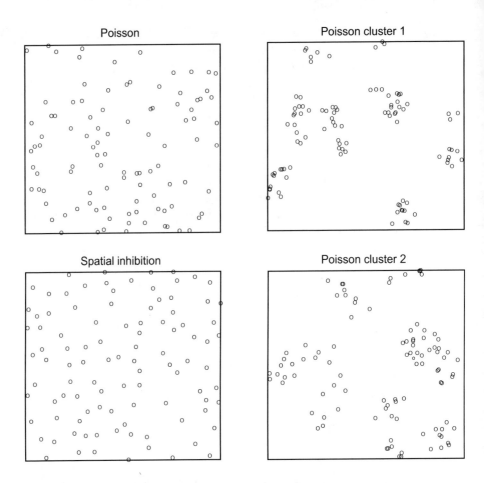

Figure 2.24 Different point pattern processes generating different types of random patterns: Poisson process (complete spatial randomness); Poisson clusters (1 and 2 using different parameter values for spacing among clusters); spatial inhibition.

- Randomize only within subsets of the locations, subregions, in which the data can be considered as spatially independent within the subregion despite overall spatial structure when all the subregions are considered. In a parkland of aspen clones, we could randomize by permuting tree diameters only within identified clonal clusters.
- Retain the spatial order that is in the data. For spatial data, this can be achieved using a two-dimensional torus constructed by connecting the map margins, north edge to south edge and east edge to west edge, and then sliding the data map on the toroidal surface. This is a good technique because it allows us to move the data (the 'toroidal shift') but maintaining most of the spatial structure, provided we are confident of stationarity (see the discussion in Section 2.6). It is especially good for testing the relationship between two variables (bivariate data) by moving the values of one variable with respect to the other, keeping the spatial structure of each variable intact. This is a recommended approach, although it can be too liberal

depending on the number of randomizations and the shape and size of the study area (Fortin et al. 1996).

One theme of this book is that one simple null hypothesis, such as complete randomness, may not be particularly interesting or useful in an ecological context. In many instances, a hierarchy of null hypotheses of increasing restriction and sophistication will be much more informative. In parallel, for randomizations, a progression of restrictions may provide more meaningful results than a single unrestricted randomization. In designing randomization procedures, it is important to think through the worst possible case and how the proposed procedure would respond to it; we may be able to resolve our concerns by finding a counter-example that shows its faults (see example in Dale & Fortin 2014, table 1.2).

The type of restriction appropriate for the randomization technique is directly related to the null hypothesis. One conclusion is that randomization and restricted randomization tests can free ecologists from parametric tests that were not designed to accommodate their novel questions and the inherent spatial structure of ecological data.

We must be careful, however, that the null hypothesis implicit in the restricted randomization is ecologically tenable. Randomization and restricted randomization tests in ecology are prone to misspecification of the null hypothesis, primarily because the null hypothesis is embedded in the randomization procedure and is not made explicit. A clear understanding of the null and alternative hypotheses is required to ensure our biological and ecological questions are correctly addressed. Furthermore, the null hypotheses and the randomization procedures that follow from them must not include the key process being tested. If all the processes are included in the null hypothesis, there is nothing left to test! In Chapter 5, we test whether the locations of two boundaries, based on two different data sets (plant species and animal species), overlap. The null hypothesis is that of no spatial relationship between the plant and animal boundaries: H_0 = no spatial association. The alternatives are: H_{1a} = the boundaries are spatially positively associated and H_{1b} = the boundaries are spatially negatively associated. The null and alternative hypotheses are at the boundary level and so the randomization procedure should be at the same level. Two options are: (a) by the toroidal shift (Fortin et al. 1996) or (b) by placing the boundaries randomly (location and orientation) within the study area. We might test which ecological processes determine the locations of the boundaries and then the randomization should be at the species level. We could examine two factors:

(1) The spatial structure (spatial dependence and spatial autocorrelation) of each species; and,
(2) The spatial interactions among species in the structure of the community.

If only the spatial structure of interest, each species can be randomized separately; if both the species and the community spatial structures are of interest, then the randomizations of species need to be linked together; this can be tricky to implement and requires a clear understanding of the species interactions (facilitation, competition, mutualism, etc.) and their spatial implications (Chapter 1).

2.11.2 Markov Chain Monte Carlo

Another approach is to develop a relatively simple model of the observed structure and then use a Monte Carlo simulation (Monte Carlo Markov Chain or MCMC) to generate artificial data sets to compare with the observed data. The idea is to determine a stochastic process that gives rise to the same spatial structure as the data (James *et al.* 2010; Remmel & Fortin 2013). This can be achieved by first estimating the structure of spatial autocorrelation in the data and then using the parameter estimates that generate simulated data with similar structures (it will not be the same pattern). Several techniques are available for that purpose, including 'conditional autoregressive' (CAR) and 'simultaneous autoregressive' (SAR) models, which are described in Chapter 6 (see also Ver Hoef *et al.* 2018). We then generate many realizations of that stochastic model and use those to generate an 'experimental' distribution of the statistic of interest to be compared with the original observed value. Such restricted randomization tests assume that the underlying process is stationary within the study area. If this is not the case, then this approach is not appropriate and other kinds of modelling of the ecological processes will be required. This is essentially a version of a Monte Carlo approach because it generates new 'data sets' from a defensible model of the process that could produce the structure of the original data; the emphasis is on 'could' because we cannot know the true underlying process. In the absence of more direct solutions, this 'model and Monte Carlo' procedure may be a good general approach. We will discuss examples of randomization and restricted randomization for more specific purposes in other sections of this book, but these comments provide useful background to those applications.

2.12 Concluding Remarks

One goal of spatial analysis is to determine the characteristics of the data's spatial structure. The possibilities include independence (spatial, temporal, phylogenetic, etc.), and several forms of lack of independence: under- or over-dispersion of events, spatial autocorrelation within a single variable or spatial dependence of one variable on another. Only when the data are spatially independent can we interpret parametric and randomization tests with any confidence. Yet, the detection of significant spatial structures (spatial aggregation, spatial dependence, spatial autocorrelation) provides meaningful insights into ecological data and their underlying processes. Space effects can be interpreted as any one of the following:

(1) A **statistical problem to be solved**: the presence of spatial autocorrelation is a problem for applying parametric and randomization tests that require independent errors (Cliff & Ord 1981).
(2) A **diagnostic indicator**: when using linear or multiple regressions, the presence of spatial autocorrelation in residuals can indicate that the variables associated with one or more processes were not included or not parameterized adequately in the model (Melles *et al.* 2011).

2.12 Concluding Remarks

(3) A **surrogate** for unmeasured processes or factors that are too difficult to measure or of which the researchers are unaware (McIntire & Fajardo 2009).
(4) A **predictor variable** that can be used in a regression or ordination framework using the relative spatial location of samples (Dray *et al.* 2006) or the spatial neighbouring effects of the environmental factors or response variables (Peres-Neto & Legendre 2010; Fortin *et al.* 2012*b*). Yet space as a predictor variable should be a last resort because it provides no insights into causal relationships.
(5) A **confounding variable** that produces confusing or spurious results when a system is analysed; it should be used as a covariable (Dray *et al.* 2006).

The degree and the sign of the spatial coefficient values can also be informative:

(1) **Negative spatial autocorrelation** may indicate that the sampling unit size or shape is inappropriate to capture the scale of the process adequately. Check the scale for unexpected cyclic or patchy structures.
(2) **Weak positive spatial autocorrelation** at small distances may indicate that some characteristic may be poorly chosen: the sampling unit size and shape, the spatial lag among the sampling locations or the spatial distance classes used to estimate spatial autocorrelation may be inappropriate to capture the scale of the process.

In the material that follows, we address different aspects of 'space' and how they affect the analysis of ecological data and its interpretation. A very broad range of ecological phenomena all have spatial components to be understood, with an equally broad range of approaches to consider. This current chapter has focused on the processes that give rise to or affect the expression of spatial structure and spatial pattern. Although we have not emphasized detailed mechanisms for the various and complex processes in ecological systems, the possibilities of factors such as landscape memory or feedback switches and the conceptual frameworks of mosaic cycles or patch-gap alternation, as described in Chapter 1, are critical background for the interpretation of the results of spatial analysis. As always, the interplay of pattern and process is the subject of further investigation and a theme that informs this book. It is an investigation that should prove both intriguing and worthwhile.

3 Spatial Analysis of Point and Quadrat Data

Introduction

Census data are frequently collected in demographic and community studies. Such data consist of the locations (x–y coordinates) of all individuals or a complete set of contiguous sample units, hence the statistical population size (N) rather than a sample of the population (n). This chapter will present methods that analyse point process data, mapped point events and quadrat data. Census data may be in one (given by x coordinate), two (x–y) or three dimensions (x–y–z). Mapped point events refer to data where locations of all 'events', such as trees, may have accompanying information, such as tree species and height or diameter (Figure 2.2). Then, quadrat data refer to either presence–absence or count data collected in contiguous quadrats. Analysing data in both space and time, such as events with coordinates x–y–t, will be presented in later chapters.

Spatial analysis of point events often begins with determining the events' neighbours and calculating measures from the distances to those neighbours. Alternatives include counting events in search windows or spatial templates. Univariate data include only one category of events, such as a single-species population of trees. Bivariate analysis considers two kinds of events, such as flowering versus non-flowering individuals. Multivariate analysis applies when there are several categories of events, such as multiple species in a community. When the events have an associated quantitative variable (such as stem diameter), versions of *marked process* analysis are used. Detailed mathematical treatment of point pattern methods can be found in Illian *et al.* (2008) and Wiegand and Moloney (2013).

The methods in this chapter include the concept of a spatial template, a set form of spatial structure, with which the data are selected or compared, or to be used to calculate the expected outcomes (Dale *et al.* 2002). We provide a summary at the end of the chapter (Box 3.1) that suggests close relationships among all these methods and with aspects of graph theory, to be discussed in Chapter 10.

3.1 Mapped Point Data

3.1.1 Introduction: Three Basic Patterns

The first set of methods analyse univariate point patterns, such as mapped flowering thistle stems. In many of these, a test statistic is calculated from the x–y coordinate data and then compared to its expected value under the null hypothesis of complete spatial randomness (CSR). By doing so, three basic patterns can be distinguished: (1) complete spatial randomness as a reference; (2) clumping or under-dispersion; or (3) regularity or over-dispersion. This trichotomy is of course too simple as the appearance and interpretation of patterns can change with the scale of study. If the events occur as clumps, and only one clump is examined, the events can appear to be over dispersed (Figure 3.1). To be most useful, the point pattern analysis must go beyond three categories and assess spatial characteristics at different spatial scales. The methods presented in this section assume the underlying process to be homogeneous (constant mean intensity) and stationary (all statistical parameters constant throughout), so that every location has an equal probability and the observed pattern at any location can be compared with CSR.

Spatial point pattern analysis has many methods available and can be applied to univariate, bivariate and multivariate data. In the latter two data types, the points in the pattern have categorical labels, and points may also have quantitative labels, giving a *marked* point pattern. Ben-Said (2021) has provided an excellent review of methods and applications of all types, with a table that summarizes applications, statistics and results; we recommend it highly. Before our detailed presentation of the range of methods, we need to emphasize one finding that must be considered for any point pattern analysis: many studies used sample sizes that were too small (see Ben-Said 2021, table 2). We should 'practice' with artificial data before designing a study to ensure that the samples will be adequate for the hypothesis and statistical approach to be used; Rajala et al. (2019) provided an instructive example.

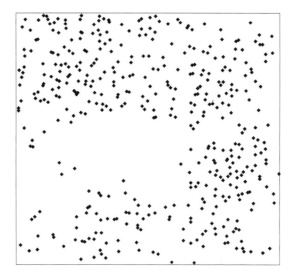

Figure 3.1 Map of living lodgepole pine at Fort Assiniboine, Alberta. The plot is 50 m × 50 m.

3.1.2 Distances to Neighbours

Given a complete census of point events in two dimensions, one approach is to calculate the distances between events that are neighbours and then to determine whether the average distance differs from that expected from CSR. If the average distance is significantly less than expected, the conclusion is that the events are clumped; if greater, they are over-dispersed.

There are, however, many ways of defining which pairs of events are the neighbours. A simple definition is each event's single nearest neighbour. Some pairs of events will be mutually nearest neighbours but not always. Spatial analysis based on the pairs of neighbouring events can be seen as equivalent to creating a spatial graph. A *graph* is a simple mathematical object consisting of a set of points, the *nodes*, joined in pairs by lines, the *edges*. In a spatial graph, the nodes have locations and the edges may or may not. Considering neighbour pairs creates a graph with the mapped events as the nodes and edges joining the neighbouring pairs but no others, like a simple 'yes or no' rule for neighbour versus not neighbour. We will have more to say about graphs, spatial graphs and networks in later chapters.

3.1.2.1 Neighbours and Distances

Given a map of all events in a study area, we can use them all or only a sample of them. For a sample of events, we can base a test statistic, Q, on the squares of the distances between neighbours. If W_{i1} is the distance between event i and its first nearest neighbour and λ is the density per unit area, consider the statistic Q:

$$Q = \pi\lambda \sum_{i}^{n} W_{i1}^2 / n. \qquad (3.1)$$

The advantage of using the distance squared is that the statistic can be compared directly to a normal distribution with a mean of 1 and a variance $1/n$. We conclude that there is clumping if Q is significantly less than 1 and that there is overdispersion if it is significantly greater. The statistic Q is one of a very large number of tests that have been proposed. Some of the other tests compare the average distance of an *event* to the nearest neighbour event (event to event) with the distance from a randomly placed *point* to its nearest event, X_{i1} (point to event). Other tests measure the distances to the first nearest, the second nearest and so on, such as W_{i3}, the distance from event i to its third nearest neighbour. Liu (2001) compared methods using the first to fifth nearest neighbours ($j = 1, 2, 3, 4, 5$) and suggested that the statistics with $j = 3, 4$ or 5 tend to perform better. An alternative is to retain and analyse the distance information from a broader range of neighbour ranks, say 1 to 10 or more, to provide a multiscale analysis (cf. Manly 2018).

3.1.3 Refined Nearest Neighbour Analysis

In using the full census of events, it is tempting to proceed with methods designed for samples (see Chapter 4). That can cause problems, however, because the values contributing to the measure will not all be independent: neighbour methods will use the location of any one event many times over. Using all the events also means that we include those that are close to the edge of the study plot, and there should be some consideration of edge effects (Chapter 2, Figure 2.10). These concerns have led to a modification, referred to by Diggle (1979) as *refined nearest neighbour* analysis.

The refined nearest neighbour analysis, G, for any given distance, v, examines all events that are farther than v from any boundary and calculates the proportion of these that are closer than v to their nearest neighbour. Under CSR, the expected value of $G(v)$, $E[G(v)]$, is:

$$E[G(v)] = 1 - e^{-\lambda \pi v^2}. \tag{3.2}$$

The test statistic is the largest difference between $G(v)$ and $E[G(v)]$:

$$d_v = \max | E[G(v)] - G(v) | \tag{3.3}$$

to be evaluated by comparison with many realizations of CSR with the same N and plot size and shape. A similar approach, $F(u)$, uses the distances from a random *point* to its nearest event, based on distance u, and its statistic is then d_u:

$$d_u = \max | E[F(u)] - F(u) | \tag{3.4}$$

to be evaluated in the same way. Because $F(u)$ is sensitive to gaps, it is called an 'empty space' function.

Insight can be gained by plotting $G(v) - E[G(v)]$ or $F(u) - E[F(u)]$ over the range of values of v or u. Diggle (1979) also suggested plotting the differences between G and F as a function of distance. For the lodgepole pine example (*Pinus contorta* Loudon) in Figure 3.1, that analysis is shown in Figure 3.2, with the difference between F and G plotted as a function of distance z. This depicts the large-scale clumping of these stems, but no significant overdispersion at smaller scales. One advantage is that the test can be modified to use any point pattern distribution for comparison, not just the Poisson associated with CSR. Illian *et al.* (2008) provided some guidance on this for different models, which may give more insight into the observed point pattern and the point process that may underlie it.

3.1.4 Second-order Point Pattern Analysis

For data from a complete census of events in the plane, the most commonly used analysis is Ripley's K (Ripley 1988). If λ is the density of events per unit area, the expected number of points in a circle radius t centred on a randomly chosen point is $\lambda K(t)$, where $K(t)$ is some function that depends on the pattern. If the points are over-dispersed, $K(t)$ will be close to 0 for small radii and increase for larger distances. The method calculates an estimate of that function $K(t)$ from the data, call it $\hat{K}(t)$:

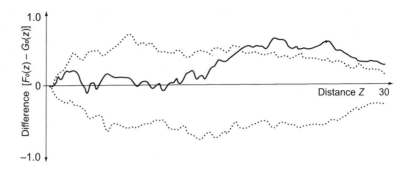

Figure 3.2 Refined neighbour analysis on living lodgepole pine trees at Fort Assiniboine, Alberta (data in Figure 3.1), where $F(z) - G(z)$ is the difference between event–event and point–event functions for a suitable chosen distance z.

$$\hat{K}(t) = A \sum_{\substack{i=1 \\ i \neq j}}^{n} \sum_{\substack{j=1 \\ j \neq i}}^{n} w_{ij} I_t(i,j)/n^2, \qquad (3.5a)$$

where A is the area of the plot; $I_t(i,j)$ is an indicator function (based on d_{ij}, the distance between points i and j) with $I_t(i,j) = 1$ if $d_{ij} \leq t$ and 0 otherwise; and weight w_{ij} corrects for edge effects. If the circle centred on i with radius d_{ij} is completely within the study plot, $w_{ij} = 1$; otherwise, it is the reciprocal of the proportion of that circle's circumference within the plot (Diggle 1983).

An alternative formulation uses a different weight: $h_i(t) = 1$ if the circle centred on i with radius t is completely within the study plot, otherwise the reciprocal of the proportion of its area within the plot:

$$\hat{K}(t) = A \sum_{\substack{i=1 \\ i \neq j}}^{n} \sum_{\substack{j=1 \\ j \neq i}}^{n} h_i(i) I_t(i,j)/n^2. \qquad (3.5b)$$

If the events resemble CSR, the number of events within a circle follows a Poisson distribution and the expected number for radius t is $n\pi t^2/A$. $\hat{K}(t)$ is compared with this expected value by subtracting the observed from the expected:

$$\hat{L}(t) = t - \sqrt{\hat{K}(t)/\pi}. \qquad (3.6)$$

Some versions use the additive inverse of Equation (3.6), hence the researcher needs to check! The square root in the transformation stabilizes the variance as a function of scale (see Illian et al. 2008 for mathematical details). $\hat{L}(t)$, from Equation (3.6), is plotted as a function of t, with negative values indicating clumping and positive values indicating overdispersion. (As a mnemonic: under the line means under-dispersed; over the line means over-dispersed.) Figure 3.3 shows the analysis

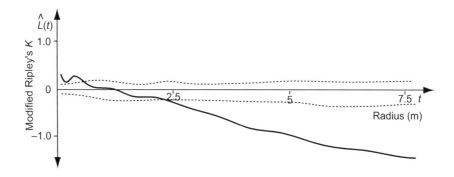

Figure 3.3 Standardized univariate Ripley's K-function analysis of living lodgepole pine at Fort Assiniboine, Alberta (data in Figure 3.1), where $\hat{L}(t)$ is Ripley's standardized statistic and t is distance.

of the data in Figure 3.1, a map of living lodgepole pines; the features of the data are clear: overdispersion at scales less than 1 metre and clumping at all larger scales.

The results can be assessed using the approximate confidence intervals for $\hat{L}(t) = 0$ or by Monte Carlo. The latter is recommended, and Figure 3.3 shows the 99% envelope generated from 100 realizations of CSR. The Monte Carlo approach allows models other than CSR for comparison. The Neyman–Scott process is often used as a null model; it produces random *clusters* of events, with events per cluster following a Poisson distribution (also called a Poisson–Poisson distribution; Pielou 1977).

We distinguish between global analysis, which evaluates the whole study area, and local analysis, which detects differences among sub-areas (see Chapter 2). The global second-order statistics discussed so far can also produce spatially explicit results. Each event, i, and distance, t, has a score comparing the observed count in the circle radius t centred on event i with the expected count based on CSR (Getis & Franklin 1987). Contour maps of those scores, drawn for ranges of radii, have clear interpretations. Getis and Franklin (1987) suggested including the scores for randomly or regularly placed points (not events) so that sparse regions can be included in the analysis, paralleling the inclusion of the 'empty space' function in the refined nearest neighbour method. The authors provided examples of this score mapping, which gives spatially explicit results for K-function analysis.

Studying the demography of a palm tree in the humid savannah of West Africa, Barot et al. (1999) explored the advantages of using several complementary methods, rather than a single one. They applied Diggle's G- and F-functions (nearest neighbour and empty space) as well as the K-function. The simultaneous application of all three found significant departures from CSR that would not have been evident otherwise. The patterns included clumping of young seedlings around maternal trees, selective mortality restricting most juveniles to nutrient-rich patches, and subsequent self-thinning due to competition in high-density patches. Their findings confirm the advantage of a combined analysis with more than one method to characterize spatial structure. Ripley's K statistics can be computed using the function *Kest* and the Ripley's L statistics with the function *Lest* of the R package *spatstat*.

3.1.5 Bivariate Data

The point patterns discussed so far have been univariate, but those methods can be easily modified for bivariate data with two kinds of events. For example, Diggle's nearest neighbour function, G, can be adapted for bivariate data by examining the distance from type 1 events to type 2 neighbours, giving G_{12}, and separately examining the distance from type 2 events to type 1 neighbours, giving G_{21}. This separation permits the detection of asymmetric associations, as between mature trees and their seedlings. Similarly, the 'empty space' function can be divided into two parts for bivariate data: F_1 describing the distance from a random point to a type 1 event and F_2 describing the distance from a random point to a type 2 event.

For the bivariate K-function, the question is: At what scales are the two kinds of events segregated from each other, and at what scales are they aggregated? We proceed by calculating:

$$\hat{K}_{12}(t) = A \sum_{\substack{i=1 \\ i \neq j}}^{n_1} \sum_{\substack{j=1 \\ j \neq i}}^{n_2} w_{ij} I_t(i,j) / n_1 n_2 \quad (3.7a)$$

and

$$\hat{K}_{21}(t) = A \sum_{\substack{i=1 \\ i \neq j}}^{n_1} \sum_{\substack{j=1 \\ j \neq i}}^{n_2} w_{ji} I_t(i,j) / n_1 n_2. \quad (3.7b)$$

The two versions are both estimates of the same function and are combined as a weighted average to compare the observed and the expected:

$$\hat{L}_{12}(t) = t - \sqrt{[n_2 \hat{K}_{12}(t) + n_1 \hat{K}_{21}(t)]/\pi(n_1 + n_2)} \quad (3.8)$$

(cf. Upton & Fingleton 1985). Values greater than 0 indicate segregation and values less than 0 indicate aggregation. Computing bivariate Ripley's K can be obtained using the function *Kcross* of the R package *spatstats*.

Figure 3.4 shows the relationship between canopy trees (mainly pine) and understorey seedlings (mainly spruce) near Grande Cache, Alberta (Dale & Powell 2001). The trees and seedlings are segregated from each other at small scales but aggregated over a range of larger scales.

Bivariate K-function analysis was also applied in a study of the Hooded Warbler, to determine the relationship between the locations of unmated males and nests (Melles et al. 2009). While the nest sites tend to be clustered, that clustering does not have a significant relationship with the locations of unmated males. In this example,

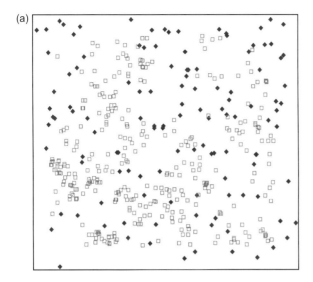

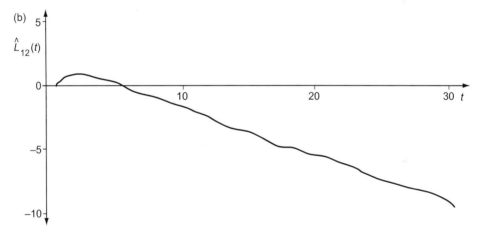

Figure 3.4 (*a*) Map of canopy trees (closed diamonds) and seedlings (open squares) in a 50 × 50 m plot near Grande Cache, Alberta. (*b*) Bivariate Ripley's K analysis of canopy trees and seedlings shown in (*a*). $\hat{L}$ is the standardized Ripley's K and t is the distance.

additional Monte Carlo tests were used to assess the spatial aggregation of females at different spatial scales.

Bivariate patterns can be tested by randomization that retains the locations but redistributes their 'labels' by shuffling or random draw (Figure 3.5). This allows us to determine whether the events of one kind, such as diseased organisms, are more clustered than can be explained by the non-randomness of the whole pattern. Dale and Powell (1994) studied plants of *Solidago canadensis* L., with or without obvious signs of insect attack, with the bivariate K-function. Based on comparisons of the observed K_{12} with CSR, the results varied among the quadrats. In one (quadrat 5), the two kinds of plants appear aggregated over a range of scales, but randomizing the labels shows that

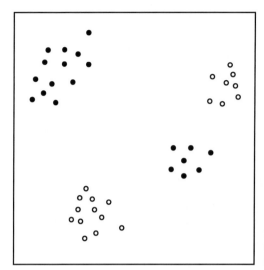 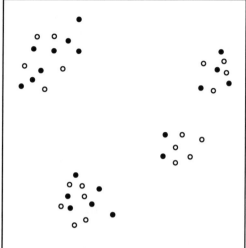

Figure 3.5 Bivariate data: re-labelling for significance testing by randomization; the events' positions are fixed but the labels are redistributed (left: original data; right: labels randomized).

this results from overall clumping (Figure 3.6a). In contrast, in another quadrat (quadrat 10) the overall result was compatible with independence, but random re-labelling shows that the two types are segregated within the overall pattern (Figure 3.6b).

Bivariate patterns can also be tested using a restricted randomization technique called a toroidal shift, applied separately to the events of each type. As locations are shifted by set amounts, say East and North, if they slide off one side, they reappear on the other (see Chapter 2). It is as though the North and South boundaries abut, and the East and West boundaries abut (Dale & Fortin 2014, figure 8.12). Perea et al. (2021) used this technique to evaluate bivariate statistics like K_{12} for saplings and adult trees in spatial networks of Mediterranean forests.

In evaluating clusters of disease incidence, there is a contrast between data in which all individuals, diseased or not, are recorded and data that include only the diseased individuals. The *Solidago* study included records of the unaffected plants but many studies concentrate on location records of disease incidence, without the whole 'at risk' population, whether from choice or because the information is not available. A common goal in epidemiology is to detect clusters of disease, but it is not always possible to distinguish the clusters of disease from the overall pattern, as we were able to do in this example. We will return to the question of how clusters can be detected and evaluated in Chapter 5.

3.1.6 Thinning or Combining

A related topic for bivariate point pattern analysis is 'thinning' *the pattern* by randomly deleting a proportion of the events. Starting with a Poisson process of intensity λ, if a proportion, p, are independently and randomly removed, the events

3.1 Mapped Point Data

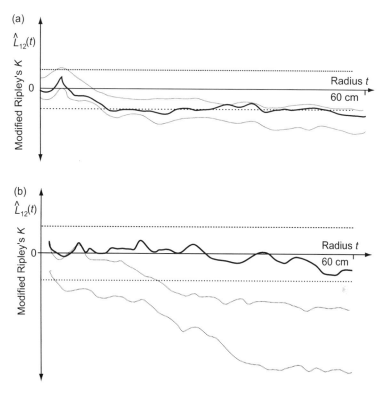

Figure 3.6 Bivariate Ripley's K analysis on infected and healthy *Solidago* plants for Quadrat 5 (*a*) and Quadrat 10 (*b*). $\hat{L}$ is the standardized Ripley's K and t is the distance.

removed form a Poisson process of intensity λp, and those that remain are a Poisson process of intensity $\lambda(1-p)$. This can be seen as splitting the original process into two subprocesses, and all three are Poisson, as would result from random labelling of a CSR with two labels, creating a bivariate division. Conversely, two independent Poisson processes of intensity λ and γ can be combined by superposition to produce a Poisson process of intensity $\lambda + \gamma$ (Baddeley *et al.* 2015).

Of greater ecological interest will be the case where some of the processes are not independent Poisson processes such that different point patterns emerge (Lewis & Shedler 1979). Andersen and Hahn (2015) used a thinning process to produce a pattern that is overdispersed at small scales and clustered at intermediate scales. These methods for combining or decombining subpatterns help in both interpreting observed patterns and modeling them (see Baddeley *et al.* 2015; Velázquez *et al.* 2015).

3.1.7 Multivariate Point Pattern Analysis Data

An obvious extension of the analysis of bivariate data is to multivariate patterns of several kinds of events. Lotwick and Silverman (1982) suggested that there were two basic approaches:

(1) Methods based on nearest neighbour and empty space functions, and
(2) Methods based on second-order analysis.

There is a close conceptual relationship between the two approaches:

(1) The first asks the question: How big can a circle centred on an event (or on a random point) grow before it encounters another event?
(2) The second asks the question: Given a circle of a given size, centred on an event, how many other events does it contain?

The first approach is based on the sizes of circles that contain no events, and the second on counting events in circles over a range of sizes.

In the univariate context, we introduced the 'event-to-nearest-event' function, $G(t)$, and the 'point-to-nearest-event' function, $F(t)$, where t is distance (Equations 3.3 and 3.4). These can be compared using their differences, or combined in a ratio to give an index of spatial interaction:

$$H(t) = \frac{1 - G(t)}{1 - F(t)}. \tag{3.9}$$

This index is close to 1 under CSR, less than 1 for clustering and greater than 1 for overdispersion.

Summary statistics for multivariate point patterns were introduced by van Lieshout and Baddeley (1999). Consider S types (like species) and revise the notation to include I and J to denote types (i and j denote events):

$G_{IJ}(t)$ is the distance function for events of type I to events of type J.
$G_{..}(t)$ is the distance function for events of any type to events of any type.
$F_J(t)$ is the empty space function from random points to events of type J.
$F_{.}(t)$ is the empty space function from random points to events of any type.

These can then be used to define two different H-functions:

$$H_{IJ}(t) = \frac{1 - G_{IJ}(t)}{1 - F_J(t)}, \quad \text{for a particular pair of types } I \text{ and } J \tag{3.10a}$$

and

$$H_{..}(t) = \frac{1 - G_{..}(t)}{1 - F_{.}(t)}, \quad \text{for of any type.} \tag{3.10b}$$

Last, with λ_I being the intensity of the Ith type, and total intensity $\lambda_{.}$, an overall index can be defined as:

$$I(t) = \sum_{i=1}^{I} \frac{\lambda_I}{\lambda_{.}} H_{II}(t) - H_{..}(t). \tag{3.11}$$

Turning to second-order methods, for a given multivariate point pattern, the choice of analysis will depend on the hypothesis of interest. There is a difference between 'Do all species tend to be segregated from any other species considered singly?' and

'Do all species tend to be segregated from all other species considered together?' It is a matter of 'partitioning' the overall pattern of events into what is attributable to single types and what is attributable to the relationships among types. Although there are insights to be gained from looking at all intraspecific statistics of form $K_{II}(t)$, there is also much to be learned from examining interspecific statistics both bivariate, $K_{IJ}(t)$, and those of form $K_{I,\sim I}(t)$ over all possible $\sim I$, call it $K_{I\cdot}(t)$. In effect, this partitions $K_{\cdot\cdot}(t)$, which includes all species pairs, into $K_{XX}(t)$, which considers all conspecific pairs, and $K_{X,\sim X}(t)$, which considers all interspecific pairs.

$K_{XX}(t)$ can be partitioned into S possible $K_{II}(t)$'s; and $K_{X,\sim X}(t)$ can be partitioned into all possible $K_{IJ}(t)$'s of which there are $S\mathcal{C}2$ ('S select 2' $= S(S-1)/2$). As in the original version of Ripley's approach to point pattern analysis, these partitioned K estimates should be transformed into the equivalent L statistics for easier interpretation. Figure 3.7a shows artificial data for four species used to illustrate this multivariate method. Figure 3.7b shows plots of the estimates of $L_{X,\sim X}(t)$ (upper), $L_{\cdot\cdot}(t)$ (middle) and $L_{XX}(t)$ (lower). These results show that, while the arrangement of events is random, events of the same type are clustered and segregated from events of a different type. The next part of the figure depicts the different scales and intensities of clumping of the individual species, with type 2 having the smallest scale of strong clumping and type 1 differing little from random (Figure 3.7c). Figure 3.7d summarizes the interspecific analyses for individual species, showing that species 3 and 4 have a high degree of negative association with events of other species over all scales. These could be further partitioned into species pairs: 1 with 2, 1 with 3, and so on. Although this analysis is more complicated than other approaches, it is also able to provide the most detailed information on the characteristics of multispecies patterns.

Rajala et al. (2019) investigated the conditions under which spatial point pattern data have the power to detect biological interactions. Although the study has implications for multispecies conclusions, they focussed on a simple two-species model. A summary of the findings is that the detection of interactions will usually require large numbers in the data and it works best when the numbers are balanced. Our conclusion echoes Ben-Said (2021), that many studies of point patterns have used sample sizes that were *much* too small.

The relationship between any two species may be changed by the presence (or absence) of a third species. Suppose species A and B are segregated at small scales in areas of preferred ecological conditions because of competition, but positively associated at larger scales because the large patches of a third species, C, exclude both. For a spatially explicit result, we can plot a Getis' style map of L-function scores by location, based on univariate, bivariate and trivariate analyses.

Condit et al. (2000) suggested multispecies analysis by modifying K for counts in rings of width Δt centred on individual events, rather than in circles. Multispecies analysis for counts in these sample rings is analogous to the versions of H_{IJ} and so on described. Counting events in rings allows the isolation of specific distance classes, rather than including the short distances with the larger, but the distance classes need to be broad to avoid erratic-appearing curves caused by zero counts. The difference

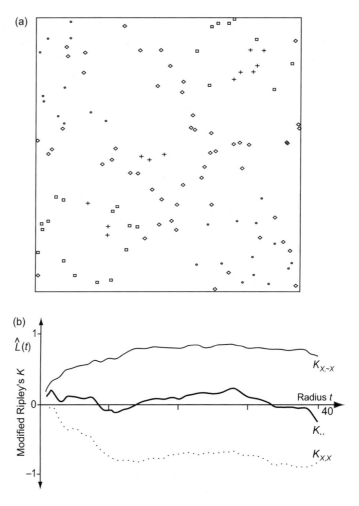

Figure 3.7 (*a*) Artificial multispecies point pattern data for illustration of the analysis. (*b*) Multivariate standardized Ripley's K analysis: conspecific pairs, $K_{X,X}$ (bottom line), all pairs, $K_{..}$ (middle line) and interspecific pairs, $K_{X,\sim X}$ (top line). (*c*) Standardized Ripley's K analysis for each species: conspecific pairs. (*d*) Multivariate standardized Ripley's K analysis interspecific pairs. $\hat{L}$ is the standardized Ripley's K and t is distance.

between using circular bands rather than full circles is that distance is the factor studied, rather than scale.

Given the range of methods available for multivariate pattern analysis, the choice may seem difficult. Methods based on Ripley's K are popular for good reason: they are easy to use and to interpret and there is a range of such techniques that cover most kinds of point data. In general, they examine scale of pattern and they can accommodate composite patterns with several scales. No single method can tell us everything we may want to know and so the use of two or three complementary methods is recommended. Many researchers, however, seem to be reluctant to do so, preferring to continue with only one method (see Velázquez *et al.* 2015).

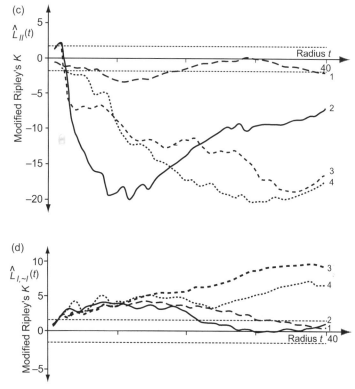

Figure 3.7 (cont.)

3.2 *K*-function Analysis for Inhomogeneous Point Patterns

Most methods for pattern analysis assume homogeneity and stationarity. What do we do if those cannot be assumed? The methods require some revision, but the modification can be applied to all data types: univariate to multivariate and quantitative marks. The major adjustment is in the calculation of the expected values, comparing local observed 'local' intensity with 'regional' values rather than the 'global' value from the whole study area. The size of the region for that estimate can be chosen by the user or determined from the data.

3.2.1 Regional Expected Values around Events

Where a non-homogeneous process is suspected over the surveyed area, comparison with CSR may be inappropriate. Illian *et al.* (2008) discussed dividing an inhomogeneous study area into subareas that meet some criterion of homogeneity. The method we describe is to derive the expected values as 'regional', based on the density of points in a subarea or region rather than in the whole study area ('global'). The *K* function uses counts centred on events and, thus, each count has a location, giving a 'local' score. Here, the same local counts are compared with expected values localized to the same event but calculated from a regional intensity.

The K-function in Section 3.1 works with λ as the density of events per unit area and the expected number of events in a circle radius t is $\lambda K(t)$, where the function $K(t)$ depends on the pattern of the events. The observed statistic $\hat{K}(t)$ is compared with its expected value. To deal with non-homogeneous structure rather than CSR, that expected value must be able to vary with location.

Changing the notation for clarity, let $c_i(t)$ be the observed count of events in a circle radius t centred on event i (excluding event i). We need to evaluate:

$$S_2 = \frac{1}{n}\sum_{i=1}^{n}\{c_i(t) - E[c_i(t)]\}. \tag{3.14}$$

Using the global expectation based on stationarity and CSR, we get a statistic that resembles the original K

$$\hat{C}(t) = \frac{A}{n}\left[\frac{1}{n}\sum_{i=1}^{n}\{c_i(t) - E_G(c_i(t))\}\right], \tag{3.15}$$

where C refers to the circle-by-circle calculation, and the subscript G refers to the global expected value for all N events and area A. In K, the edge correction appears in the counting procedure, the 'observed', whereas, in C, it is in the calculation of the expected value.

3.2.2 Regional Values for Subareas of the Study Plot

For a regional version (a subplot of the full area), the expected value is $E_R(c_i(t))$, based on the number of point events, n, and the area of that subplot; using a circle of radius $T > t$ centred on event I,

$$\hat{C}_R(t) = \frac{A}{n}\left[\frac{1}{n}\sum_{i=1}^{n}\{c_i(t) - E_R(c_i(t))\}\right] \tag{3.16}$$

becomes

$$\hat{C}_R(t) = \frac{A}{n}\left[\frac{1}{n}\sum_{i=1}^{n}\left\{c_i(t) - c_i(T)\frac{a_i(t)}{a_i(T)}\right\}\right], \tag{3.17}$$

with regional intensity estimated for a circle of radius $T > t$, centred on point event i, with the circle's area within the study area designated as $a_i(r)$, for any radius r.

In the original analysis, a square root transformation was applied to make the statistic, L, linear with radius, t. The same logic suggests that the comparable statistic for a non-stationary point pattern should be:

$$\hat{L}_R(t) = \sqrt{\frac{A}{n}\left[\frac{1}{n}\sum_{i=1}^{n}\{\sqrt{c_i(t)} - \sqrt{E_R(c_i(t))}\}\right]}. \tag{3.18}$$

The standard analysis detects a tendency to clumping at all scales, but the inhomogeneous approach shows that the pattern is approximately random once the effects of the trend are removed. Computing Ripley's K for the inhomogeneous cases can be obtained using the function *Linhom* of the R package *spatstats*.

Non-stationarity is one violation of the assumptions of Ripley's K but anisotropy should also be considered. Illian *et al.* (2008) included the analysis of point patterns' orientation, based either on the directions of nearest neighbours or on a subdivision of the circular templates for counting in a K-function analysis. The first, using directions, resembles features of the fibre analysis we describe in Chapter 10, where the graph edges between the nearest neighbour events (nodes) are the fibres that can be analysed for direction (see Figure 10.2). The second approach, subdividing the full 360° into equal segments (say 4 or 8), is described in Dale and Fortin (2014, figure 3.14).

3.3 Mark Correlation Function

These methods were designed to investigate the interactions of neighbouring forest trees based on measured characteristics such as height or diameter (Penttinen *et al.* 1992; Stoyan & Penttinen 2000). The approach modifies the methods in Section 3.1 to include a quantitative characteristic (the *mark*) for each event, m_i (Illian *et al.* 2008).

The K function for mark data is

$$\hat{K}_m(t) = \sum_{i}^{n} \sum_{\neq j}^{n} w_{ij} I_t(i,j) m_i m_j, \qquad (3.19)$$

and the observed and expected values are compared by calculating

$$\hat{L}_m(t) = t - \sqrt{\hat{K}_m(t)/\pi\mu^2}, \qquad (3.20)$$

where μ is the mean value of m_i. When $\hat{L}$ is plotted as a function of t, large positive values indicate overdispersion of the marks (equivalent to negative autocorrelation at scale t) and large negative values indicate their aggregation (positive autocorrelation at scale t). Parallels with the interpretation of the univariate K-function analysis are obvious.

Goulard *et al.* (1995) provided an excellent example of the usefulness of this approach in a study of the clumps of sprouts of sweet chestnut, *Castanea sativa* Mill., in the Limousin region of France. In addition to the locations of coppiced clumps, four variables were measured: diameter, number of shoots before cutting, height at 1 year after cutting, and height at 3 years after cutting. These were the 'marks' used in the analysis. The authors also measured soil depth to the granite beneath, at 120 locations, and produced an interpolated (Kriged) estimated surface (see Chapter 4) of soil depth for the whole study area. They found that the clumps were regularly dispersed, with diameter and number of shoots displaying negative autocorrelation at smaller distances. Heights were not strongly correlated. The analysis showed that small clumps were aggregated in gaps between larger clumps and that heights could be related to their spatial relationship with local soil variables. This study provides a good example of the sophisticated and detailed analysis that the marked point process approach can offer, especially when used in combination with other forms of analysis.

Ledo et al. (2011) have extended the method further in describing an intertype mark correlation function that can be used to characterize relationships between different categories, such as among tree diameters for different tree species in a forest. Darsanj et al. (2021) used the technique on four tree characteristics: height, diameter at breast height (DBH), crown diameter and light crown height. As examples of their results, in a pure *Populus* stand, DBH values were negatively correlated at distances of less than 5 m and greater than 40 m, but crown diameters were positively correlated from 5 m to 40 m. Ballani et al. (2019) recommended the use of mark–mark scatterplots to improve the interpretation of mark correlation analysis and provided three good examples of this application.

3.4 Point Patterns in One and Three Dimensions

3.4.1 One Dimension

Patterns of events in one dimension often arise in ecology: the positions of species boundaries on an environmental gradient, encounters with individuals or objects along a line transect, and so on. Assessing boundaries on gradients, Dale (1999) described several statistics characterizing the one-dimensional arrangements of n events along some given length.

Standardizing the total length to 1.0, the positions of n events, the x's, divide the interval 0 to 1 into $n + 1$ pieces with lengths μ_i (see Figure 3.8a). Two statistics can help to characterize different arrangements of the events:

$$W_n = \sum_{i=1}^{n+1} \mu_i^2 \tag{3.21a}$$

and

$$h_n = \sum_{i=1}^{n} \mu_i \mu_{i+1}. \tag{3.21b}$$

The statistic W_n detects the clumping of boundaries which produces short segments, with larger values indicating a greater inequality of segment lengths. The statistic h_n measures serial autocorrelation because large values indicate that short segments are adjacent to short segments and large segments adjacent to large segments. Tables of approximate critical values and guidance for applying these statistics are provided in Dale (1999).

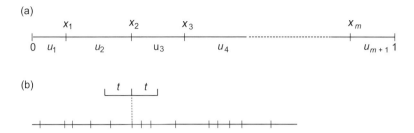

Figure 3.8 (*a*) Illustration of standardized one-dimensional transect data, with *n* events. (*b*) A line segment with *n* events, and the '*t*-bar' used for calculating the one-dimensional version of Ripley's *K*-function.

We have emphasized methods related to *K*-function analysis and there is a one-dimensional version, illustrated by Figure 3.8*b*: consider a line of length *B* in which *n* events occur. For a range of values, *t*, the number of events within distance *t* of each event is counted:

$$\hat{K}(t) = B \sum_{\substack{i=1 \\ i \neq j}}^{n} \sum_{\substack{j=1 \\ j \neq i}}^{n} w_i(t) I_t(i,j) n^2, \qquad (3.22)$$

where $w_i(t)$ corrects for edge effects. If a bar of length $2t$ centred on event *i* is completely within the line, $w_i(t) = 1$, otherwise, it is the reciprocal of the proportion within the line. The number of random events expected inside a bar of length $2t$ is $2nt/B$; therefore plot

$$\hat{L}(t) = t - \hat{K}(t)/2, \qquad (3.23)$$

as a function of *t*. Positive values indicate overdispersion and negative values indicate clumping. An artificial example is given in Figure 3.9*a* and *b*. A physical example of rapids on the Morris/Pipestone River in Northern Ontario (52° 15′ N, 90° 45′ W) is in Figure 3.9*c* and *d*, in which $W_m = 0.089$ and $h_m = 0.027$. The first is significant, indicating clumping, but the second is not, indicating no tendency for clusters of short inter-rapid river sections.

An obvious modification of Equation (3.22) produces a bivariate version of the *K*-function for one dimension. Methods for one dimension can also be modified for networks of one-dimensional structures. Okabe and Yamada (2001) provided univariate and bivariate forms of *K*-function analysis for events constrained to the edges of a two-dimensional spatial network (see also Baddeley *et al.* 2021). This analysis can be applied to ecological studies such as the occurrence of aquatic invertebrates in water channels. Spooner *et al.* (2004) have analysed the spatial arrangement of *Acacia* trees on a road network in Australia, providing a model for future studies of the spatial relationships of organisms and anthropogenic linear structures. This approach ties in well with our discussion of the analysis of linear structures with point events in Chapter 10.

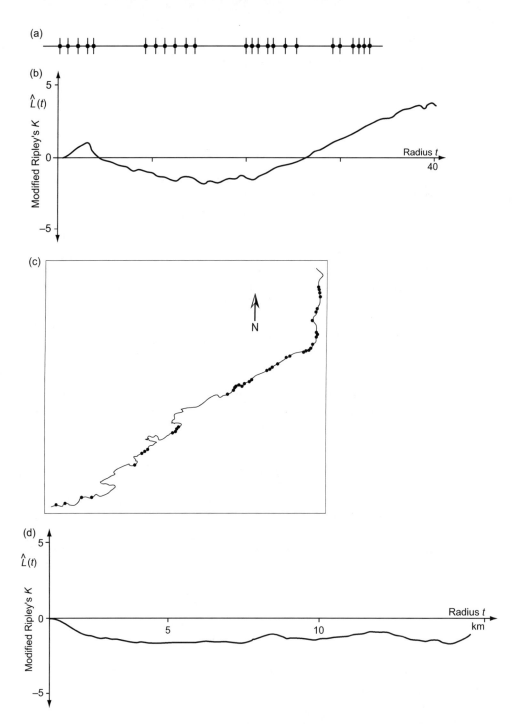

Figure 3.9 (*a*) Artificial one-dimensional transect data (clumps). (*b*) Standardized Ripley's K analysis on one-dimensional transect data. There is overdispersion at both very small and very large scales. $\hat{L}$ is the standardized Ripley's K and t is distance. (*c*) Rapids on the Morris/Pipestone River. (*d*) Standardized Ripley's K analysis of the Morris/Pipestone River rapids data.

3.4.2 Three Dimensions

The analysis of spatial patterns in three dimensions has received less attention, in part because such data used to be encountered less. Many of the methods described for two dimensions can be adapted for three. K-function analysis for three dimensions is a straightforward modification from 2-D (Baddeley *et al.* 1987; König *et al.* 1991). With the usual notation, and V being volume, calculate:

$$\hat{K}(t) = V \sum_{\substack{i=1 \\ i \neq j}}^{n} \sum_{\substack{j=1 \\ j \neq i}}^{n} w_i(t) I_t(i,j) n^2. \tag{3.24}$$

Count the number of events within a sphere of radius t centred on each event. The edge correction factor $w_i(t)$ is 1 if the sphere, centred on i and radius t, is completely within the study volume; otherwise, it is the reciprocal of the proportion of the sphere that lies within the study volume. Because the volume of the sphere is $4\pi t^3/3$, to make the response linear, calculate:

$$\hat{L}(t) = t - \sqrt[3]{3\hat{K}(t)/4\pi}. \tag{3.25}$$

Under CSR, the expected value is 0 and significant departures from 0 are interpreted in the usual way. Extensions of this approach to bivariate and multivariate analysis will proceed with the modifications described for two dimensions. Computing 3-D, Ripley's K can be obtained using the function *K3est* of the R package *spatstats*.

In addition to 3-D *K-function* analysis, König et al. (1991) described three-dimensional versions of the 'event-to-nearest-event' and 'random-point-to-nearest event' statistics, versions of the G-function and F-function, as well as an analysis of a three-dimensional marked point process. These three-dimensional approaches have been applied only rarely to ecological examples.

In the second edition of this book, we wrote 'Now that it is well-known that methods for three-dimensional pattern analysis are available and easy to apply, we expect to see further applications of these methods in ecological studies.' How true! The title of Jackson et al. (2020) tells it all: 'Three-dimensional digital mapping of ecosystems: a new era in spatial ecology.' Three-dimensional data can be obtained using LiDAR (Light Detection and Ranging). LiDAR pulsed laser remote sensing technology can be deployed from satellites, planes or drones to detect and evaluate the three-dimensional structure of features like forests by recording a point cloud of the heights of signal returns. Because there can be multiple returns at the same location, the data forms natural layers of return ranks: first returns are the top of the canopy and last returns are the ground. Clusters in the point cloud reveal the shapes of individual trees or shrubs which can then be classified further, for example as deciduous or coniferous. The process can be automated through cluster detection and trained machine learning to characterize the structural composition of different areas of the three-dimensional map of canopy types, densities, heights, and so on.

3.5 Circumcircle Methods

Counting point pattern events in circles can be extended to consider new ways of locating the circles. Each triplet of points defines a triangle and a circle that goes through all three of its corners, the circumcircle. Counting the events in those circles follows Ripley's approach, but also relates to the Delaunay triangulation, in which the edges are determined by circumcircles containing no other point events (Okabe et al. 2000; Chapter 2).

For a mapped plot, area A with n events, there are approximately $n^3/6$ circumcircles. For the kth circumcircle, there are n_k events inside it (excluding the three that define it) and a_k is its area within the plot. The expected count is $e_k = (n-3)a_k/A$, based on CSR. The Freeman–Tukey standardized residual compares the observed and expected:

$$z_k = \sqrt{n_k} + \sqrt{n_k + 1} - \sqrt{4e_k + 1}. \tag{3.26}$$

Values less than -1.96 indicate gaps and greater than 1.96 indicate patches, but low-density areas may produce overlapping 'gap' circles and high density may produce overlapping 'patch' circles. To distinguish among competing overlapping circles, we can consider the 'best' to be those that contrast most with their surroundings. To find these, count the number of events in a ring around each circle; for circle radius r_k, ring's width is $(\sqrt{2}-1)r_k$ to give the same area as the circle. The ring count is p_k with an expected value f_k, and its standardized residual is:

$$\varsigma_k = \sqrt{p_k} + \sqrt{p_k + 1} - \sqrt{4f_k + 1}. \tag{3.27}$$

The contrast between circle and ring densities is the statistic:

$$Z_k = (z_k - \varsigma_k)/\sqrt{2}. \tag{3.28}$$

Because Z is calculated on a double circle template, it can be considered as a wavelet, the 'boater' wavelet (Figure 3.10, Section 3.8; Dale & Powell 2001). Z measures how well the data match the shape of the template. Therefore, plot Z^2 as a function of the circle radius to determine the peaks that reflect the sizes of patches and gaps (see Dale & Powell 2001). The results can be made spatially explicit by associating the z or Z score in a size class with the centre of the circle. In that way, a contour map of the scores could be produced for each of several size classes or scales.

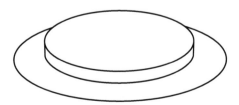

Figure 3.10. The 'boater' wavelet used in circumcircle analysis.

Bivariate and multivariate versions of this approach are straightforward adaptations of the univariate version, as described in Dale and Fortin (2014). An additional and interesting analysis for multispecies maps would be to calculate species diversity and species evenness for each circle. A spatially explicit result could be derived from the diversity scores, given a threshold value, to be superimposed on the original map of events. More discussion of diversity and of local measures of diversity for spatial analysis is provided in Chapter 9.

3.6 Areal Unit Analysis

Data are often collected in units that have two-dimensional area, whether they are sample units (scattered, systematically placed or contiguous), study areas (defined by natural processes or created) or experimental units (naturally or deliberately demarcated). Such circumstances require approaches different from those that start with dimensionless points with locations because the areal units do have locations but they also have sizes and shapes, and usually boundaries and neighbours that may or may not share boundaries between them.

3.6.1 Quadrat Variance Methods

We will begin with data in one dimension, a series of values, x_1 to x_n, representing density, counts or presence–absence data in a transect of n contiguous units. The spatial pattern of a single species can be studied with methods that examine how variance changes with scale: low values indicate similarity and high values indicate dissimilarity. The distances that produce high and low variance are related to the scales of patterns and the sizes of patches and gaps in the data.

These methods can be extended to bivariate and multivariate data and to two or more spatial dimensions. To compare with variance-based approaches, we describe techniques based on wave forms such as spectral and wavelet analysis. It turns out these approaches are very similar to the quadrat variance methods and we conclude by comparing these methods.

The variance methods originated from examining counts in sampling units for evidence of spatial non-randomness. The underlying argument is that, if the events occur randomly, the counts per unit should follow a Poisson distribution, which has the property that the variance and the mean are equal. The recommendation was then to calculate the variance-to-mean ratio of count data as an index of dispersion:

$$I_D = s_c^2 / m_c, \qquad (3.29)$$

where s_c^2 is the sample variance of the counts and m_c is their sample mean. The interpretation was that values around 1 indicate randomness, greater than 1 indicate clumping and less than 1 indicate overdispersion. Of course, the logic of this recommendation is imperfect (see Dale 1999, figure 7.16), but the use of the variance among counts for pattern analysis became well-established.

To improve the imperfect variance-to-mean index, one modification is to make the calculation for a range of unit sizes, to determine how scale (as grain) affects the outcome because plants can be overdispersed at small scales but clumped at larger scales. However, there are limitations to the sizes that can be sampled and using scattered quadrats loses the spatial information of their relative positions. A second improvement is to sample many small contiguous units to be combined into larger blocks; spatial patterns can then be assessed for a range of block sizes.

Two 'blocked quadrat' methods used contiguous quadrats from a transect and combined them into blocks for a range of sizes (Dale 1999). The first, *two term local quadrat variance* or TTLQV, has a two-part window of $b + b$ units, as the calculation template. The block size, b, affects the length of the window and the distance between its halves. It is:

$$V_2(b) = \sum_{i=1}^{n+1-2b} \left(\sum_{j=i}^{i+b-1} x_j - \sum_{j=i+b}^{i+2b-1} x_j \right)^2 \bigg/ 2b(n+1-2b). \qquad (3.30)$$

The sums within the brackets are the block totals and the outer sum creates the average over all window positions. This variance is calculated for $b = 1, 2, 3, \ldots$, and, when plotted against b, its peaks are interpreted as indicating scales of pattern in the data (Dale 1999), as illustrated in Figure 3.11. This method is often applied to density or cover data, but it can also be applied to counts or binary data.

An alternative to blocking is a two-part template with each part a single unit. This gives *paired quadrat variance*, PQV, and only the spacing between the halves of the template change, d, not their size:

$$V_p(d) = \sum_{i=1}^{n-d} (x_i - x_{i+d})^2 / 2(n-d). \qquad (3.31)$$

The terms within the bracket are single quadrat values and the outer sum creates the average over all positions of the window for spacing d. As for TTLQV, peaks in the plot of V_p as a function of d are interpreted as indicating scales of pattern in the data. TTLQV and PQV are similar in that they both use a two-part window but, in PQV, the size of the window does not change, only the spacing; in TTLQV, both the size and distance between the centres of the two parts change.

Both TTLQV and PQV can be extended to three-parts, *three term local quadrat variance*, 3TLQV (Hill 1973), and *triplet quadrat variance*, tQV (Dale 1999). 3TLQV is:

$$V_3(b) = \sum_{i=1}^{n+1-3b} \left(\sum_{j=i}^{i+b-1} x_j - 2 \sum_{j=i+b}^{i+2b-1} x_j + \sum_{j=i+2b}^{i+3b-1} x_j \right)^2 \bigg/ 8b(n+1-3b). \qquad (3.32)$$

For tQV, calculate:

$$V_t(d) = \sum_{i=1}^{n-2d} (x_i - 2x_{i+d} + x_{i+2d})^2 / 4(n-2d). \qquad (3.33)$$

3.6 Areal Unit Analysis

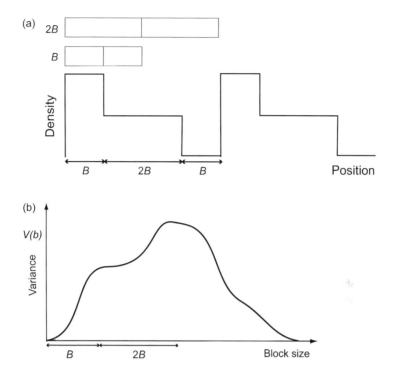

Figure 3.11 (*a*) The test data are created by adding together two square wave patterns, with scales of *B* and 2*B*. (*b*) The variance (TTLQV) is plotted as a function of block size; peaks (at scales *B* and 2*B*) are interpreted as corresponding to the scales of pattern in the data.

The contents of the main brackets are collected by the window templates and the outer sums generate averages over positions. Peaks in the variances again indicate scales of pattern and the three-part windows are less sensitive to trends in the data (Dale 1999).

In many forms of data analysis, a key step is a test of statistical significance. In spatial analysis, the purpose is usually exploratory, with a focus on hypothesis development. Nevertheless, statistical significance is definitely of interest (and could be very useful!) but is made difficult by several forms of lack of independence in the data, and in how the methods use the data.

First, the values found in adjacent quadrats will tend to be more similar than those at some distance from each other. That fact is part of the underlying logic of spatial pattern analysis, but it also represents spatial autocorrelation in the data (see Chapter 4) which can complicate the evaluation of statistical tests. Positive spatial autocorrelation tends to make statistical tests too liberal: they give more apparently significant results than the data actually justify.

The second form of lack of independence lies in the fact that each piece of data is used more than once, in calculating both the variance at a single block size and the variances at different block sizes. This lack of independence makes it difficult to provide statistical tests to evaluate the results of a single analysis or to compare results.

Randomization procedures can be used to help evaluate the significance of detected patterns in data but, because that kind of assessment requires both the data and a considerable number of reanalyses, it is not always feasible. In some cases, restricted randomizations in which the spatial structure is preserved are possible, as described in Chapters 1 and 2.

In the previous edition of this book (Dale & Fortin 2014), we pursued the derivation of parameters on which to base significance testing but concluded that it was not worthwhile. The values derived for the mean and variance depend strongly on the underlying distribution of the data and we never know the underlying distribution or whether the distribution is stationary. For that reason, notwithstanding analytical complexities, these methods will remain techniques for data exploration, not for statistical testing, except when a meaningful restricted randomization approach is available and perhaps not even then.

3.6.2 Two or More Species

An obvious extension of single-species analysis is to examine the associations of pairs of species, looking at the effect of scale on covariance (Greig-Smith 1961). Adapting the quadrat variance methods is based on the relationship between the covariance and the variances:

$$\text{Cov}(A, B) = [\text{Var}(A + B) - \text{Var}(A) - \text{Var}(B)]/2. \tag{3.34}$$

This formula can be applied to any of the methods to produce corresponding covariances TTLQC, tQC, and so on. Figure 3.12 shows the application of 3TLQC to binary data from a study of sedge meadows on Ellesmere Island (Young et al. 1999). The relationship between *Eriophorum triste* and *Carex aquatilis* changes with block size from negative to positive. The negative association at small scales can be attributed to the divergent ecological 'preferences', one more mesic and one semi-aquatic, but coexisting in a wet hummock–hollow habitat.

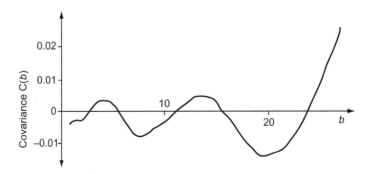

Figure 3.12 Covariance analysis (3TLQC) for the Ellesmere Island sedge meadow data, where b is the block size. The association of *Carex aquatilis* and *Eriophorum triste* cycles between positive and negative.

3.6 Areal Unit Analysis

In many cases, we want to evaluate the spatial pattern of a whole community, but the analysis of all possible pairs of species may be difficult to interpret and so methods for evaluating all species together are preferred: multispecies analysis. The recommended method for evaluating multispecies patterns is the one introduced by Noy-Meir and Anderson (1971) based on a combination of the quadrat (co)variance calculations described in Equation (3.34) and the ordination method of principal components analysis (PCA; Legendre & Legendre 2012). This is usually called multiscale ordination (MSO; Ver Hoef & Glenn-Lewin 1989; Wagner 2004).

From k species' densities, x_1 to x_k, in a string of n quadrats, calculate the $k \times k$ variance–covariance matrix for each block size, b, from 1 to some maximum, B: call the matrices $\mathbf{C}(1), \mathbf{C}(2), \ldots, \mathbf{C}(B)$. The variances and covariances can be from any of the methods described in Section 3.6.1, but we recommend 3TLQV and 3TLQC because they are less affected by trends (Dale & Zbigniewicz 1995). The matrices, $\mathbf{C}(b)$, are summed for $b = 1, 2, \ldots, B$, and that sum matrix is eigenanalysed, as in principal components analysis. Eigenanalysis creates linear combinations of the original variables (the species densities, x_1 to x_k) that are mutually orthogonal and the linear combinations, y_1 to y_k, in order, must explain as much of the total variance in the data as they can. For each y_i, its eigenvalue, λ_i, is the proportion of the total variance that it explains and its eigenvector is the vector of the weights in the linear combination of the x's producing y_i.

In this method, the largest eigenvalues are each partitioned into the amounts of variance contributed by each block size, using the weights in the eigenvectors, and these are plotted as a function of scale. Peaks or plateaus in each of these variance plots are interpreted as scales of patterns. Because larger block sizes tend to dominate the analysis, the covariance matrices should be weighted prior to summing and Dale and Zbigniewicz (1995) suggested weighting of the variance–covariance matrix for block size, b, by $6b^2/(b+2)$. One feature of interest is the relative strength of the contributions of the various species to the pattern. If one species overly dominates an eigenvector, the pattern detected is not truly multispecies. A measure of species' contributions can be based on the weights from the eigenanalysis that provide the linear combination of the original species densities and is given by the eigenvector:

$$y_i = \sum_{j=1}^{k} u_{ij} x_j, \quad \text{with} \quad \sum_{j=1}^{k} u_{ij}^2 = 1. \tag{3.35}$$

How evenly the species contribute to the pattern can be measured using the variance of the absolute values of the weights, u_{ij}. If all species have equal weights, the variance will be 0; and if one weight is 1.0 and the rest are 0, then the variance is $(k-1)/k^2$. Consider the coefficient of variation, which is the square root of the sample variance, $s_i(u)$, divided by the mean, $m_i(u)$. Its maximum value is the $k^{0.5}$, and a measure of the evenness of the ith eigenvector is therefore:

$$E_i = 1 - s_i(u) / \left[m_i(u) k^{0.5} \right]. \tag{3.36}$$

The measure may be clarified by an example. We established six 50 m transects at a site near Fort Assiniboine, Alberta, which was dominated by Jack Pine and Aspen, and we sampled the understorey in each transect with 200 contiguous 25 × 25 cm quadrats. The species list is typical of the Canadian boreal forest, including vascular plants, such as *Linnaea borealis*, and feather mosses, such as *Ptilium crista-castrensis*. We analysed the data with multiscale ordination based on 3TLQV and 3TLQC. Using the southern east–west transect as an example, we find that, for the 12 most common species, the first three axes explained 22.6, 16.8 and 15.2% of the variance (55.6% in total). The evenness of the eigenvector weights was also high, 0.868, 0.801 and 0.819, indicating the true multispecies pattern, not domination by any single species. Figure 3.13a shows the data for this transect and Figure 3.13b shows the partitioned variances as a function of block size for the first three eigenvalues. The three axes provide clear evidence of a pattern in the range of 17–27 units (8.5–13.5 m), which accords with a visual assessment of the data.

3.6.3 Two or More Dimensions

The basic concepts of the quadrat variance methods, described for a single dimension, can also be extended to two-dimensional data collected on a grid of contiguous units. Again, any of the methods can be adapted for this purpose.

When the one-dimensional formula for tQV (Equation 3.33) is expanded to two dimensions, it becomes:

$$V_5(d) = \sum_{i=1}^{n-2d} \sum_{j=1}^{m-2d} \left(x_{i,j} + x_{i+d,j} - 4x_{i+d,j+d} + x_{i+2d,j} + x_{i,j+2d}\right)^2 / 20(n-2d)(m-2d).$$

(3.37)

The other extensions to two dimensions are simple in concept, but the equivalents of TTLQV and 3TLQV, $V_5(d)$ and $V_9(b)$, require long equations (see Dale 1999). They are understood more intuitively from the templates for their calculation, shown in Figure 3.14. Relaxing the concept of $V_5(d)$ to compare any sample unit with all units that are approximately distance d away, as in Figure 3.14c, comes very close to the estimation of the omnidirectional variogram, as described in Chapter 4.

Two-dimensional grid data are often studied using a lacunarity analysis based on the variance-to-mean ratio in units of different sizes. The technique is popular for two-dimensional grid arrays such as pixels of satellite images and has been used for the study of fragmentation in tropical landscapes and similar applications (McIntyre & Wiens 2000; Wu & Qi 2000; Swetnam *et al.* 2015; Newman *et al.* 2019).

The extension to three dimensions is possible, and the units in a three-dimensional array are called 'voxels', to parallel the term 'pixel'. Voxel analysis is commonly applied in medical imagery and also in computer graphics and computer games. Applications of voxels are still rare in ecological studies but Kükenbrink *et al.* (2017) and Wang *et al.* (2008a) used LiDAR point clouds as the basis for modelling the canopy layer of a forest or the crowns of individual forest trees. Among the

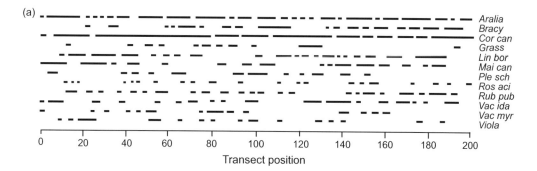

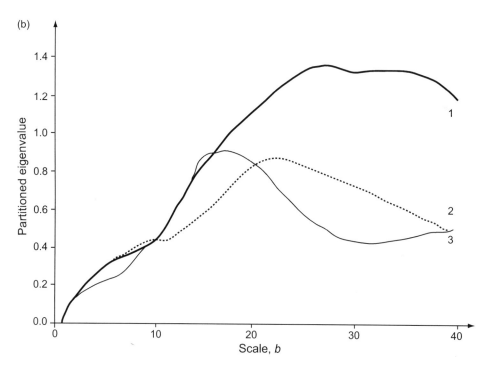

Figure 3.13 (*a*) Presence–absence of 12 understorey species along a 50 m transect at Fort Assiniboine, Alberta. (*b*) Partitioned variances as a function of block size for the first three eigenvalues in multiscale ordination of the data in (*a*). Key to species abbreviations: Aralia, *Aralia nudicaulis* L.; Brachy, *Brachythecium* sp.; Cor can, *Cornus canadensis* L.; Grass, Poaceae; Lin bor, *Linnaea borealis* L.; Mai can, *Maianthemum canadense* Desf.; Ple sch, *Pleurozium schreberi* (Brid) Mitt.; Ros aci, *Rosa acicularis* Lindl.; Rub pub, *Rubus pubescens* Raf.; Vac ida, *Vaccinium vitis-idaea* L.; Vac myr, *Vaccinium myrtilloides* Michy; Viola, *Viola* sp.

ecological applications of LiDAR, we note its application to relate the structural complexity of forest plots (Shannon entropy for 1 m 'bins' of data) to the observed diversity of small mammals (Schooler & Zald 2019).

As the three-dimensional methods become better known, ecologists will see their usefulness for their own studies and take advantage of them more frequently, especially as the data for this analysis becomes more and more readily available.

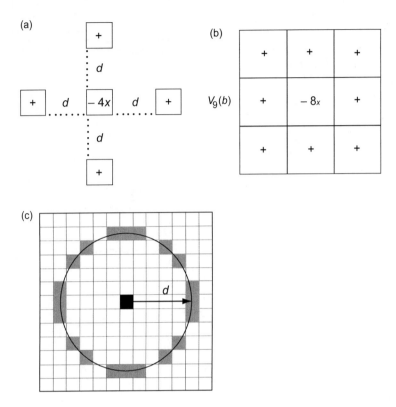

Figure 3.14 (*a*) Template used for the calculation of $V_5(d)$. (*b*) Template used for the calculation of $V_9(b)$. (*c*) An illustration of extending the $V_5(d)$ concept to compare a sample unit to all others at a distance of approximately d.

3.7 Spectral Analysis and Related Techniques

Spectral analysis is designed to detect a repeating pattern in quantitative series by fitting sine and cosine functions to the data, thus determining which frequencies or wavelengths fit best (Ripley 1978). The data are measured variables in continuous or evenly spaced series. The approach works best for 'rich' data sets, with large numbers of observations, and where assuming stationarity is justified. One well-known technique for spectral analysis is the Fourier transform, which decomposes the 'signal' (the data) into sine and cosine waves of various frequencies and starting positions (Figure 3.15*a*; see Legendre & Legendre 2012). A closely related technique is the Walsh transform which decomposes the signal into square waves instead of sine waves, of various frequencies and positions (Figure 3.15*b*; see Ripley 1978). Spectral analysis can be applied to two-dimensional ecological data. It was originally developed for continuous signals but it can also be applied to point pattern data (see Mugglestone & Renshaw 1996). As a bridge to the next section, we can comment that spectral analysis is usually associated with the global analysis of long data series, rather than working on local detail.

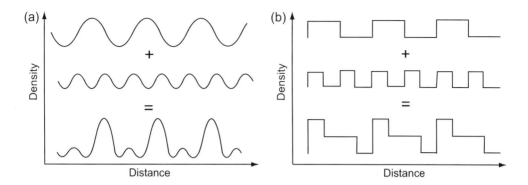

Figure 3.15 Wavelet template shape. (*a*) The concept of Fourier transform: the combined pattern of sine and cosine waves is resolved back into its original components. (*b*) A Walsh transform resolves a combination of different frequencies of square waves into its components.

3.8 Wavelets

Wavelet analysis is closely related to spectral analysis, but its wave template is finite for localized comparison with the data. This method provides both a local evaluation of the spatial structure and a global measure of fit using the wavelet variance, which gives variance peaks at scales of pattern in the data, as in the quadrat variance methods. The analysis evaluates how well the wavelet template, at different sizes, matches the data over a range of positions. The data used for this approach are typically quantitative in a continuous or evenly spaced series. Where g is the wavelet function, its transform, T, is a function of the wavelet size and position:

$$T(b, u_i) = \frac{1}{b} \sum_{j=1}^{n} y(u_j) g\left[(u_j - u_i)/b\right], \tag{3.38}$$

where b is the wavelet's relative width and $y(u_j)$ is the data value at position u_j. The transform evaluates the inner product of $y(u)$ with the wavelet function, localized in size and position (Daubechies 1992). $T(b, u_i)$ takes large positive values when the match of the wavelet and the data centred at u_i is very good and large negative values when the match is bad (see Dale 1999, figure 9.8, or Dale *et al.* 2002, figure 6). As a technical note, the wavelet transform given in Equation (3.38) is the discrete form, using summation, whereas the continuous wavelet transform uses integration. Discrete wavelet transforms can be computed using the *dwt* function of the R package *waveslim*.

Wavelet analysis can employ many different templates, but they must have an integral of zero; four are shown in Figure 3.16: 'Mexican hat' (probably the most common), the Haar, the French top hat (FTH) and the Morlet.

A wavelet variance can be defined based on the transform, as:

$$V_W(b) = \sum_{i=1}^{n} T^2(b, u_i)/n. \tag{3.39}$$

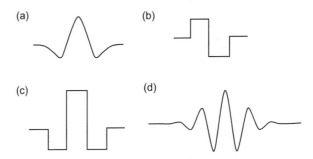

Figure 3.16 Wavelet templates of different shapes: (*a*) Mexican hat, (*b*) Haar, (*c*) French top hat, and (*d*) Morlet.

The wavelet variance based on the Haar wavelet is equivalent to TTLQV and the French top hat is equivalent to 3TLQV (Dale & Mah 1998). Both are also related to spectral analysis using the Walsh transform, which decomposes a signal into combinations of square waves of different frequencies (cf. Ripley 1978). The FTH wavelet is also similar in form to the 'boater wavelet' introduced in the discussion of circumcircle methods.

Whatever wavelet is used, wavelet variance analysis can be modified to give a wavelet covariance for bivariate data using Equation (3.39), thus leading to multivariate analysis. Ding and Ma (2021) chose the complex-valued Morlet to examine interspecific associations among soil invertebrates in a transect of 119 10 × 10 cm quadrats. The wavelet correlation method produced networks of species associations, allowing the assessment of how those change along the gradient at a range of scales.

Wavelet analysis can also be applied to two-dimensional data such as densities measured in a plane, such as vegetation cover in a grassland (Csillag & Kabos 2002). It can also be combined with scale-specific regression to assess the relative importance of environmental variables on the spring phenology of European vegetation (Carl et al. 2016). As in other aspects of spatial analysis, there is a trend toward extending methods to spatiotemporal analysis and that is true also of wavelet analysis. As one example among many, Dastour et al. (2022) highlighted the advantages of wavelet algorithms in a spatiotemporal analysis of climate and vegetation in the Athabasca River basin in Alberta, Canada.

Wavelets have many applications in many fields of research from geology to medicine and in technology, including image compression and analysis, and we will briefly discuss related properties of wavelets as they emerge in a description of boundary detection (see Chapter 5).

3.9 Concluding Remarks

The range of data types and analysis methods in this chapter is wide but the coverage is not exhaustive. The practising ecologist may encounter different kinds of data or

3.9 Concluding Remarks

requirements for different analysis. These situations may be completely novel (that is a challenge!) or they may have been encountered before, but are not easy to find in the literature. We will describe a few such situations.

Many of the methods used for the analysis of the positions of events in two dimensions are based on the assumption that the events can be adequately represented as dimensionless points in a plane. In the analysis of a particular set of data, it may become clear that the assumption is not sufficiently realistic and the events should be treated as circles of non-zero radius: what then? There is even a discussion of solving the same problem in three dimensions, considering the dispersion of spheres, instead of circles. Wiegand *et al.* (2006) considered alternatives to points as the units of interest for the spatial analysis of discrete objects.

For our recommendations on methods, it is clear that those based on the *K*-function cover many different kinds of data and a range of situations. That set of methods is highly recommended, although there may be some debate about the relative advantages of whole circles for counts (Ripley's original method) or rings of different widths. The decision on the method to use will depend, of course, on the data and the question being asked, but we suggest using two or more complementary approaches (Goulard *et al.* 1995 is a good example), so as not to miss important features of the data.

In considering the material in this chapter, we can identify three themes:

The first is the 'relatedness' of the various approaches used, both conceptually and mathematically. The basic concept of wavelets, the comparison of the data with some kind or shape of templates, recurs and unifies diverse approaches. Not just wavelets, but the *K*-function, TTLQV, and many of the other techniques (almost all!) can be considered in this way. Box 3.1 illustrates this fact. Mathematically this same fact can be expressed as a cross-product (see Getis 1991; Dale *et al.* 2002). Some of the methods included may decline in popularity, as TTLQV may be superseded by wavelet analysis (but see Karami *et al.* 2015; Yang & Zhou 2017). We have retained some treatment of the older techniques, not for historical context but because the concepts and the mechanics are basic and provide the conceptual and mathematical basis for the methods that followed. Methodological development continues, with the greatest occurring, it seems, in the application of wavelets (see Floryan & Graham 2021, Dastour *et al.* 2022) and refinement of multivariate and marked point pattern methods (see Ballani *et al.* 2019; Rajala *et al.* 2019; Ben-Said 2021).

Many of the methods can also be explained in the context of graph theory for spatial graphs. Each instance of the PQV template is a spatial graph with two nodes, one in the centre of each quadrat of the pair, one with weight $+1$ and one with weight -1, joined by an edge of length equal to the scale being tested (not a very interesting graph, we admit!); $V_5(d)$ has a five node template with nodes weights of -4 for the central node and $+1$ for the four 'arms' with edges length d (Figure 3.14a). Similarly, the template for scale b in TTLQV is a spatial graph of $2b$ nodes, the first b nodes with weight $+1$ and the second b nodes with weight -1, joined in a string by edges all of length 1 (again not very exciting ...). These small graph templates are reminiscent of the small subgraph structures called motifs (Milo *et al.* 2002) that are used in the analysis of complex network structures (Dale & Fortin 2021). Table 3.1 and Box 3.1

Table 3.1 Summary of the spatial analysis methods presented in Chapter 3

Spatial analysis method	Template		
Points			
Nearest neighbour	Expanding circle based on event		Simple linkage
Refined nearest neighbour	Expanding circle based on event or random point		Simple linkage
Univariate K	Expanding circle based on event		
Getis	Expanding circle based on event or random point		
Condit	Rings centred on events		
Bivariate K	Expanding circle		

Table 3.1 (cont.)

Multivariate K	Expanding circle	
Mark correlation	Expanding circle	
Events networks	Links	
One-dimensional Ripley's K	"t - bar"	
One-dimensional lacunarity	Moving window/gliding box	
Three-dimensional nearest neighbour	Expanding sphere	Simple link
Three-dimensional Ripley's K	Expanding sphere	
Contiguous units		
TTLQV and TTLQC		
PQV and PQC		
3TLQV and 3TLCV		
tQV and tQC		
Lacunarity	One point window	

3.9 Concluding Remarks

Table 3.1 (*cont.*)

2DtQV = 5QV	$V_5(d)$ — cross pattern with center $-4x$, four $+$ arms at distance d	
3DtQV = 9TLQV	$V_9(b)$ — 3×3 grid with center $-8x$ and eight $+$ cells	
Two-dimensional lacunarity	(box with arrows)	Gliding box
Spectral analysis: Fourier	(sine wave)	Sine wave
Walsh transform	(square wave)	Square wave
Wavelets		
One-dimensional data	(Mexican hat shape)	Mexican Hat
	(Haar shape)	Haar
	(French top hat shape)	French Top Hat
	(Morlet shape)	Morlet
	(sine shape)	Sine
Two-dimensional data	(sombrero 3D shape)	Sombrero
Circumcircle	(triangle in circle)	Simple
	(triangle in circle with $-/+$; boater hat)	Double = boater wavelet

Box 3.1 Comparison of Spatial Analysis Methods with Rules for Spatial Graphs
In this chapter, we have described several methods for the analysis of patterns in transects or grids of sample units like quadrats. We also reviewed different methods for point pattern analysis and, in Chapter 2, we described different rules for the creation of edges in a spatial graph. These included rules based on distance thresholds, topological considerations and functional relationships between nodes. All the methods for point pattern analysis rely on, or are closely related to, one of the spatial graph edge rules. The methods for quadrat transects and grids can also be described by spatial graphs, although their graphs are necessarily more uniform and regular in structure. The purpose of Table 3.2 is to make those relationships explicit.

Table 3.2 The relationships between point pattern analysis methods and spatial graphs

Point Pattern Analysis Method	Spatial Graph Edge Rule and Conditions
Univariate	
1st nearest neighbour distances (1° analysis)	nearest neighbour rule (NN network)
ith nearest neighbours	ith nearest neighbour graph
refined nearest neighbour distance	nearest neighbour and random points
Ripley's K (2° analysis) & Getis map	distance threshold
Condit's Ω	two-threshold distance graph (upper and lower threshold distances)
circumcircle and 3° analysis	Delaunay triangulation
Bivariate (categorical node labels)	
Ripley's K (2° analysis)	distance threshold & node labels
Condit's Ω	two-threshold graph & node labels (upper and lower distance limits)
circumcircle and 3° analysis	Delaunay triangulation & node labels
Multivariate	
Ripley's K & Condit's Ω	threshold distance graphs & node labels
circumcircle and 3° analysis	Delaunay triangulation & node labels
Quantitative node labels	
mark correlation	distance thresholds & node labels
Quadrat Variance Analysis Method	**Edge Rule and Other Graph Conditions**
PQV, etc.	two nodes at a time: weights $+1$ and -1 edge length = scale investigated
TTLQV, etc.	$2b$ nodes at a time: weights $b + 1$'s and $b - 1$'s edges all unit length
3TLQV	$3b$ nodes: weights $2b + 1$'s and $b - 2$'s edges all unit length
PQV, TTLQV, etc. in 2D	unit length becomes $\sqrt{2}$ for diagonals
Wavelets	As for TTLQV; weights determined by wavelets' function

provide summaries of the methods in relation to the spatial graphs that are closely related.

The second theme in this chapter is the concern for using null hypotheses other than complete randomness for the evaluation of spatial structure. The most commonly used null model for point pattern is CSR, but we noted some examples where the Poisson–Poisson distribution or a Markov inhibition model seemed more appropriate for comparison. The null hypothesis of complete randomness is not really expected to be true and so other hypotheses should be considered. More work needs to be done on this topic, especially in areas other than univariate point pattern analysis.

The third theme is the usefulness of Monte Carlo and randomization techniques to solve or circumvent problems with analytical approaches. It is often most useful to test the observed results against something that does not destroy all the spatial structures or assume its absence (see Chapters 1 and 2). Again, more work should follow, but this approach has found its place in our age of fast and easy computation.

In recommending methods to use, for point patterns, the set of methods based on Ripley's K-function covers many kinds of data and a range of situations. That set of methods has much to recommend it. On the other hand, this chapter shows that the set of methods based on wavelets combines flexibility, in template choice, with a conceptual sophistication that is very appealing. Those methods can also be adapted to a wide range of data types. The decision on the method will depend on the data and the question being asked, but using two or more complementary approaches may help in not missing important features.

Concluding this chapter, Box 3.1 gives a summary of the methods described, emphasizing similarities and differences by focusing on the template or window function used in the calculation of the statistic.

For novel data types or combinations of different objects, such as points and circles, some ingenuity may be required. Chapter 10 covers various combinations of points and lines. For greater complexity, say a mixture of points, lines, and circles, there may be little choice but to switch from object-oriented analysis to pixels and treat the complete spatial data set as a multilayer raster array.

4 Spatial Analysis of Sample Data

Introduction

Point pattern processes generate spatial distributions of point events, and their spatial structure can be evaluated by methods presented in Chapter 3 ('point pattern analysis') based on the events' locations in a study area. Point pattern processes are common but other ecological processes like surface pattern processes generate spatially continuous quantitative data (e.g. surface air temperature) that should be analysed by 'surface pattern' methods, as presented in this chapter.

The dichotomy between point and surface patterns is not absolute, however. Indeed, point data can be converted to surface data by counting events in sampling units to create density values. When the entire study area is surveyed using contiguous units, these quantitative data represent can be analysed using methods presented in Chapter 3. Yet, qualitative data could also be sampled using contiguous units and their spatial pattern can be analysed using join count statistics (presented in Section 4.1).

Sampling data using contiguous units requires, however, a high sampling effort in terms of time and cost. This is why most ecological studies are based on sampled data where variables are measured in units according to some sampling design (Chapter 2), based on the researchers' assumptions about the process under investigation. Then, the spatial pattern of the data can be inferred using a range of spatial statistics presented in this chapter. Most of these spatial methods share the notion that nearby values of a variable are more likely to be similar than distant ones: 'Tobler's first law of geography' (Tobler 1970); yet this may not be universally true.

An often-forgotten limitation of statistical inference is that the smallest spatial scale at which a variable is observed is the smallest grain size and that no information can be obtained below this resolution. Hence the measured value is assigned to the entire sampling unit assuming it is uniform within each sampling unit. Furthermore, the spatial layout of samples (Chapter 2) can affect the assessment of spatial patterns due to confounding (Fortin 1999*a*). The potential biases in evaluating spatial patterns will also be influenced by the spatial statistics chosen according to their assumptions and sensitivities to extreme values, as will be detailed in what follows.

4.1 Join Count Statistics

The occurrence of the same characteristics (binary, qualitative or categorical data) in adjacent, contiguous, sampled locations (e.g. lattice data, regions, counties) can be used to estimate and test spatial association or aggregation using 'join count statistics'. Join count statistics are based on the frequencies of co-occurrences of matches and mismatches in the characteristics and test whether the observed frequencies can be accounted for by randomness alone (complete spatial randomness).

Join count statistics were developed by geographers to test whether adjacent counties showed significant aggregation of diseases (Moran 1948; Cliff & Ord 1981). In such analyses, each county is given a binary value (black indicating disease incidence, white indicating none) as if the nominal attribute prevailed for the entire administrative polygon. With contiguous polygons (Figure 4.1), the 'join' adjacency is defined as polygons sharing part of a boundary, that is, first-order neighbours.

For binary data (e.g. black and white), there are three join count statistics: two that count matched pairs, J_{BB} (for black–black) and J_{WW} (for white–white), and one that counts mismatched pairs, J_{BW} (black–white). The J_{BB} and J_{WW} measure positive spatial association and J_{BW} measures repulsion (negative spatial association). The total number of joins (J) is determined by the spatial arrangement of the sampled locations and the connectivity algorithm used:

$$J = J_{BB} + J_{WW} + J_{WB}. \tag{4.1}$$

The observed join count statistic, J_{BB}, is the number of joins of adjacent sampling units, regions, in the same category of black:

$$J_{BB} = \frac{1}{2}\left(\sum_{\substack{i=1\\i\neq j}}^{n}\sum_{\substack{j=1\\j\neq i}}^{n}\delta_{ij}x_ix_j\right), \tag{4.2}$$

where i and j are the two sampling units, x_i is the attribute of unit i (black = 1 and white = 0) and δ_{ij} is the connectivity matrix entry for units i and j: 1 when i and j are

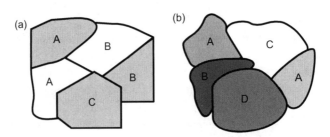

Figure 4.1 Spatial adjacency of polygons. (*a*) Arbitrary political boundaries of counties make it possible to have adjacent counties with the same category (here either white or grey). (*b*) Natural forest stands have, by definition, different types of adjacent forest surrounding them (different grey shades).

adjacent and 0 otherwise. The statistic J_{BW} counts the pairs of adjacent sampling units with unlike categories:

$$J_{BW} = \frac{1}{2} \left[\sum_{\substack{i=1 \\ i \neq j}}^{n} \sum_{\substack{j=1 \\ j \neq i}}^{n} \delta_{ij}(x_i - x_j)^2 \right], \qquad (4.3)$$

and J_{WW} can be computed using the two previous statistics:

$$J_{WW} = \frac{1}{2} \left(\sum_{\substack{i=1 \\ i \neq j}}^{n} \sum_{\substack{j=1 \\ j \neq i}}^{n} \delta_{ij} \right) - (J_{BB} + J_{BW}). \qquad (4.4)$$

This means that only two of the statistics are independent.

The null hypothesis of complete spatial randomness (CSR) of categories can be tested, assuming the spatial stationarity of the generating process, by computing the expected values using the observed and expected values for comparison with a Normal distribution, $N(0, 1)$. To establish the probability of each category:

- It is assumed that the assignments are independent for each unit (sampling with replacement) with $p = q = 0.5$; or
- It is treating the categories as dependent: the category in any one unit affects the categories in others (sampling without replacement), with $p = n_B/n$ and $q = n_W/n = (1 - p)$, with n_B black units and n_W white units.

Mathematical details are given in Cliff and Ord (1981).

Join count statistics can be assessed from lattice data (Figure 4.2) where the adjacency can be based using three neighbour rules: rook, bishop and queen. When sampling units are not contiguous, the units' centroids can be treated as nodes with adjacencies determined by any of the network algorithms presented in Chapter 2 (e.g. Delaunay triangulation). In such a network, join count statistics can be computed for higher-order neighbours: second-order, third-order and so on. An alternative is to compute the Euclidean distances between sampling units to create distance classes to determine adjacencies (Chapter 2).

Using simulated data with known spatial patterns (random, uniform and patchy), Figure 4.2 illustrates the results for rook, bishop and queen connectivity assessing spatial association. There are 100 units (50 black and 50 white) where the number of joins varies with the connectivity rule (rook = 180, bishop = 162 and queen = 342). For the random data, the rook joins suggest a significant difference from random but the other join rules' statistics are not significant. For the uniform data, both the rook and bishop connectivity rules give significant results while the queen rule does not. For patchy data, all three connectivity rules produced significant join count statistics. These examples highlight how sensitive these join count statistics may be for detecting spatial aggregation under different adjacency rules. These examples stress

Spatial Analysis of Sample Data

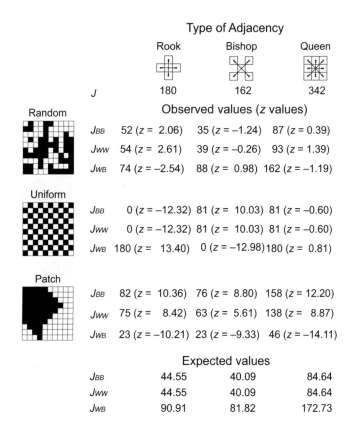

Figure 4.2 Join count statistics (J_{BB}, J_{WW} and J_{WB}). The expected values are based on the number of joins (rook = 180, bishop = 162 and queen = 342) and the number of black and white values ($n_B = 50$; $n_W = 50$). The observed values of the three statistics (and corresponding z values) are computed according to the three chess moves (rook, bishop and queen) based on binary data from a 10 × 10 lattice from three types of spatial patterns (random, uniform and patchy). This example stresses how the numbers of joins (i.e. the connectivity rule) affect the observed statistics and their significance. For example, in the case of the random pattern, the observed statistics based on the rook move are significant (values > |−1.96|), while these statistics are not significant with the bishop and the queen moves.

that there is a close relationship between join counts and detecting and locating spatial clusters (Anselin & Li 2019).

4.1.1 Join Count Statistics for *k*-categories

Join count statistics have been extended beyond the binary case to *k* categories where $k > 2$ (Cliff & Ord 1981). They have also been extended to consider neighbours beyond first order, defined by the neighbour network topology, or measured distances among sampling units as *d*-neighbours. Combining these two extensions, Equation (4.2) becomes:

$$J_{rr}(d) = \frac{1}{2}\left[\sum_{\substack{i=1\\i\neq j}}^{n}\sum_{\substack{j=1\\j\neq i}}^{n}\delta_{ij}(d)x_{ri}x_{rj}\right], \tag{4.5}$$

and Equation (4.3) becomes:

$$J_{rs}(d) = \frac{1}{2}\left[\sum_{\substack{i=1\\i\neq j}}^{n}\sum_{\substack{j=1\\j\neq i}}^{n}\delta_{ij}(d)x_{ri}x_{sj}\right], \tag{4.6}$$

where r and s indicate categories, $\delta_{ij}(d)$ indicates whether units i and j are adjacent in the distance class d (by d-neighbours or d-Euclidean distance), and x_{ri} is an indicator function that takes the value 1 when unit i belongs to category r, and 0 otherwise.

The spatial units we have considered so far have been arranged so that any pair of units can belong to the same category (i.e. having both matches and mismatches; Figure 4.1a) but, in some cases, this is not so where the units were delineated to obtain only mismatches (Figure 4.1b). Lowell (1997) demonstrated, based on simulations, that, when there is a minimum of five or so categories, the join-count statistic, J_{rs}, is robust and provides unbiased results. Join count statistics can be computed using the functions *joincount.test* and *joincount.multi* and their significance with the function *joincount.mc* of the R package *spdep*.

4.2 Global Spatial Statistics

When quantitative data are measured at sampling locations, the spatial pattern is estimated using second-order spatial statistics as described here.

4.2.1 Spatial Covariance

Ecologists are well-accustomed to the notion of covariance and correlation between two variables and of determining whether they covary positively, negatively, or not at all, by testing the null hypothesis of absence of correlation (i.e. 0). Linear correlation between quantitative variables can be estimated using Pearson's correlation coefficient. This coefficient is the standardized sample covariance between two variables (x and y) which measures the deviations of the variables from their respective means:

$$\hat{\rho}(x,y) = \frac{\sum_{i=1}^{n}(x_i - \bar{x})(y_i - \bar{y})}{\sqrt{\sum_{i=1}^{n}(x_i - \bar{x})^2 \sum_{i=1}^{n}(y_i - \bar{y})^2}}, \tag{4.7}$$

and which estimates their correlation,

$$\rho(x,y) = \frac{\text{Cov}(x,y)}{\sqrt{\text{Var}(x)\text{Var}(y)}}. \tag{4.8}$$

Stemming from this notion of covariation between two variables, the estimates of the covariance or the correlation of a single variable with itself (giving the prefix 'auto-') can be computed for all pairs of units that are separated by the same given spatial lag (spatial autocorrelation). The spatial autocovariance, $C(d)$, of x for distance d, can be estimated as the product of the deviation from the expected value, $E(x_i)$, at location i and the deviation from the expected value, $E(x_i)$, at location $i + d$:

$$C(d) = E\{[x_i - E(x_i)][x_{i+d} - E(x_i)]\}. \tag{4.9}$$

The spatial autocorrelation, $\rho(d)$, of x for distance d is the autocovariance at d divided by the variance, which is $C(d)$ for $d = 0$:

$$\rho(d) = \frac{C(d)}{C(0)}, \tag{4.10}$$

with

$$C(0) = \text{Var}(x) = \sigma^2. \tag{4.11}$$

The spatial structure of a variable can be estimated by various statistics derived from correlation (see Equation 4.8); all require the assumption of stationarity. These spatial statistics are 'global' because they estimate the spatial pattern for the entire study area producing a single value for each distance d.

Spatial pattern is the lack of independence among values due to location and results from endogenous and exogenous processes (Chapter 1), but it is not usually possible to disentangle the contributions of these two kinds of processes based on spatial statistics. The only way to determine their contributions is to combine prior knowledge of the processes and the factors acting on the data with an appropriate experimental design (see Chapters 2 and 8); spatial autocorrelation coefficients cannot discriminate between spatial structure inherent in the variable itself and that induced by dependence on other factors that have spatial structure of their own.

4.2.2 Spatial Autocorrelation Coefficients for One Variable

4.2.2.1 Moran's *I* Coefficient

Moran (1948) proposed a coefficient of spatial autocorrelation, Moran's *I*, computed for each distance class, d as:

$$I(d) = \left(\frac{n}{W(d)}\right) \frac{\sum_{\substack{i=1 \\ i \neq j}}^{n} \sum_{\substack{j=1 \\ j \neq i}}^{n} w_{ij}(d)(x_i - \bar{x})(x_j - \bar{x})}{\sum_{i=1}^{n}(x_i - \bar{x})^2}, \tag{4.12}$$

where $w_{ij}(d)$ is the distance class connectivity matrix that indicates whether a pair of sampling locations, i and j, are in distance class d; x_i and x_j are the values of x at i and j; and $W(d)$ is the sum of $w_{ij}(d)$, here the number of pairs of locations in the distance class, d. As with Pearson's coefficient, positive autocorrelation is indicated by positive values (0 to 1), negative autocorrelation by negative values (0 to −1) and the expected value with no autocorrelation is close to 0: $E(I) = -1/(n-1)$ (Cliff & Ord 1981). Moran's I may be favoured over other spatial statistics presented in the rest of Section 4.2 because it behaves like Pearson's correlation coefficient, facilitating its interpretation. The estimate from Equation (4.12) is the average value of autocorrelation at distance d in all directions and for the entire study area: a *global average isotropic* estimate.

As Moran's I values are computed as the deviation of each value to the arithmetic mean, $\bar{x}$, the estimate can be biased when the data are not normally distributed. A skewed distribution resulting from a few outliers may bias the estimate of the arithmetic mean, which will then produce under- or over-estimation. Furthermore, with too few pairs of locations in distance class d, the estimated value can fall outside the range of −1 to +1, most often at the largest distances that have the fewest pairs (Figure 4.3a). DeWitt et al. (2021) proposed two ways to rectify Moran's I to remove its lack of consistency while maintaining its intuitive value for interpretation.

To avoid some of the inconsistencies resulting from computing second-order spatial statistics from deviations to the mean, $\bar{x}$, Geary's c and semi-variance are computed using deviations based on the values between sampling locations according to distance classes.

4.2.2.2 Geary's c Coefficient

To avoid measures based on the arithmetic mean, Geary (1954) proposed to estimate spatial autocorrelation based on the difference between the values at location pairs in distance class d:

$$c(d) = \left(\frac{n-1}{2W(d)}\right) \frac{\sum_{\substack{i=1 \\ i \neq j}}^{n} \sum_{\substack{j=1 \\ i \neq j}}^{n} w_{ij}(d)(x_i - x_j)^2}{\sum_{i=1}^{n}(x_i - \bar{x})^2}. \quad (4.13)$$

Geary's c behaves like a dissimilarity, distance, measure that varies from 0 (positive autocorrelation) to 2 or more (negative autocorrelation) where the expected value of the absence of spatial autocorrelation is 1. When distance d is zero, Geary's c is 0. As for Moran's I, Geary's c based on too few pairs will give strange values, often greater than 2. Also, the estimated values of Geary's c will be biased in the presence of skewed data because the differences between adjacent locations are squared (Equation 4.13), which amplifies the effect of an extreme value and may, thus, distort the autocorrelation estimate.

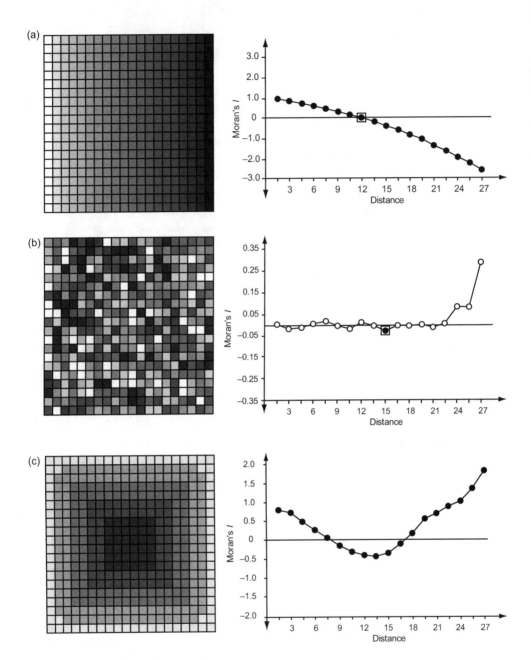

Figure 4.3 Moran's *I* spatial correlograms corresponding to simulated spatial patterns (a 20 × 20 lattice where the values increase from 0 to 10 from white to black). (*a*) Gradient: the correlogram shows the corresponding characteristic trend of significant positive values at short distances to negative ones at large distances. (*b*) Random: the values oscillate along the zero value (i.e. the absence of significant spatial autocorrelation). (*c*) One big patch: the values are all significant and positive at short and large distances, while negative at intermediate. The spatial range (zone of influence, patch size) is around 7.5 units, a distance at which the sign of the values changes from positive to negative. (*d*) Sixteen patches: a first change of sign from positive to negative values occurs around 2.0 units, which corresponds to the spatial range of the patches. Then the correlogram repeats this oscillation in decreasing amplitude with distance

4.2 Global Spatial Statistics

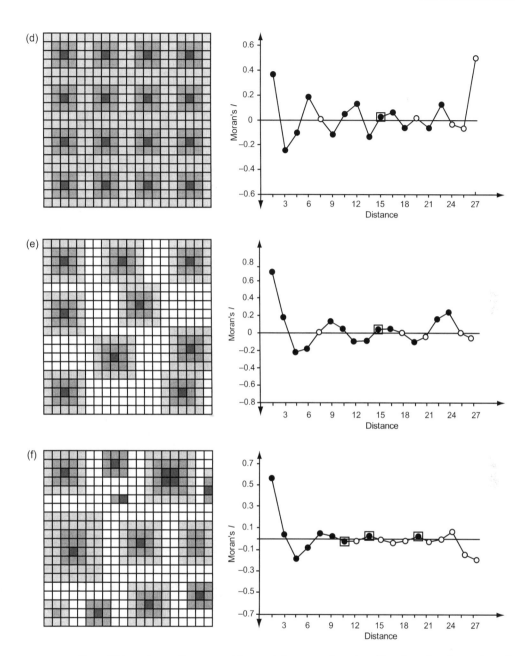

Figure 4.3 (cont.) revealing a repetitive spatial pattern of patches. Note here that the patches all have the same size and distance among them. (*e*) Nine patches: as described previously, the sign change of the values indicates that the patch size is around 3.0 units and that there is a repetitive pattern of patches. Note here that the patches have the same size but not the same distance among them. (*f*) Twelve patches: the sign change of the values indicates that the patch size is around 2.5 units. Here the repetitive pattern of patches is not detected by the correlogram as the patch size and distance among the patches vary. Solid circles indicate significant coefficient values at $\alpha = 0.05$; open circles indicate non-significant coefficient values; open squares indicate coefficient values that are non-significant after progressive Bonferroni correction.

4.2.2.3 Significance of Coefficient Values

Significance testing allows us to determine which spatial autocorrelation coefficient values can be interpreted legitimately for understanding the spatial structure. The significance of spatial autocorrelated value at a given distance class can be tested by a randomization procedure or by Normal approximation. Mathematical details can be found in Cliff and Ord (1981). Yet, the significance testing for several distance classes is problematic due to the lack of independence in the data used to estimate coefficients for many distance classes (Fortin & Dale 2009). This problem is inherent to all multiple testing procedures (e.g. quadrat variance methods; Chapter 3). Several procedures have been proposed to address this issue. The most widely used (Oden 1984) is the Bonferroni correction, which adjusts the probability level for significance, α, by dividing it by the number of distance classes, k:

$$\alpha' = \alpha/k. \tag{4.14}$$

For $\alpha = 0.05$ and $k = 10$, we get $\alpha' = 0.005$, and the probability threshold for each coefficient is, thus, $\alpha' = 0.005$. The Bonferroni-corrected value is affected by the number of distance classes selected, which may be somewhat arbitrary. To diminish this effect, a progressive correction test can be used where the Bonferroni-corrected level is computed for each distance class, one by one (Legendre & Legendre 2012). Hence, with the progressive Bonferroni tests at $\alpha = 0.05$ with a maximum of $k = 10$ $\alpha'(d = 1) = 0.05$, several α' can be used in sequence: $\alpha'(d = 2) = 0.025$ for the first comparison, $\alpha'(d = 3) = 0.0167$ for the third comparison, and so on up to $\alpha'(d = 10) = 0.005$ for the 10th comparison. For $d = 1$ to 10, Equation (4.14) is rewritten as:

$$\alpha'(d) = \alpha/d. \tag{4.15}$$

4.2.2.4 Spatial Correlograms

The characteristics of spatial structure include its magnitude (degree), spatial range (zone of influence) and shape (isotropic/anisotropic). To assess these characteristics, we plot the values of autocorrelation against distance d as a spatial correlogram, like Figure 4.3, where significance is indicated by symbols, such as solid when significant ($\alpha = 0.05$) and open when not. At distance zero, Moran's I is 1 and Geary's c is 0. Most ecological data show some positive autocorrelation, at short distances, the first values are usually positive; with the strongest autocorrelation in the first distance class. As the strength of the process decreases with distance, as in seed dispersal where more seeds fall closer to the tree, autocorrelation decreases with distance. How this appears in the correlogram can help characterize the pattern: a trend in the correlogram, from positive to negative autocorrelation, indicates a gradient in the data. A plot (Figure 4.3b) that levels off around the expected value of no autocorrelation indicates the absence of spatial pattern. When positive values at short distances are followed by somewhat regular fluctuation around the expected value, this can be interpreted as patchiness (Figure 4.3*c–f*) and a repeated alternation of values, positive and negative, can reflect the patch structure in the study area (Figure 4.3*d* and *e*). The distance at which autocorrelation first reaches the expected value is the spatial range, zone of influence or patch size of the spatial pattern, depending on the phenomenon being studied.

Interpreting the shape of the correlogram as indicating characteristics of the spatial pattern requires consideration of the significance of the entire correlogram using the Bonferroni or progressive Bonferroni correction described in Section 4.2.2.3. The progressive Bonferroni correction facilitates the interpretation of the correlogram as the non-significant values from the progressive correction at higher $\alpha'(d)$ levels are not considered.

Another approach that reduces the effect of multiple tests is to compute a *partial spatial correlogram* instead of a spatial correlogram. The technique corrects the autocorrelation at lag *d* for the lack of independence caused by observations intervening between the pairs of values separated by *d*. This is the spatial equivalent of the partial time series analysis using Durbin's autoregressive (or autoregressive – moving average) procedure. The mathematical details are briefly presented in Cliff and Ord (1981) but autoregressive and moving average models are discussed more fully in Chapter 6.

The interpretation of Geary's *c* correlogram (Figure 4.4) differs from Moran's *I* correlogram in that the values at short distances are close to zero and then, at a given distance, show fluctuation around the expected value, which is 1. When the assumptions of the spatial statistics are met then Geary's *c* correlograms are the upside-down versions of Moran's *I* correlograms. Again, somewhat regular fluctuations in the correlogram can be interpreted as indicating patchiness in the data (Figure 4.3*c*).

Moran's *I* and Geary's *c* spatial autocorrelation statistics can be computed using the respective functions *moran.test* and *geary.test* and their significance with the functions *moran.mc* and *geary.mc* while spatial correlation correlograms can be plotted using the function *sp.correlogram* of the R package *spdep*. Other R packages are also computing spatial autocorrelation statistics using the function *correlog* in *ncf* and *pgirmess* packages.

4.2.2.5 Anisotropic Spatial Correlograms

Up to this point, the calculation of spatial autocorrelation assumed implicitly that the underlying process is *isotropic* (the same in all directions, Chapter 2); these spatial correlograms are called omnidirectional or all directions (Figure 4.5*a*). Many ecological processes generate patterns that vary with direction (consider winds and water currents), resulting in elongated patches; these patterns are *anisotropic*. To determine the degree of anisotropy, spatial autocorrelation can be estimated by considering both distance (a distance class matrix) and direction using an angle class matrix (Oden & Sokal 1986). The angle class matrix indicates pairs of samples that have the

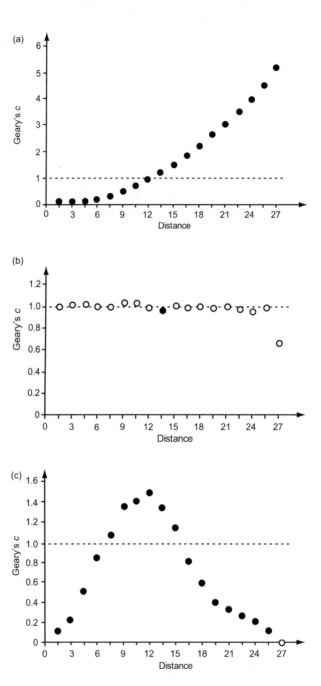

Figure 4.4 Geary's c spatial correlograms based on the same simulated spatial patterns from Figure 4.3. (*a*) Gradient: the correlogram shows the corresponding characteristic trend of significant positive spatial autocorrelation values (so between 0 and 1) at short distances to negative spatial autocorrelation values (so larger than 1) than ones at large distances. (*b*) Random: the values are oscillating along the expected value of 1. (*c*) One big patch: the values are all significant and positive (between 0 and 1) at short and large distances, while negative at intermediate (greater than 1). Solid circles indicate significant coefficient values at $\alpha = 0.05$; open circles indicate non-significant ones.

orientation within the specified search angle (the angular separation). To determine spatial structure related to wind direction, we can specify search sectors of 45°, centred on directions of interest (Figure 4.5b). Spatial autocorrelation is computed in each direction, producing directional correlograms (Figure 4.5b), and anisotropy is determined by comparing the shapes and ranges of these. In Figure 4.5b, two directional correlograms (45° and 135°) match in shape and range, but two do not, reflecting the strong anisotropy of the pattern. Estimating autocorrelations based on distance and direction implies fewer pairs of locations than the omnidirectional case. Therefore, we recommend using fewer distance classes (Figure 4.5) and no more than four direction classes. Anisotropy in the degree of spatial autocorrelation can be computed using the function *Sncf2D* in the R package *ncf*.

4.2.3 Variography

Parallel to the development of spatial statistics by geographers, mining engineers developed a family of statistics to estimate spatial patterns, known as geostatistics (Matheron 1970). Geostatistics uses the spatial variance, estimated from sampled data by computing the spatial variance to predict values at unsampled locations by modelling the spatial structure using techniques known as Kriging (Section 4.7.4). Here, we provide a brief overview of the estimation of this variance, referred to as variography. More information about the wide range of geostatistical techniques can be found in numerous textbooks (among others, Cressie 1993; Chilès & Delfiner 2012).

Geostatistics is based on the 'regionalized variable theory' assuming that the value of a variable z at location s is a realization of a random function $Z(x)$. The value, $z(x)$, is composed of three components:

$$z(s) = m(s) + \varepsilon(s) + \varepsilon, \tag{4.16}$$

where $m(s)$ is the deterministic structural function of the variable at location s; $\varepsilon(s)$ is the spatially dependent residual of $m(s)$, the spatial variance component; and ε is the spatially independent normally distributed residual component. Under the assumption of stationarity of the process, $m(s)$ is the average value of the variable within the study area. The expected difference between the values at two locations that are distance h apart is 0 such that the variance of this difference depends only on h. True stationarity over a study area is rare, and we can still proceed when only quasi-stationarity (weak stationarity) prevails: the first-order moment, $E(Z(s))$, and second-order moment, Var$(Z(s))$, are stationary within a relatively small neighbourhood (an isotropic moving window).

Notation in geostatistics differs from the notation in spatial statistics. For example, the lag is h rather than d, and Equation (4.10) is rewritten as:

$$\rho(h) = C(h)/\sigma^2. \tag{4.17}$$

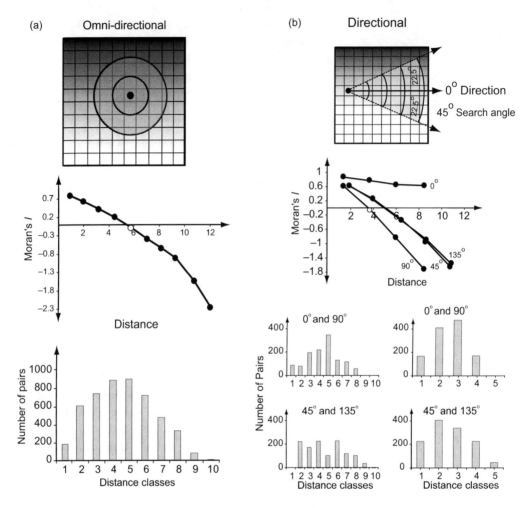

Figure 4.5 Moran's *I* spatial correlograms. (*a*) Omnidirectional correlogram based on 10 isotropic equidistant classes. The trend in the 10 × 10 lattice data is detected by the characteristic shape of the correlogram. The histogram of the pairs of sampling locations per distance class shows the edge effects at short and large distances, but, except for the last distance class, there are more than 20 pairs. (*b*) Directional correlograms were computed in four directions (0°, 45°, 90° and 135°) using a search angle of 45° (22.5° on each side). The directional correlograms are based on five equidistant classes; because, with 10 equidistant classes, the number of pairs of locations per distance and angle class is too low. With fewer classes, the number of pairs per distance and angle class increases, making the coefficient estimates and the comparisons more reliable. Here, the four directional correlograms allow us to detect anisotropy because the values of Moran's *I* do not all overlap one another (although the 45° and 135° ones are identical) and the correlograms have different spatial ranges (0°: no spatial range, 45° and 135°: around 4.0 units and 90°: 3.8 units).

4.2 Global Spatial Statistics

A direct relationship between spatial covariance, $C(h)$, and spatial variance, $\gamma(h)$, and the variance of the variable, σ^2, is this:

$$\gamma(h) = \sigma^2 - C(h). \tag{4.18}$$

Under the assumption of stationarity, the values of spatial autocovariance and spatial variance are mirror images (flipped). Similarly, the relationship between spatial autocorrelation and spatial variance is:

$$1 - \rho(h) = \gamma(h)/\sigma^2. \tag{4.19}$$

The spatial variance of the quantitative variable, z, is then estimated by the semi-variance function:

$$\hat{\gamma}(h) = \frac{1}{2n(h)} \sum_{i=1}^{n(h)} (z(s_i) - z(s_i + h))^2, \tag{4.20}$$

where z is the value of the variable at the ith sampling location, s_i, and $n(h)$ is the number of pairs of sampling locations separated by distance h. (Some texts use **h** instead of h to denote a vector of distance and direction). Because the sum ranges from 1 to n, each pair is included twice; hence, because of the division by 2, the spatial variance is called 'semi-variance'. The plot of the semi-variance values against the lag h is a *semi-variogram* but it is simply referred to as a 'variogram'. When the variogram is computed using empirical data, it is an *experimental variogram* (or *sample variogram*; Figure 4.6a). Then the experimental variogram is fitted to a *variogram model* or *theoretical variogram* (Figure 4.7).

The equation for semi-variance (Equation 4.20) is comparable to Geary's c equation (Equation 4.13) but without the division by the standard deviation, which standardizes the autocorrelation value. Hence, the semi-variance function is in the same units as the observed data, hence it is not bounded like Geary's c (0 to 2). At short lags, the semi-variance is small (close to zero), indicating that the spatial structure is at its strongest. As the lag increases, the semi-variance rises and levels off at a plateau called the sill (Figure 4.5a).

The shape of the variogram can provide insights into the spatial pattern that produced it. The experimental variograms in Figure 4.6, calculated from the simulated data in Figure 4.3, are similar to Geary's c correlograms in Figure 4.4 and are flipped compared to those for Moran's I in Figure 4.3.

Three parameters are usually estimated from an experimental variogram: the nugget effect, C_0, the range, a, and the sill, C_1 (Figure 4.6a). The nugget is the intercept when it is greater than zero. Theoretically, at $h = 0$, the semi-variance is also equal to 0 but it can be unrealistic sometimes to force the theoretical variogram to go through 0 and the nugget parameter accounts for the observed variability at short distances due to local random effects or measurement errors. The spatial range indicates the distance up to which the structure varies, the greatest distance for spatial autocorrelation. Beyond the

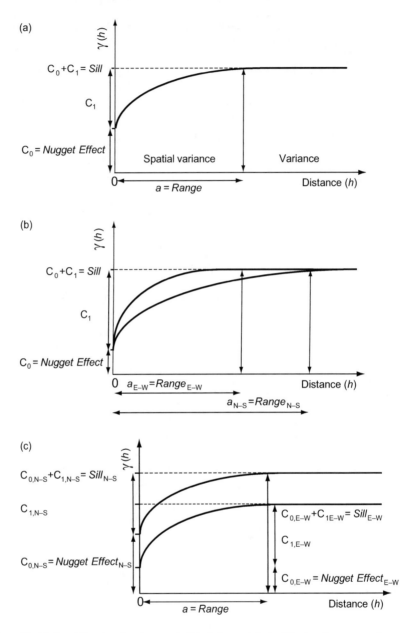

Figure 4.6 (*a*) Omnidirectional variogram (solid circles and dashed line) and corresponding theoretical spherical variogram (solid line) and its parameters: the range, the nugget and the sill (*h* indicates the spatial lag; $\gamma(h)$ is the semi-variance). Directional variograms (here in the north–south, N–S, and east–west, E–W, directions) allow the identification of anisotropic spatial variance in the data). (*b*) Geometric anisotropy is depicted as the range of values differed, but the sill values are the same. (*c*) Zonal anisotropy is depicted as the range values are the same, but the sill values differ.

range, the semi-variance values level off, forming the sill. One implication of a variogram that reaches a sill at a finite range is that pairs separated by distances greater than that range are effectively independent of each other. In some cases, the experimental variogram does not have a sill. This implies that the range is undetermined (Figure 4.7a) or that there is none. This may occur when the extent is smaller than the scale of the pattern or if differences in the underlying processes continue to accumulate.

The next step is to find the best variogram model to fit the experimental variogram. The most common unbounded variogram models (with no sill, Figure 4.7a) are:

- nugget:

$$\gamma(h) = C_0, \qquad (4.21)$$

- linear:

$$\gamma(h) = C_0 + bh, \qquad (4.22)$$

where b is the slope,

- exponential:

$$\gamma(h) = C_0 + C_1 \left\{ 1 - \exp\left(-\frac{h}{a}\right) \right\}. \qquad (4.23)$$

Bounded variogram models (with sill, Figure 4.7b) include:

- spherical:

$$\gamma(h) = C_0 + C_1 \left\{ \frac{3h}{2a} - \frac{1}{2}\left(\frac{h}{a}\right)^3 \right\} \quad \text{for } 0 < h < a, \qquad (4.24)$$

$$\gamma(h) = C_0 + C_1, \quad \text{for } h \geq a,$$

- Gaussian:

$$\gamma(h) = C_0 + C_1 \left\{ 1 - \exp\left(-3\frac{h^2}{a^2}\right) \right\}, \qquad (4.25)$$

- linear with a sill:

$$\gamma(h) = C_0 + bh \quad \text{for } 0 < h < a, \qquad (4.26)$$

$$\gamma(h) = C_0 + ah \quad \text{for } h \geq a,$$

- hole-effect:

$$\gamma(h) = C_0 + C_1 \left\{ 1 - \frac{a \sin(h/a)}{h} \right\}. \qquad (4.27)$$

Because the strongest values of spatial variance (and spatial autocorrelation) occur at short distances, while fitting an experimental variogram to a theoretical model, the

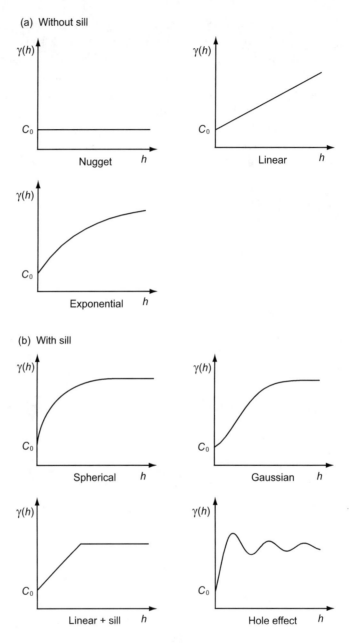

Figure 4.7 Theoretical variograms: (*a*) without a sill: showing the pure nugget effect, linear, and exponential models; (*b*) with a sill: showing spherical, Gaussian and nested models (linear and a sill) and the hole effect. See the text for the equations that describe these models.

parameters should match the shape of the variogram as much as possible over the distances shorter than the range a (Figure 4.7). Then, the estimated parameters of the theoretical variogram are used for the spatial interpolation at unsampled locations using Kriging algorithms (see Section 4.7). Inappropriate parameter estimations can result in

4.2 Global Spatial Statistics 115

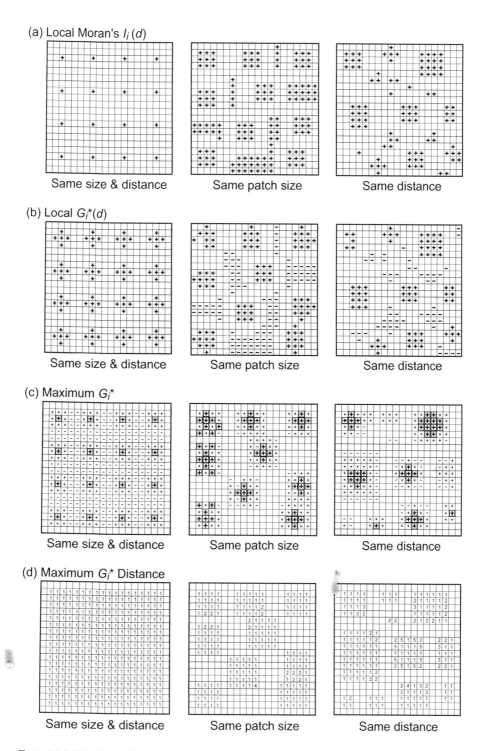

Figure 4.8 LISA (Local Indicators of Spatial Association) using the same simulated data shown in Figure 4.3: 16 patches (Figure 4.3d); 9 patches (Figure 4.3e) and 12 patches (Figure 4.3f). (a) Local Moran's I_i estimated using a distance interval class of 1.5 units (+ indicates significant values at 0.05, positive or negative). (b) Local G_i^* estimated using a distance interval class

very different, and possibly very misleading, estimated values. If the range is too small, the structure will be modelled for only small distances generating small patches and, if the range is too large, the structure will be overly smoothed. Selection of the best variogram model and its parameters used to be determined 'by eye' based on experience. Now, we use generalized least-squares and maximum likelihood measures like Akaike's information criteria to select the parameters (Chilès & Delfiner 2012).

Like Geary's c, the semi-variance is affected by outlier values from skewed data because the estimation is based on the squared differences. It is therefore recommended to:

- Plot the data at different lags to identify outliers using h scattergrams; and,
- Transform the data, as appropriate, to reduce skewness.

Furthermore, the number of pairs contributing to the statistic distances decreases as distance increases and so one rule of thumb is to interpret only the first two-thirds of a variogram. Because **h** denotes a vector including both distance and direction, anisotropic patterns can be determined by comparing the shape and parameters of directional variograms (Figure 4.6). In geostatistics, three types of anisotropy are distinguished:

(1) the ranges differ but the sills are the same (geometric anisotropy; Figure 4.6b),
(2) the ranges are the same, but the sills differ (zonal anisotropy; Figure 4.6c), and
(3) both the ranges and sills differ.

Geometric anisotropy can be modelled, using a linear relationship, to predict the values of a variable at unsampled locations. It is not as straightforward for the two other types of anisotropy where more than one variogram model is needed to model spatial structure. This can be achieved by determining different models as a function of distance, as illustrated in Figure 4.9, where a spherical model is used up to the range and the sill is used beyond the range.

Ecological data are rarely stationary and often show some trend or gradient, called 'drift' in geostatistics. The presence of drift implies that $m(s)$ is not a good estimate of the average over the entire study area. Drift can be detected by comparing the values of $m(s)$ computed using only the values at the beginning of vector **h** (at $z(s_i)$, 'head' locations) with those at the end of vector **h** (at $z(s_i + \mathbf{h})$, 'tail' locations). When these averages differ, the covariance is non-ergodic because ergodic covariance requires $m_{-\mathbf{h}}(\text{head}) = m_{+\mathbf{h}}(\text{tail}) = m(s)$ (Cressie 1996). In the presence of drift, the generalized random intrinsic functions of order k should be used to estimate the spatial variance, rather than the experimental variogram (for mathematical details see Journel & Huijbregts 1978; Webster & Oliver 2007; Chilès & Delfiner 2012).

Ecological data are often binary as presence–absence data or as indicator data for threshold responses: the spatial variance of qualitative or categorical data can be estimated using an indicator function with a threshold value to determine the response

Figure 4.8 (cont.) of 1.5 units (+ indicates positive and − negative significant values at 0.05). (c) Maximum local $G_i{}^*$ estimated using five distance classes of 1.5 units each (+ in a square indicates positive and − negative significant values at 0.05; + indicates positive values but nonsignificant). (d) Maximum local $G_i{}^*$ distance (number indicates the distance class at which the maximum local $G_i{}^*$ was estimated). See text for more details.

as 0 or 1. Continuous variables can also be transformed into binary indicator variables $i(s; z)$. The semi-variance for an indicator function is:

$$\hat{\gamma}_{i(s;z)}(h) = \frac{1}{2n(h)} \sum_{i=1}^{n(h)} (i(s_i; z) - i(s_i + h; z))^2. \quad (4.28)$$

The indicator function can be extended to include several thresholds, corresponding to multiple-state variables, or to explore the sensitivity of the selected thresholds. The indicator function is then used to perform *indicator* Kriging analysis (Webster & Oliver 2007; Polakowska *et al.* 2012).

To conclude this overview of variography, we stress that variograms aim to determine the parameters used to model spatial patterns. Hence, to maintain the same measurement units as the original data, semi-variance values are not standardized. Furthermore, while no significance tests have been developed for semi-variance functions, randomization and bootstrap procedures can assess their significance. Lastly, we note that a much wider range of geostatistical techniques is available than presented here but this is beyond the scope of this book. We refer the reader to more advanced textbooks for greater detail (Cressie 1996; Webster & Oliver 2007; Chilès & Delfiner 2012; among others). Semi-variance values and anisotropy can be computed using the function *variogram* and the estimations of the parameters for the theoretical variogram models using the function *fit.variogram* of the R package *gstat*.

4.3 Sampling Design Effects on the Estimation of Spatial Pattern

As presented in Chapter 2, regardless of the type of spatial analysis used, the sampling design (including the three components of sample size, sampling unit size and their layout in space) determines the power of detecting significant signals from the sampled data. Furthermore, the determination of one of the components of the design affects the other two. The first component is the sample size because it is directly related to the sampling effort or ability (usually limited by cost and time). In time series analysis and geostatistics, it is recommended that at least 100 data points be sampled to detect patterns. In ecology, sample sizes are usually smaller when fieldwork is involved and, if a signal is strong, fewer, perhaps 50, can detect it. Moreover, if the spacing among units is within the pattern's spatial range, then even 20–30 may be able to capture its essence but the pattern needs to be clear for small numbers to work (Fortin 1999a). Especially with small numbers, failure to detect any significant patterns does not necessarily imply that there are none and a revised sampling design may be required to detect a pattern if present (Fortin *et al.* 1989; Fortin 1999a).

The sampling design can be revised in two ways: (1) change the spacing of the units and (2) change the size of the units. The spacing of the sample locations can be arranged to maximize, as much as possible given the sample size and number of locations within the range of the variable. Spatial layouts with more than one lag seem to be more efficient at capturing spatial patterns with relatively small sample sizes (Fortin *et al.* 1989).

Interestingly, a random sampling design can detect significant spatial patterns because the range of spatial distances among the sampling locations is large (Fortin *et al.* 1989).

The other consideration is the choice of unit size. When spatial autocorrelation at the first distance class is not significantly high, this can indicate that the unit size is larger than the range or that more than one spatial pattern is present (Fortin 1999a); a smaller unit size should be used. How small? When no prior knowledge is available about a system, we recommend the smallest unit that can include more than one object of interest (Chapter 2). With this fine spatial resolution, the data can be aggregated for coarser resolution as needed (Fortin 1999a). The values of autocorrelation change with the sample unit size, as is well-known in spatial analysis (Dungan *et al.* 2002). This fact is referred to as the 'modifiable area unit problem' (MAUP; Openshaw 1984; Perry *et al.* 2002) or as the 'change of support' in geostatistics (Journel & Huijbregts 1978; Dungan *et al.* 2002). The problem is like making inferences about individual objects based on aggregated information; consider trying to use stand-level data to make conclusions about individual trees when no information about individual trees was recorded. These problems are inherent in almost any spatial analysis, and they are the subject of active research in geography, especially for satellite image analysis; geostatistics (Cressie 1996); and landscape ecology (Perry *et al.* 2002; Wu *et al.* 2002; Wu 2004).

4.4 Spatial Relationship between Two Variables

So far, we have presented methods that quantify the spatial structure of only one variable at a time. In many cases, it is of interest to analyse the spatial interaction between two variables, as for Ripley's bivariate K (Chapter 3), or to predict the spatial pattern of one variable based on its correlation with another. To do this, autocorrelation is modified to estimate cross-correlation between two variables:

$$I_{xy}(d) = \left(\frac{n}{W(d)}\right) \frac{\sum_{i=1}^{n} \sum_{\substack{j=1 \\ i \neq j \\ j \neq i}}^{n} w_{ij}(d)(x_i - \bar{x})(y_j - \bar{y})}{\left(\sqrt{\sum_{i=1}^{n}(x_i - \bar{x})^2}\right)\left(\sqrt{\sum_{i=1}^{n}(y_i - \bar{y})^2}\right)}. \quad (4.29)$$

Similarly, the semi-variance function is modified to estimate the cross-covariance:

$$\hat{\gamma}_{uv}(\mathbf{h}) = \frac{1}{2n(\mathbf{h})} \sum_{i=1}^{n(\mathbf{h})} (z_u(s_i) - z_u(s_i + \mathbf{h}))(z_v(s_i) - z_v(s_i + \mathbf{h})). \quad (4.30)$$

Of special interest in bivariate geostatistics is the concept of *spatial confounding* in studies of the relationship between a covariate x and an outcome variable y using a regression model. Spatial confounding occurs when an unmeasured variable with its spatial structure may be affecting either x, y or both (Gilbert *et al.* 2022). Several methods have been proposed to deal with this concern (Urdangarin *et al.* 2023) and we will discuss the topic further in Chapter 7, in our treatment of spatial regression.

4.5 Local Spatial Statistics

As more and more ecologists engage in ecological studies with large spatial extents, the likelihood that their data sets violate the assumption of stationarity is very high. When using global statistics with these data sets, local and small areas of spatial heterogeneity are masked by the fact that the statistics are summaries for the entire study area, with a single average value of spatial autocorrelation or a series of average values for different distances as in a correlogram or variogram. In such cases, a global assessment of spatial dependence can be misleading because average values provide information about neither the range of variability in the degree of dependence nor the exact localization of patterns. For example, in a sloping study area, tree abundance may vary from the top of the slope to the bottom, with some small openings here and there, creating localized patches with lower tree abundance. Global statistics may detect the large-scale trend in abundance values, but they may miss local patterns because their algorithm lumps together all the local deviations by summation and averaging. While the average value of spatial dependence is meaningful where only one process occurs (either induced or inherent dependence), it is misleading when several processes act at various intensities in different parts of the study area.

These limitations and the misapplication of global statistics have been acknowledged for decades (Getis & Ord 1992, 1996) and recognition that local assessment of spatial autocorrelation may provide insights led to the development of local statistics (Getis & Ord 1992; Kabos & Csillag 2002). Anselin (1995) proposed 'LISA' (local indicator of spatial association) as an acronym for these local statistics and we will use that acronym.

As presented in Section 4.2, the calculation of global statistics involves a computation at each sampling location which are then all combined, averaging local variation over the study area. The local spatial statistics first proposed were the components of global statistics, calculated at each sampling location i but not accumulated into an overall average. Consequently, the local Moran's I_i is:

$$I_i(d) = \frac{(x_i - \bar{x})}{\frac{1}{n}\sum_{i=1}^{n}(x_i - \bar{x})^2} \sum_{\substack{j=1 \\ j \neq i}}^{n} w_{ij}(d)(x_j - \bar{x}), \qquad (4.31)$$

where $w_{ij}(d)$ is the weight matrix for distance d. It can be binary, stressing only the connectivity among locations, or with unequal weights such as inverse distance to emphasize neighbourhood effects.

Assuming completely uniform randomization, the expected value of I_i is:

$$E(I_i) = \frac{-1}{n-1} \sum_{j=1}^{n} w_{ij}. \qquad (4.32)$$

Consequently, unlike global Moran's *I*, which has the same expected value for all locations and distances, the expected value of local Moran's *I* varies by location. The significance of local I_i can be tested by comparison to the standard Normal distribution once I_i is standardized to $z(I_i)$:

$$z(I_i) = \frac{[I_i - E(I_i)]}{\sqrt{\text{Var}(I_i)}}. \tag{4.33}$$

Here, $\text{Var}(I_i)$ is derived by assuming either complete randomization or conditional randomization. As for global Moran's *I*, local Moran's I_i can be computed with different distances, *d*. Any type of Bonferroni correction that adjusts for both the number of locations and the number of neighbour search distances, which results in a very large number of multiple comparisons, would be too conservative, and only rarely would I_i be considered significant. In general, we recommend using local spatial statistics as exploratory tools to detect localized structures, especially when a lack of stationarity is suspected.

As with global Moran's *I*, the local version computes the deviations from the average value of the variable, *x*, over the entire study area. Positive I_i values occur when the value at location *i* is similar to those of its neighbours in their deviation from the average ($\bar{x}$). In other words, positive values of I_i indicate that the values in the vicinity of location *i* and at location *i* are either all larger (positive deviation) or smaller (negative deviation) than the average. Negative values of I_i also indicate that deviation from the average is either larger or smaller than the average but where the value at location *i* is of a different sign from its neighbours. When the value of I_i is close to zero, the deviation from the average is small and no local structure can be detected. This can occur because spatial patterns are either absent or so subtle that local values are too similar to the overall average. Figure 4.8a shows significant local I_i at $\alpha = 0.05$ based on the same simulated data used to compute the omnidirectional correlograms in Figure 4.3d (16 regularly spaced patches of equal size), Figure 4.3e (9 irregularly spaced patches of equal size) and Figure 4.3f (12 regularly spaced patches of different size). These maps of I_i help to identify the position, size, shape and layout of local structures. In the case of the 16 regularly spaced patches (Figure 4.3d), the centroid of each patch is identified as having the most positive autocorrelation with its neighbours (Figure 4.8a). The maps of local I_i values for the two other cases (Figure 4.8a) are not as informative, because we cannot discriminate between clusters of high and low spatial structure. Indeed, positive values of I_i (as indicated by +) can result from clusters of either high or low values. Without looking at the maps of the raw data (Figure 4.3e and f), we cannot determine which clusters have high or low values. Other local statistics may be more appropriate, as described in the rest of this section.

Global Geary's *c* can also be modified to obtain a local statistic, c_i:

$$c_i(d) = \frac{1}{\frac{1}{n}\sum_{i=1}^{n}(x_i - \bar{x})^2} \sum_{\substack{j=1 \\ j \neq i}}^{n} w_{ij}(d)(x_i - x_j)^2. \tag{4.34}$$

4.5 Local Spatial Statistics

The difference between the local I_i and the local c_i is in the numerator, where, for local I_i it is the deviation from the value at location i and the overall average of the variable, while for local c_i it is the value of the variable at location i. Similarly, under the assumption of complete randomness, the expected value of c_i is proportional to the number of local neighbours:

$$E(c_i) = \frac{2n}{n-1} \sum_{j=1}^{n} w_{ij}. \qquad (4.35)$$

The equations for $\text{Var}(c_i)$ under complete and conditional randomness can be found in Boots (2002). Positive values of local Geary's c_i correspond to cases where the value at location i is like its neighbours, while negative values indicate a difference in sign from its neighbours.

Getis and Ord (1992) proposed two new local spatial statistics: local G_i, in which the value at location i is excluded from the computation; and local G_i^*, in which the value at location i is included. Local G_i is computed as follows:

$$G_i(d) = \frac{\sum_{\substack{j=1 \\ j \neq i}}^{n} w_{ij}(d) x_j}{\sum_{\substack{j=1 \\ j \neq i}}^{n} x_j}, \qquad (4.36)$$

where the expected value, under the assumption of complete randomness, depends on the number of local neighbours:

$$E(G_i) = \frac{1}{n-1} \sum_{\substack{j=1 \\ j \neq i}}^{n} w_{ij}. \qquad (4.36)$$

Similarly, the local G_i^* is computed as:

$$G_i^*(d) = \frac{\sum_{j=1}^{n} w_{ij}(d) x_j}{\sum_{j=1}^{n} x_j}, \qquad (4.37)$$

where the expected value, under the assumption of complete randomness, also depends on the number of local neighbours:

$$E(G_i^*) = \frac{1}{n} \sum_{j=1}^{n} w_{ij}. \qquad (4.39)$$

These two statistics are, in essence, the ratio to the global average of local averages (a) *around* location i and (b) *at* location i. They detect clusters of high or low values: 'hot spots' and 'cold spots'. As with local Moran's I_i, the local G statistics cannot differentiate between the absence of spatial structure and where local averages equal the global average.

As the number of neighbours increases with search distance, the G statistics are asymptotically normal and can be standardized to facilitate interpretation (Getis & Ord 1996): positive G values indicate clusters with high values (hot spots) and negative ones indicate clusters with low values (cold spots). Figure 4.8b illustrates how, unlike local Moran's I_i, G_i^* can discriminate between locations of significant hot spots (+) and cold spots (−) in different arrangements of patches.

Although these local statistics, G_i and G_i^*, are more informative than local Moran's I_i, they are estimated relative to the global average. They are therefore sensitive to overall global structure (e.g. trends). Kabos and Csillag (2002) proposed the 'H Moran statistic', which is a local statistic for qualitative data that is not affected by the presence of a global pattern in both its computation and its significance tests. So far, no local statistics for quantitative data have been developed that can account for global patterns and detect local patterns. Local Moran's I, local Geary's c and local G statistics can be computed using the respective following functions of the R package *spdep*: *localmoran*, *localC*, and *localG*.

4.6 Spatial Scan Statistics

Referred to as 'scan statistics', a set of statistical techniques has been designed to detect clusters of events. The name is logical because they scan the data for clusters and usually involve the calculation of associated probabilities to indicate the most unexpected clusters (Glaz *et al.* 2001). Scan statistics can be formulated in spatial, temporal or spatio-temporal versions, depending on the situation and purpose. Typical examples of application include examining health data for spatial clusters of a particular type of cancer, scanning time series for clumps of rare events or studying epidemiological data for groupings of disease onset in time and space (Grillet *et al.* 2010a). The reliability of the probabilistic calculations that determines how unexpected the detected clusters are depends on the assumption of stationarity, as do so many of the spatial methods.

The key to scan statistics is the scan template and this fits our theme of the importance of templates for the calculation of spatial statistics. Spatial scans make use of localized windows that sweep over the entire study area, with a probability or likelihood ratio associated with each location. The shape of the template used may interact with the shape of the clusters detected (cf. Duczmal *et al.* 2006). The second key to interpreting scan statistics is to remember that low-probability occurrences are still possible and, given enough positions of a scan template, some will have significant results, even when the system is truly random. The larger the number of template positions used, the more likely it is that some will appear to be significant. (Some coincidences are just coincidental!) Instead of relying on the raw probabilities for each template and position combination, the maximum of a likelihood ratio is used (Kulldorff 1997) and, in many cases, the number of events of an 'at risk' population must be considered as well as the number of those in the category of interest.

When the numbers are large, the required probability calculations become unwieldy, and much of the work in developing scan statistics has been in providing useful and accurate approximations (cf. Glaz *et al.* 2001), as well as determining the power of the tests. Most applications of scan statistics are two-dimensional, dealing with geographic data such as locations of disease or other kinds of critical events. The choice for the template shape is usually the isodiametric circle or square, especially when the phenomenon is isotropic or there is no reason to expect otherwise (cf. Glaz *et al.* 2001). When anisotropic structures are known or expected, anisodiametric templates such as ellipses will work for mildly irregular clusters, but for highly irregular clusters a simulated annealing method may be preferable (Duczmal *et al.* 2006).

In considering circles as templates for two-dimensional scans (Figure 4.9), it is obvious that the technique is closely related to Ripley's *K*-function analysis and the circumcircle methods, because they are based on counts of events in circles. The concept of the calculations is also similar, comparing the observed counts with the expected in some standardized way.

Applications of scan statistics in three spatial dimensions are not common but they do exist. One commonly cited application concerns the clustering of galaxies in space, for which the three dimensions are reduced to two by using scan statistics on astronomical photographs (Glaz *et al.* 2001). Scan statistics are often applied in the spatio-temporal analyses of epidemiological data (Grillet *et al.* 2010a; Read *et al.*

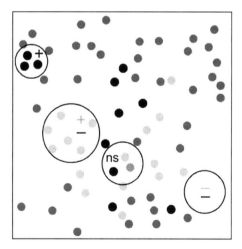

Figure 4.9 Scan statistics based on simulated data where the grey tone corresponds to the scan statistic values: darker grey indicates a higher probability of being part of a significant cluster. The circle indicates the zone of influence of the circle being either positive (+), negative (−) or not significant (ns).

2010) but scan statistics have appeared only rarely in ecological studies. The potential for greater use of these techniques in ecology is clear, particularly as the use of large geographic databases continues to grow. More work may need to be done on developing those approaches that are best suited to ecological spatial data and on making them available in a readily applied format. Scan statistics can be computed using the function *scan* of the R package *base*.

4.7 Spatial Interpolation

In ecological studies, information of interest for a target population over the study area could be species abundance, species behaviour and so on. Without a complete census or survey, we need methods for the estimation and prediction of the values of a variable at unsampled locations. This is achieved by modelling the spatial pattern with interpolation techniques, assuming an underlying process that is stationary. The model attempts to summarize the spatial pattern using as few parameters as possible. In fitting these parameters to the data, the interpolation techniques are modelling the main spatial signal while minimizing the error, resulting in smoothed estimated values at unsampled locations. Several interpolation methods are available, each having advantages and limitations; these can be classified into broad categories:

- Global: A single interpolation function for the entire study area. The resulting map of the interpolated data is usually a smooth surface (e.g. trend surface analysis); changing one value will affect the function and, thus, the predicted values everywhere.
- Local: The interpolation function is applied locally to a limited number of locations. The resulting map of the interpolated data is smooth but includes both global and local patterns. Changing one value will affect only neighbouring locations (e.g. proximity polygons, inverse distance weighting, Kriging).
- Approximate: At the sampling locations, the predicted values will not be the same as the observed ones (e.g. trend surface analysis).
- Exact: At the sampling locations the predicted values will be exactly the observed values (e.g. proximity polygons, inverse distance weighting, Kriging).

Other properties can also be used to characterize these interpolation techniques: deterministic, with only one possible predicted value at a given location (all methods except Kriging, which can be stochastic with a conditional annealing procedure), point interpolators (all methods except the proximity polygons), and areal (only proximity polygons and Kriging). Here, we provide a brief overview of four spatial interpolation methods and the readers to texts that present more detail (Chilès & Delfiner 2012).

Spatial models are available (moving average, MA; simultaneous autoregressive, SAR; conditional autoregressive, CAR) to simulate data with a known degree of spatial dependence (Ver Hoef *et al.* 2018). These spatial autoregressive models are mostly used to simulate data rather than to interpolate and, therefore, we will present them in Chapter 6.

4.7.1 Proximity Polygons

Without realizing it, we often interpolate sampled data by assuming that the value or characteristic is the same throughout the entire sampling unit. For point locations, we associate a polygon (named for Dirichlet, Thiessen or Voronoi; Chapter 2) with each point, consisting of all parts of the plane closer to that point than to any other (Okabe et al. 2000). The characteristic at the x–y coordinate of the reference point is assigned to the whole polygon; when the characteristic is qualitative, this creates a categorical attribute; when it is quantitative, the value that is assumed to apply to the entire polygon. These Voronoi polygons can be seen as defining the area of influence around each point and the technique results in abrupt changes when crossing from one polygon to another (Figure 4.10).

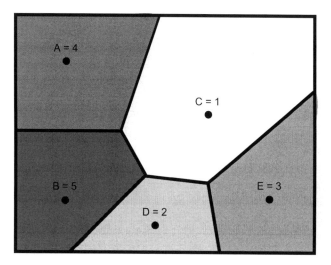

Figure 4.10 Proximity polygon interpolation using Voronoi polygons. The five sampling locations (A–E) were first linked using a Delaunay tessellation algorithm and then the Voronoi polygons (solid lines) were determined. Each polygon indicates the zone of influence of each sampling location and the value at the sampling location is assigned to the entire polygon (here indicated by a different grey tone corresponding to the sampling value).

4.7.2 Trend Surface Analysis

If we ignore the spatial context, the best prediction for a variable's value at an unsampled location is its average value over the entire study area. When information is available about the relationship between an independent variable and the variable of interest, the most common interpolation is by regression. In a spatial context, the x–y coordinates of sampling units can be used as independent variables in a regression and, when the pattern is a linear trend, interpolation can proceed by multiple linear regression (omitting interaction between x and y):

$$\hat{z}(\mathbf{s_0}) = b_0 + b_1 x + b_2 y, \tag{4.42}$$

where $\hat{z}(\mathbf{s_0})$ is the predicted value at location $\mathbf{s_0}$, b_0 is the intercept, and b_1 and b_2 are coefficients of the surface's slope (Figure 4.11a). This multiple regression approach is useful because a general equation can model large-scale patterns over the whole study area.

When the pattern is a non-linear trend, the values can be approximated by polynomial regression of various orders such as quadratic (x^2 or y^2), cubic (x^3 or y^3) or higher. When the pattern is a smooth, monotonic, curved surface, a second-order polynomial can be used (Figure 4.11b):

$$\hat{z}(\mathbf{s_0}) = b_0 + b_1 x + b_2 y + b_3 x^2 + b_4 xy + b_5 y^2. \tag{4.43}$$

When the pattern has a saddle in its shape, a third-order polynomial may be appropriate (Figure 4.11c):

$$\hat{z}(\mathbf{s_0}) = b_0 + b_1 x + b_2 y + b_3 x^2 + b_4 xy + b_5 y^2 + b_6 x^3 + b_7 x^2 y + b_8 xy^2 + b_9 y^3. \tag{4.44}$$

We could continue with higher-order polynomials (Figure 4.11d) but it is not recommended because the simplicity gained by polynomial regression is lost by the need to estimate many regression coefficients. More importantly, improving the fit in some areas may decrease the fit in others. Trend surface analysis is a global interpolator and should not be used to model local patterns. In addition, polynomial regression is not the best way to fit a smooth surface to a more complex spatial structure; the use of splines deserves exploration (Hayes 1974). Trend surface analysis can be computed using polynomial regression (e.g. the function *poly* of the R package *stats*) using the x and y coordinates as predictors.

The key advantage of using a trend surface analysis for interpolation is that no prior knowledge of the spatial pattern is needed because the interpolation is based on empirical data. The disadvantages are that (a) it should not be used when the pattern consists of several small patches and (b) the predicted values at sampled locations are not the observed ones as it is an approximate interpolator, not an exact one.

4.7.3 Inverse Distance Weighting

Following the concept of trend surface analysis, linear interpolation can be based on the data from sampled locations in a restricted neighbourhood. The underlying premise is that nearby locations are better predictors and the linear interpolator weights the interpolation at unsampled locations, s_0, according to the proximity of sampled locations:

$$\hat{z}(\mathbf{s_0}) = \sum_{j=1}^{m} \left(w_j z(\mathbf{s_j}) \right). \tag{4.45}$$

4.7 Spatial Interpolation 127

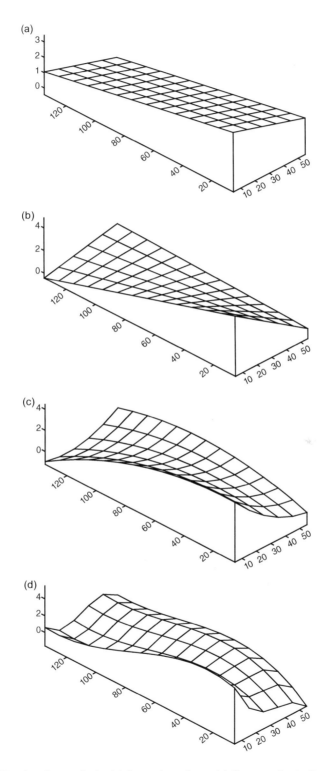

Figure 4.11 Trend surface analysis: (*a*) first-order polynomial (linear surface), (*b*) second-order polynomial (non-linear surface: valley type), (*c*) third-order polynomial (non-linear surface: saddle surface) and (*d*) fourth-order polynomial (non-linear surface: wavy surface).

Here $z(s_j)$ is the value of variable z at the sampled location j, m is the number of sampling locations within the search radius and w_j is the weight according to the distances with the weights totalling 1. The most common inverse distance weight is:

$$\hat{z}(\mathbf{s_0}) = \sum_{j=1}^{m}\left(d_{ij}^{-k}z(\mathbf{s_j})\right) \div \sum_{j=1}^{m}d_{ij}^{-k}, \tag{4.46}$$

where k is a value between 0 and 1, and d_{ij} is the Euclidean distance between the unsampled location i and sampled location j. More weight can be put to nearby locations by varying the value of the exponent k. When the distance between the sampled and the unsampled locations is zero, the interpolated value will be the observed one. The advantage of this type of linear interpolator is that it is weighted locally around each location, therefore preserving more of the complexity of local patterns (Figure 4.12) than trend surface analysis. It is also very easy to use and does not require prior knowledge about the data. When a map of the study area is needed for illustration purposes alone, this linear interpolation technique is quite useful. Interpolated values using the inverse distance weighting method can be computed using the function *idw* of the R package *spatstat*. It does not, however, provide any information about the discrepancies between the interpolated values and the 'real' spatial pattern at the unsampled locations. This is why Kriging is often preferred over the inverse distance weighting method.

4.7.4 Kriging

Stemming from the two previous interpolation methods, Krige (1966) proposed an interpolation method using a system of linear equations, based on prior knowledge of the degree of spatial dependence. This technique is, in essence, a weighted moving average called Kriging. Kriging is a geostatistical technique that uses the parameters of range, nugget and sill estimated by the experimental variogram (Journel & Huijbregts 1978).

Its origins are in application to mining questions, and Kriging was developed for the prediction of ore by interpolating from either punctual (point) or block (area) samples over a two-dimensional region and also predicting the values of the ore for a given volume (three-dimensional).

Kriging involves a set of linear regressions that determine the best combination of weights for interpolation by minimizing the variance as derived from the spatial covariance in the data. Here, the weights, w_i, are based on the spatial parameters of the variogram model, derived from an experimental variogram, so that sampling locations within the range have more influence on the predicted value. To solve this system of equations, the sum of the weights is constrained to be 1, giving more equations than unknown parameters to estimate:

$$\hat{z}(\mathbf{s_0}) = \sum_{j=1}^{m}(w_j z(\mathbf{s_j})) \quad \text{with} \quad \sum_{j=1}^{m}w_j = 1. \tag{4.47}$$

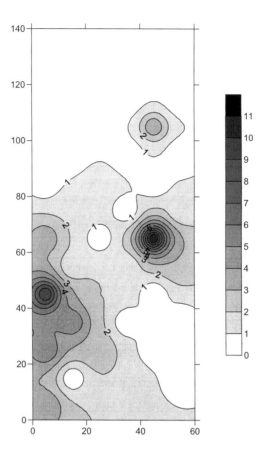

Figure 4.12 Inverse weighted distance. Interpolated values of sassafras abundance data ($n = 84$) based on 42 sampling locations (as shown in Dale & Fortin 2014, figure 6.13*b*). The estimated values are illustrated using contour lines.

where *m* is the number of samples in the search neighbourhood. To minimize the estimator error, the Lagrangian multiplier (λ) is added as a constant (see Journel & Huijbregts 1978; Cressie 1996):

$$\sigma_E^2(\mathbf{s_0}) = \sum_{j=1}^{m} \left(w_j \gamma(\mathbf{s_j}, \mathbf{s_0}) \right) + \lambda. \tag{4.48}$$

The estimation error is also called the Kriging variance or Kriging error. Equation (4.47) can be written in matrix notation:

$$\mathbf{Cw} = \mathbf{c}, \tag{4.49}$$

where the predicted value at an unsampled location, s_0, is a vector **c** that is obtained by starting with the variance–covariance matrix, **C**, between known locations, *i* and *j*, as estimated by the theoretical variogram model selected and multiplying by the unknown vector of weights, **w**. Then **c** is the vector of covariances between the sampled locations

i and the unsampled location. The covariance matrix and vector values are given by the variogram model, where the vector of weights is to be estimated:

$$\begin{bmatrix} \gamma(d_{11}) & \gamma(d_{12}) & \cdots & \gamma(d_{1m}) & 1 \\ \vdots & \vdots & \ddots & \vdots & \vdots \\ \gamma(d_{m1}) & \gamma(d_{m2}) & \cdots & \gamma(d_{mm}) & 1 \\ 1 & 1 & \cdots & 1 & 0 \end{bmatrix} \cdot \begin{bmatrix} w_1 \\ \vdots \\ w_m \\ \lambda \end{bmatrix} = \begin{bmatrix} \gamma(d_{10}) \\ \vdots \\ \gamma(d_{m0}) \\ 1 \end{bmatrix}. \quad (4.50)$$

This is achieved by left-multiplying both sides of Equation (4.50) with the inverse covariance matrix, $\mathbf{C}^{-1}$, solving for vector $\mathbf{w}$:

$$\mathbf{Cw} = \mathbf{c},$$

$$\mathbf{C}^{-1}\mathbf{Cw} = \mathbf{C}^{-1}\mathbf{c},$$

$$\mathbf{Iw} = \mathbf{C}^{-1}\mathbf{c},$$

$$\mathbf{w} = \mathbf{C}^{-1}\mathbf{c}, \quad (4.51)$$

because ($\mathbf{C}^{-1}\mathbf{C}$) is the identity matrix $\mathbf{I}$. The determination of weights is therefore related to both the variogram model and the number of locations considered. Most geostatistical software packages offer a choice of rules to determine the search neighbourhood. The first method is to define a search distance. Given that most of the spatial dependence occurs within the range, the search distance should not usually exceed the range. If the data are sparse, the search radius should increase until a minimum number of locations is reached, usually set at 15–20. In these circumstances, the values at sampled locations beyond the range contribute little to the Kriged values. The search neighbourhood does not need to be isotropic and can be more elliptic or even a volume in the case of three-dimensional Kriging.

Kriging resembles trend surface analysis in that only one model is used for the entire study area and also the inverse weighted distance method in that the interpolation is performed locally. Thus, given that the weights are proportional to both the spatial variance and the distance among locations, Kriging is an exact interpolator because the Kriged values at sampling locations are equal to the observed ones. However, the selected variogram model and its parameters may not be the best ones, which is why cross-validation is proposed to evaluate the overall robustness of the model. Cross-validation consists of removing each sampled location, one at a time, and then Kriging that location, comparing how reliably this Kriged value matches the observed one. This procedure was especially important before maximum likelihood methods were available to facilitate model selection. The effectiveness of the Kriging depends on how well the selected model fits the data. Spatially interpolated values based on Kriging can be estimated using the function *krige* of the R package *gstat*.

Another validation procedure for Kriged values is to map both the Kriged values and their associated estimation error (Figure 4.13). Indeed, given that the Kriging

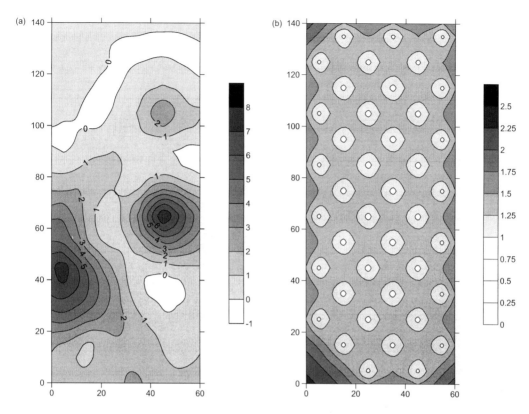

Figure 4.13 Kriged values of sassafras abundance data ($n = 84$) based on 42 sampling locations (as shown in Dale & Fortin 2014, figure 6.13b). The estimated values are illustrated using contour lines. (a) An isotropic spherical model using a range of 32 metres. (b) The associated Kriging errors with the Kriged values based on the isotropic spherical model. The Kriged errors are higher at the unsampled locations than at the sampled ones but, given the uniform spatial sampling design, the errors are comparable in terms of values.

variance is in the same units as the Kriged data, areas where the errors are higher than others can be identified. These areas of high variance can be due to:

(1) too few sampled locations within the range in those areas or
(2) an inappropriate variogram model or incorrect parameter values (Figure 4.14).

Note that the Kriged errors are relative to the variogram model used. If the selected theoretical model seems to be the best one, areas of high error may suggest places for more sampling effort and this procedure can be used to determine the optimal sampling design.

Building from this system of linear equations, several Kriging variants have been developed to account for the particular characteristics of environmental data that the property of unbiased predictors can hold. First, interpolation can be performed for specific x–y coordinates (*punctual* Kriging) or for an area (*blocked* Kriging). When the

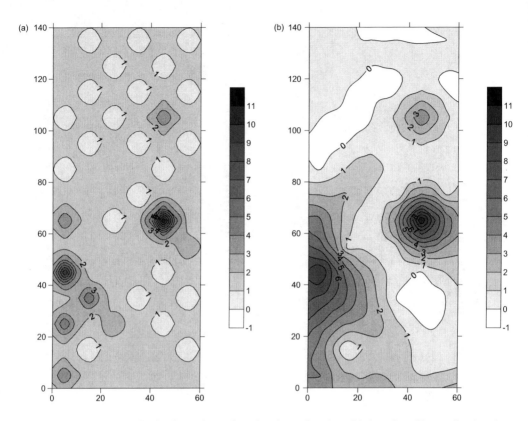

Figure 4.14 Kriged values of sassafras abundance data ($n = 84$) based on 42 sampling locations (as shown in Dale & Fortin 2014, figure 6.13b). The estimated values are illustrated using contour lines. (a) An isotropic spherical model based on a range of 10 metres, which is too small (compared to the fixed 32-metre range of Figure 4.13) and creates small patches around each sampled location. (b) An isotropic spherical model using a range of 60 metres, which is too large (compared to the fixed 32-metre range of Figure 4.13a) creates smoothed values between sampled locations.

mean is known, second-order stationarity applies; this system of linear equations is referred to as *simple* Kriging. In most cases, the mean is not known and only the weak stationarity assumption prevails, thus *ordinary* Kriging should be used. However, especially with environmental data, a large-scale trend could be present as well as local patterns in the data. This implies a 'drift' in the mean and that the simple system of linear equations cannot be used. *Universal* Kriging was developed to model such large-scale trends and then the residuals are Kriged after the trend is removed. As with trend surface analysis, the trend may not be linear, so an 'intrinsic random function of order k' should be used. These functions are the equivalent of the k-order polynomial functions in trend surface analysis and account for non-linear trends in the data before Kriging. When the spatial pattern shows non-linearity that cannot be fitted by a polynomial function, *non-linear* Kriging can be used.

A spatial pattern can be anisotropic where the variance changes with direction. When anisotropy is such that the sill is constant and only the range varies, it is geometric anisotropy and the distance matrix can be adjusted to account for it. When the anisotropy is 'zonal', with the range constant but the sill varying, the adjustment is made in the covariance matrix by adding nested terms. A nested procedure can also be implemented by using different variogram models as a function of distance. Stemming from this nested property of adding more terms in the system of linear equations, *stratified* Kriging offers a way to interpolate over regions that have different spatial variances due to changes in stratum type. In forested landscapes, for example, it is inappropriate to Krige deciduous and coniferous stands using the same variogram. Using forest type as a stratification criterion, stratified Kriging can be used based on a different variogram for each type (Wallerman *et al.* 2002).

In ecological and environmental studies, the variable of interest, z_1, is often too costly to sample. When we know how it correlates with a more accessible variable, z_2, *co-Kriging* can interpolate the value of z_1 given z_2 based on their cross-variogram. This is an appealing method, but it assumes that the relationship between the two variables holds even at locations where z_1 is not sampled. Often, the resulting Kriged map of z_1 looks like a mirror image of z_2, which may or may not reflect the real spatial pattern of z_1.

We are often interested in mapping the structure of an assemblage of species as a community and this can be done using *multivariate* Kriging (Wackernagel 2013). Lastly, given threshold responses or presence–absence data, we may want to use our quantitative variables as qualitative and determine their spatial structure in that format. An indicator variogram can be estimated (as presented in Section 4.2.3) and the *indicator* Kriging can be used for that purpose (Polakowska *et al.* 2012).

Here, we have presented a very succinct overview of the dynamic field of geostatistics. Many more geostatistical methods are available in advanced geostatistical textbooks. One last aspect of geostatistics that is very useful to ecological studies deserves to be mentioned: stochastic simulation based on conditional annealing. Spatial stochastic simulations generate a series of data with a given degree of dependence in order to evaluate whether or not observed sample data show significant spatial patterns (Fortin *et al.* 2012a). To do so, the parameters for the variogram model are derived from an experimental variogram and can be used to generate stochastic simulations having the same degree of spatial variance. This approach was proposed to generate maps that have more variability than the Kriged ones, resulting in more realistic spatially simulated values. Such simulated data are generated by an iteration process where the values at the sampled locations are kept as anchor points from which the annealing algorithm iteratively spreads data values around them while ensuring that the overall structure is maintained through the range, nugget, and sill values. This spatial stochastic simulation approach, based on a theoretical variogram, allows us to address significance testing of spatially autocorrelated data.

4.8 Concluding Remarks

The first issue to be decided is whether the spatial analysis of the data should be global, summarizing the structure for the entire study area, or local, measuring the structure in neighbourhoods. The choice should be guided by the goal(s) of the study and by knowledge about the stationarity of the processes of interest. When no prior information is available, we recommend that more than one method is applied and their results compared to identify whether the assumption of stationarity holds.

For global analysis, Moran's I is usually favoured, if only because of its direct correspondence in meaning with Pearson's linear correlation. Users should be aware of its sensitivity to extreme outliers, which influence the average used to estimate the spatial deviation. This is why many prefer Geary's c or the semi-variance γ because outlier values affect only the deviations (differences) computed with them. Unfortunately, these differences have more weight in the final measure because they are squared.

Similarly, for local analysis, some spatial statistics are affected by the presence of a global pattern which may result in biased estimates. In the absence of overall structure, local G_i^* statistics in their standardized versions are easier to interpret as indicating local areas of high (hot spots) or low (cold spots) values. Both global and local statistics provide information only about the spatial structure.

Interpolation techniques can be used for informal evaluation, in mapping spatial patterns where the simplest method (inverse weighted distance) provides good results. Such maps can be modified by using smoothing algorithms, such as a spline, to give a more visually pleasing or more intuitive product. On the other hand, when information about the values and the associated errors at unsampled locations is required, one of the various Kriging techniques should be used. There is no magical recipe to determine which methods and which parameters should be favoured under all circumstances. The ability to achieve meaningful interpolation using Kriging comes with experience where the general rule of thumb is to capture the intensity and range of the spatial variance at short distances as the priority. Last, keep in mind that Kriging errors associated with Kriged values are a function of the theoretical model selected and the parameter values provided, not of the data themselves. The results depend crucially on the appropriateness of the model on which Kriging is based.

Chapters 3 and 4 are distinguished by organizing complete data including sets of contiguous sample units versus sample data. While the distinction is clear and has clear consequences, the question may arise of how to proceed when the data from an array of contiguous units are incomplete, whether by structural constraints or by error. Dale and Fortin (2014) provided a good discussion of this problem, focussing on the example of incomplete video transects across coral reefs to determine the density and distribution patterns of a cold-water coral *Lophelia pertusa* and of the mega-fauna associated with it. The conclusion is that censored data presents a problem, but it is not insurmountable. Both the standard methods for sample data and the methods designed for spatially complete data sets assisted by numerical approaches can be used to provide the looked-for insights.

Table 4.1 Summary of the spatial analysis methods presented in Chapter 4

Spatial analysis method	Template
Join count statistics (Topology: network)	
Join count statistics (Lattice: chess moves)	
Global spatial statistics (Isotropic spatial lag)	
Global spatial statistics (Anisotropic spatial lag)	
Local spatial statistics (Topology: network)	
Local spatial statistics (Lattice: chess moves)	

Returning specifically to sample data, the focus of this chapter, several statistics can estimate spatial dependence for such data (Table 4.1). They have a common root in determining the spatial covariance among the values of variable(s) at different sampling locations. Hence the question becomes: which one is best for the purpose?

Lastly, the unresolved issue is that of significance testing while estimating spatial dependence at several distances based on the same data. Although spatial statistics traditionally have significance tests and even progressive Bonferroni corrections, applied to them, these do not fully account for the dependence of values from one distance class to another. We will revisit this problem in Chapter 6.

5 Spatial Partitioning
Spatial Clusters and Boundary Detection

Introduction

To understand ecological processes at multiple scales, ecological studies may need to be carried out over large regions. In doing so, such larger study areas are likely to include several ecological processes acting at different spatial and temporal scales (Stein et al. 2014; Urban et al. 2020). In such cases, the standard assumption of stationarity of the process (constant mean and variance; isotropy; see Chapter 2) is unlikely to be valid over space and time. To analyse ecological data over large regions properly, it is therefore necessary to first partition them into smaller, more homogeneous areas (patches) that are generated by the same process. Stratifying a region can also help in the monitoring and management of resources locally. There are two main families of approaches to spatially partitioning ecological data:

(1) grouping adjacent locations that have similar values of the indicator variable(s), generating spatial clusters (Figure 5.1a) [joining the most similar ones to create clusters] or
(2) dividing the whole region into subareas, based on their degree of dissimilarity, by delineating boundaries between areas (Figure 5.1b) [separating the ones that are different with boundaries].

In theory, these two approaches should give the same spatial partitions but, in practice, there may be misalignments due to sampling locations, data accuracy and not accounting for the spatial embedding of multiple drivers. Furthermore, some partitioning approaches are more descriptive than others by not having any significant tests for the identified partitions, leaving the users to subjective interpretations. In this chapter, we present the spatial partitioning methods that are most relevant and most appropriate for delineating spatially cohesive areas for ecological data (Figure 5.2).

5.1 Patch Identification

5.1.1 Patch Properties

A patch is a spatially homogeneous area with at least one variable having consistent attributes throughout, either categorical (e.g. forest rather than crop) or quantitative

5.1 Patch Identification

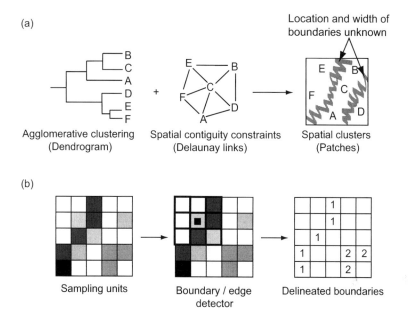

Figure 5.1 Spatial partitioning. (*a*) Spatial clustering where clusters are formed only when the degree of similarity between sampling locations (A, B, C, D, E and F), as determined by the algorithm (here an agglomerative one), is high, and where the sampling locations are adjacent to one another (e.g. A is adjacent to F, C and D) in a chosen spatial connectivity structure (here Delaunay links). The exact location and width of the boundaries between the spatial clusters are not determined by this method, as illustrated by the grey zigzags. The spatial clustering algorithm only identifies the membership of each sampling location to a spatial cluster.
(*b*) Boundary detection using a kernel filter approach where the 1's indicate the locations where boundaries are the most pronounced (sharp) and the 2's indicate the second most pronounced boundaries. The grey shades correspond to quantitative values of a variable (low, black; high, white).

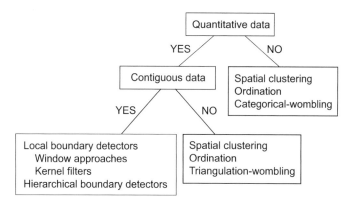

Figure 5.2 Decision tree to guide the selection of boundary detection methods for ecological data.

(e.g. tree age). Consequently, adjacent patches differ for at least one variable. The juxtaposition of patches creates a mosaic in which each patch can be characterized by its structural properties such as area (e.g. small, large), shape (elliptic, square, sinuous, etc.), boundary properties (e.g. sharp, gradual) and contrast between adjacent patches (e.g. low between forest types; high between forest and crop). This definition of 'patch' is data-driven based on a targeted variable being more-or-less uniform spatially. Not all the variables in any one patch are expected to be spatially uniform. When patches are arbitrarily delineated, like administrative units, the within-patch spatial structure can range from weak trends to strong spatial autocorrelation and can reduce our ability to delineate boundaries accurately (Philibert et al. 2008).

5.1.2 Spatial Clustering

Patches can be seen as spatial clusters based on both spatial adjacency and similarity in values of a given variable. Spatial clusters can be determined by grouping sampling locations having similar values using a clustering algorithm (see Legendre & Legendre 2012 for a review and mathematical details) and then imposing a spatial constraint (e.g. Delaunay triangulation algorithm; Chapter 2; Figure 5.3) to keep only groups that have spatially adjacent sampling locations. Pang et al. (2023) provided guidance on how to choose the appropriate clustering algorithms according to the objective(s) of the study. The most common algorithms are various 'linkage' versions of hierarchical agglomerative methods. These methods start with each location recognized as a separate group and then merge groups into larger groups based on the degree of similarity among them. To address the particular nature of ecological data (e.g. presence–absence data, rare species, the double-zeros of dual absences, ...), many similarity and dissimilarity metrics are explained in more detail in Legendre and Legendre (2012). The researcher can define a priori the degree of similarity at which samples are merged into an existing cluster. The 'linkage' family

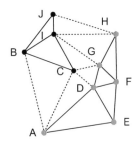

Figure 5.3 Delaunay connection links (the solid lines) among 16 sampling locations.
If some large-scale process occurs (e.g. two climatic zones indicated by black dots and grey dots), then the dashed links between sampling locations should be removed, creating two spatial clusters.

of methods uses a series of criteria that need to be met before groups are merged; for example, in single linkage, two groups merge based on the minimum distance (or maximum similarity) between them; in intermediate linkage, the similarity of two groups is compared using all possible pairs in both before merging; and, in complete linkage, the merging of groups occurs when the maximum distance (or minimum similarity) between groups is reached. The advantage of agglomerative clustering procedures is that the final groups are non-overlapping clusters. The most important drawbacks are the subjectivity involved in the selection of the measure of similarity for the creation of clusters and then the threshold to determine the number of clusters to consider.

Clusters can also be determined by an algorithm known as *k*-means partitioning, which requires the user to decide in advance the number of clusters to be formed. Then the algorithm uses an iterative process to optimize the clusters, minimizing the within-cluster sum of squares error and maximizing the similarity of each sample to its cluster's centroid. *k*-means algorithm has some subjectivity in the choice of the number of clusters and, when changing the number of clusters, *k*, the clusters are overlapping from one number of clusters to the next one.

To create *spatial* clusters, spatial constraints need to be added to the clustering algorithm (Legendre & Fortin 1989). Spatial adjacency among sampling locations can be determined from any of the connectivity algorithms presented in Chapter 2. Then spatial clusters are formed by merging adjacent locations that have comparable values (Figure 5.1*a*). A by-product of forming spatial clusters is that there are divisions or 'boundaries' between them (Figure 5.1*a*) but their exact location and width are unknown. This is the deficiency of these spatial clustering methods: the locations have known cluster membership but the boundaries between clusters are not well defined.

Spatial clusters can be obtained for any data type (qualitative or quantitative, univariate or multivariate) and any sampling design but they have two major problems:

Problem 1: Without prior knowledge or independent information about the data, the researcher needs to select either the level of similarity for the agglomerative algorithms or the number of spatial clusters for *k*-means partitioning. To achieve this, Gordon (1999) developed a goodness-of-fit index, *G*, that indicates how *k* contributes to minimizing the sum of square error of the between-cluster variability, *B*, when compared to the within-clusters variability, *W*:

$$G = [B/(k-1)] \div [W/(n-k)], \qquad (5.1)$$

where *k* is the number of clusters and *n* is the number of sampling locations. The value of this goodness-of-fit index is a guide to the appropriate number of clusters. It is important to have an underlying ecological question to provide an upper limit to the number of clusters. The choice of the appropriate number of clusters should be guided by the goal of the research and knowledge of the study area.

As an example, in a study of forest canopy composition in 84 10 × 10 m samples of 26 tree species (Fortin 1997) varying the number of spatial clusters reveals very different spatial arrangements of tree assemblages. For comparison, we selected three numbers of clusters: 5, 10 and 20. The differences in the layout of spatial clusters based on two clustering algorithms (hierarchical agglomerative centroid and k-means) combined with spatially constrained (Delaunay algorithm) are depicted in Figure 5.4. Both algorithms provide similar clusters, but the centroid agglomerative algorithm creates more singletons (4 of 5; 6 of 10; 15 of 20) than the k-means method (2 of 5; 5 of 10; 12 of 20). Here, the spatial partition based on 20 clusters divides the study area into patches showing the spatial heterogeneity of the lower area and the greater homogeneity of the upper area. The advantage of the hierarchical agglomerative centroid algorithm is that it is hierarchical when increasing the number of clusters: the spatial clusters are kept when the number of clusters increases. This is not true for the k-means algorithm: increasing the number does not maintain the clusters of the previous partition, due to the random assignment of locations to clusters to begin each iteration. To reduce this problem, spatial partitioning based on a hierarchical clustering algorithm could provide the assignment input for the k-means procedure.

Problem 2: Both spatial clustering (Section 5.1.2) and boundary detection algorithms (Section 5.2) will provide spatial clusters or boundaries even when none exist. An obvious example is the case where there is a gentle gradient across a region: any boundaries identified will not reflect discontinuities but may respond most to local noise in the data. Furthermore, in some circumstances, spatial clusters may include sampling locations that are highly similar to adjacent sites but other spatial clusters may have a high degree of dissimilarity. Hence, applying spatial constraints in clustering locations may not create spatial clusters with strong similarity among locations. For example, two non-adjacent sampling locations may have comparable values because they are in the same climatic regime or the same topography. Then, it may be appropriate to customize the spatial constraints among the sampling locations by adding or removing links accordingly.

5.1.3 Fuzzy Classification

The clear dichotomy of membership by which every location is either a member or not a member of any given group (as presented in Figure 5.1) may not be appropriate in all circumstances either because the quantitative data cannot be accurately classified into discrete classes (e.g. mixed forest dominated by deciduous or by coniferous species), their spatial location is uncertain or inaccurate (e.g. telemetry data) or the data measurement is only approximate (e.g. percentage cover reported in classes). Fuzzy classification and fuzzy k-means have been suggested to be more appropriate spatial clustering methods in such cases (Jacquez *et al.* 2000). These methods are based on fuzzy set theory where the membership function is not a discrete dichotomy (0 or 1) but rather a real number ranging from 0 to 1, called a 'possibility'. The

5.1 Patch Identification

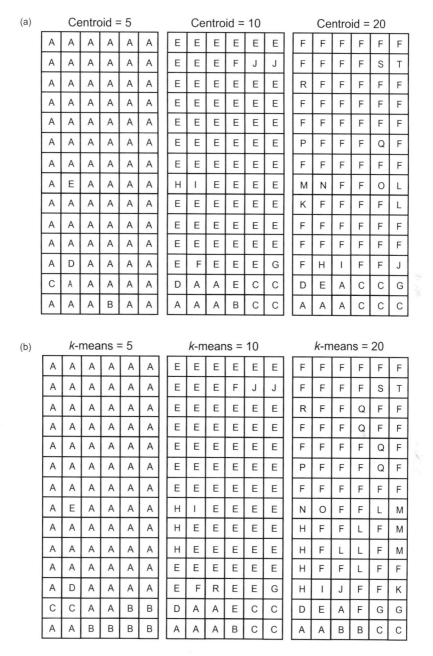

Figure 5.4 Spatial clusters on the tree abundance data of 84 sampling locations based on Delaunay connection links and (a) the centroid agglomerative algorithm and (b) the k-means method. Spatial partitions based on five spatial clusters identify that most of the spatial heterogeneity is concentrated in the lower left part of the study area and there is some heterogeneity in the middle of the plot. Spatial partitions based on 10 and 20 spatial clusters divide the large spatial cluster A, based on five spatial clusters, into smaller spatial clusters stressing much spatial heterogeneity over the entire plot. The k-means clustering algorithm creates larger spatial clusters than the agglomerative one.

possibility of being a member of a cluster is based either on expert knowledge or on spatial location uncertainty. The membership function can take several different shapes (e.g. linear, sigmoidal, symmetric, asymmetric) defined over a range of values of a variable, called transition zones. The advantage of the fuzzy classification approach is that it may more adequately reflect ecological processes and species' responses to environmental conditions. The drawback is that it requires more knowledge about the processes and any user-defined decisions can introduce a lot of subjectivity, resulting in non-optimal spatial clusters. To reduce the amount of subjectivity, the researcher can employ the fuzzy k-means. This method is also an iterative procedure that minimizes the within-cluster variability but a fuzzy exponent is added to allow the overlap of clusters. When this fuzzy exponent is set to zero, it is equivalent to the k-means algorithm; when it is too high, all the clusters overlap. McBratney and de Gruijter (1992) suggested starting with a value of two for the fuzzy exponent. Using the same tree species data, the fuzzy k-means (five classes and exponent $= 2$) produces a different spatial partition from the equivalent spatial k-means clustering example (Figure 5.5). Here, the five crisp spatial clusters are smaller than previously but are surrounded by a gradient of fuzzy membership values. The advantage of the fuzzy k-means method is that it allows the determination of fuzzy boundary zones between spatial clusters. Fuzzy set theory has been applied to detect fuzzy boundaries (Hufkens *et al.* 2009; Fiorentino *et al.* 2018).

5.2 Boundary Delineation

Several disciplines have developed analytic tools to detect boundaries (Figure 5.2). In ecology, the detection of edges, ecotones and boundaries between vegetation types are all of interest (Fortin 1992; Hufkens *et al.* 2009).

5.2.1 Ecological Boundaries

In ecology, boundary detection methods have a long tradition associated with ecotone delineation. Ecotones occur at the interface between two communities or ecosystems, where the exchange of nutrients and other forms of 'information' occurs. Ecotones have distinct structural and functional properties that differ from the adjacent systems (Hansen & di Castri 1992; Ferro & Morrone 2014; Kolasa 2014) and their structural properties are directly related to the underlying processes that generate or maintain them. Alpine treeline ecotones, for example, have been a focus of studies for decades and they exhibit a range of morphological characteristics (abrupt versus gradual; tree islands versus Krumholtz) which can be related to underlying processes (Bader *et al.* 2021). Thus, ecotones, as ecological boundaries, represent linear or non-linear responses (such as thresholds) to environmental gradients (Table 5.1). Our ability to detect ecological boundaries depends on the ecological process(es) under investigation and on the sampling design and analysis chosen (Fortin & Drapeau 1995; Fortin 1997, 1999*b*; Fortin *et al.* 2000). Sampling designs for detecting boundaries should

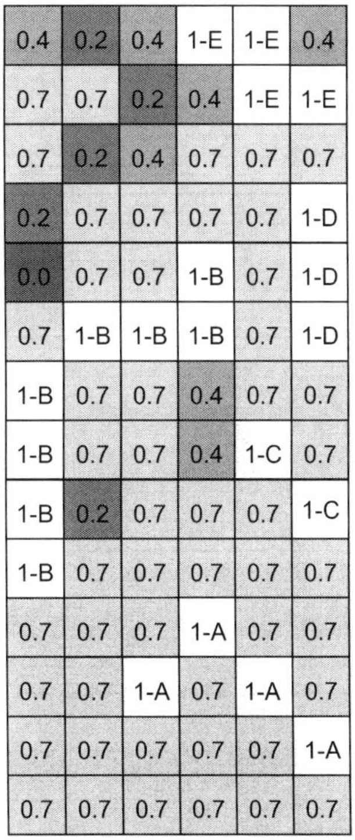

0.4	0.2	0.4	1-E	1-E	0.4
0.7	0.7	0.2	0.4	1-E	1-E
0.7	0.2	0.4	0.7	0.7	0.7
0.2	0.7	0.7	0.7	0.7	1-D
0.0	0.7	0.7	1-B	0.7	1-D
0.7	1-B	1-B	1-B	0.7	1-D
1-B	0.7	0.7	0.4	0.7	0.7
1-B	0.7	0.7	0.4	1-C	0.7
1-B	0.2	0.7	0.7	0.7	1-C
1-B	0.7	0.7	0.7	0.7	0.7
0.7	0.7	0.7	1-A	0.7	0.7
0.7	0.7	1-A	0.7	1-A	0.7
0.7	0.7	0.7	0.7	0.7	1-A
0.7	0.7	0.7	0.7	0.7	0.7

Figure 5.5 Fuzzy k-means spatial clusters of the tree abundance data of 84 sampling locations using $k = 5$. The sampling locations parts of a spatial cluster have a membership possibility of 1 and are indicated by 1-A, 1-B, 1-C, 1-D and 1-E, where A, B, C, D and E signify the five spatial clusters. The other sampling locations have a membership possibility of belonging to a spatial cluster varying from 0 to 0.9999. For illustration purposes, the membership possibilities were classed into four categories: 0.7 (from 0.5 to 0.9), 0.4 (from 0.3 to 0.49), 0.2 (from 0.1 to 0.29) and 0.0 for 0. The five spatial clusters each contain very few sampling locations. Most of the sampling locations almost belong to a spatial cluster (as indicated by 0.7) and illustrate the spatial extent (width, area) of the boundary zones between the spatial clusters.

include sufficient sampling locations over a transect, or an area, so that adjacent patches as well as the boundary itself are covered sufficiently. There is an obvious advantage in trying to understand boundary processes when historical data are available, such as air photos to document spatial shifts over time and to be interpreted relative to ecological processes (Birre et al. 2023).

5.2.2 Boundary Properties

To start, we should define some terms. The term *edge* in boundary detection and image segmentation refers to a sharp demarcation that separates areas (not the graph theory meaning of a join between nodes!). In image segmentation, there are three major types of edges: the *step edge*, the *roof edge*, and the *spike edge*. The step edge is where two well-defined and almost uniform patches of different types meet (e.g. forest

Table 5.1 Processes and environmental factors creating and maintaining ecological boundaries

Environment and landscape structure changes	Processes and factors creating or maintaining boundaries	Type of boundary	Ability to detect edges and the underlying processes and factors
Sharp environmental changes	Geomorphology, topography, biogeochemistry, climate	Sharp, narrow, persistent	Possible to detect abrupt changes in diversity or abundance
Gradual environmental changes	Geography, climate, species' ranges (species physiological limits), species interactions	Blurred, wide, persistent, or transient	Difficult to detect changes in biomass and abundances; possible to detect compositional changes
Spatial heterogeneity within large disturbances	Fire, storm, drought, species interactions, succession	Sharp to smooth, transient	Possible to difficult depending on the intensity of the disturbance
Spatial heterogeneity within small gaps	Treefall, species interactions, succession	Blurred, transient	Difficult to detect due to qualitative and quantitative noise
No environmental changes	Species interactions, dispersal ability	Sharp, persistent	Possible to difficult depending on species interactions

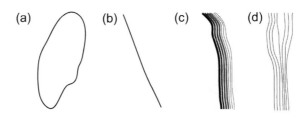

Figure 5.6 Boundary properties: sharp (crisp, step edge or line) boundaries (*a* and *b*); gradual (intermediate, fuzzy or transition) boundaries (*c* and *d*), open (difference) boundaries (*b*, *c* and *d*) and closed (area) boundary (*a*).

and agricultural land), while the roof edge occurs when either or both patches are spatially autocorrelated (Philibert *et al.* 2008). The spike edge, where an abrupt change in intensity occurs only locally, is rarer in ecology than in image processing. In ecology, the term *edge* often refers to an anthropogenic step edge, as in *edge effect* in landscape dynamics. Several similar terms refer to step edges, such as sharp, crisp or line boundaries (Figure 5.6*a* and *b*). The opposite of step edges includes gradual, intermediate or fuzzy boundaries or transition zones (Figure 5.6*c* and *d*). The term *boundary* can apply to either sharp (line) or gradual (zone) demarcations.

Another characteristic of a boundary is width: narrow or broad. This width may vary asymmetrically along the boundary as well as from one side to another (Figure 5.6*c* and *d*). Locations where ecological processes are sharper than others may have localized boundaries that do not enclose an area. These are *difference* or *open* boundaries (Figure 5.6*b–d*), whereas those that enclose an area completely are *area* or *closed* boundaries (Figure 5.6*a*). Anthropogenic boundaries tend to be straight, but not always, and boundaries originating from natural processes are often sinuous or recurving.

Lastly, all the structural and functional properties of boundaries are scale-dependent (Fortin 1999b; Csillag et al. 2001; James et al. 2011). Boundary studies need to acknowledge the scale of the ecological processes being studied as well as the effects of spatial resolution, of both the sampling unit and the extent of the study area, on the accuracy of boundary detection (James et al. 2011).

5.2.3 Boundary Detection and Analysis for One-Dimensional Transect Data

This section focuses on the analysis of how organisms are arranged on simple one-dimensional environmental gradients. Specifically, environmental gradients are considered to have monotonic changes in the intensity of a single environmental factor with physical distance. The gradient is the spatially explicit version of the kinds of monotonic changes in a factor that may be detected by any ordination technique (e.g. PCA, RDA, CCA; see Legendre & Legendre 2012). The concept of spatial structure is somewhat different in the context of a gradient but it still refers to non-randomness in space that has a certain amount of predictability. A gradient produces predictable and directional non-stationarity and the resulting spatial structure is in the appearance of species where they were previously absent and their disappearance where they were previously present. The predictability of the spatial pattern is in how the species enter and leave along the gradient and in the relationships between the ranges and densities of the species where they are present. For any analysis of this kind of spatial pattern, a key element is the detection of multispecies boundaries along the gradient, as discussed in Section 5.2.4.

For the analysis of gradients, including the detection of boundaries, both the sampling design and data type need to be considered. The obvious kinds of data are density measures and presence–absence records; the two designs are continuous transects over distance and spaced samples such as quadrats or points.

The simplest and most effective way to detect ecological boundaries from quantitative data is to apply a moving split-window technique (Webster 1973; Johnston et al. 1992). This technique consists of computing the difference between two halves of the window. The window size can vary upward from a single sampling location per half. Various metrics can be used to measure the differences between the two adjacent window-halves like discriminant functions, Mahalanobis distances and squared Euclidean distance (Ludwig & Cornelius 1987). The squared Euclidean distance, D_E^2, is the most common metric:

$$D_E^2(x_1, x_2) = \sum_{i=1}^{p} (z_{1i} - z_{2i})^2, \tag{5.2}$$

where x_1 and x_2 are the two locations to compare and z_{1i} and z_{2i} are the values of the p variables at these two locations. The window then slides along the entire transect, one sampling location at a time, so that all adjacent locations can be compared. In the

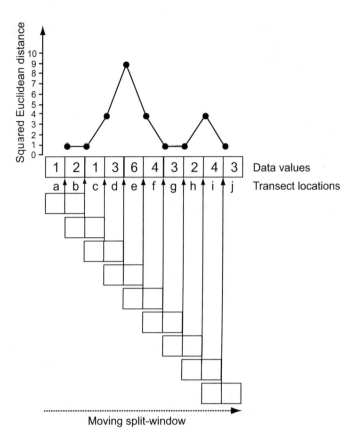

Figure 5.7 Moving split-window along a transect of 10 contiguous sampling locations. The split-window size is two sampling units (one sampling unit in each half). The squared Euclidean distance is computed for each pair of sampling units, resulting in nine values; the strongest peak (the best boundary) is between the sampling locations 'd' and 'e' and the second weaker peak is between sampling locations 'h' and 'i'.

example illustrated in Figure 5.7, each half contains the data from one location and the resulting measures are located between the locations for a total of $n - 1$ difference values. Sharp boundaries occur where high and narrow peaks identify the location of ecological boundaries, whereas gradual boundaries occur where peaks are low and wide. By computing differences based on adjacency, the moving split window is in essence a local boundary detector. The drawback of all local boundary detectors is that they are sensitive to local noise in the data. To minimize the undesirable effects of local noise in the data, computation of the differences can be performed using windows of increasing sizes. The results can then be drawn on the same plot where peaks corresponding to ecologically meaningful boundaries will persist while peaks due to local noise will be smoothed out. Ecological boundaries will produce peaks at the same locations. The similarity to analysis with a Haar wavelet is obvious (Dale *et al.* 2002; James *et al.* 2011; James & Fortin 2012).

Comparable methods are available for presence–absence data collected at contiguous sampling locations. The observed number of species present at a given location is compared against a random distribution derived from a Monte Carlo procedure: Dale's (1986, 1988) method computes the amount of spatial overlap at each location, while the McCoy *et al.* (1986) technique is based on the probabilistic similarity between pairs of locations. A different approach can be used in which the presences and absences of the individual species are initially dealt with as a separate source of data for analysis, usually summarized as the species' range on the gradient, with its endpoints defined by the first and last occurrences of the species. There are some interesting questions about the meaning and importance of absences within any one species' range, but there are some advantages to using this kind of data to study the boundaries of species in one dimension, as we now describe.

Many studies of the pattern of species on environmental gradients have used contiguous or spaced quadrats and recorded density or presence–absence in them. Quadrat data can be used for many of the boundary analyses described, such as overlap, gap size, intermingling of boundary types and clustering of boundaries. If the quadrats are small enough that there is never more than one boundary in a quadrat, then the methods described for continuous ranges can be used directly, with the quadrat position in the sequence being the equivalent of distance along the transect. The problem comes when the quadrats are larger, producing ties in the ranking of boundary order. These ties obviously represent a loss of information. If the relationships among the ranges and boundaries of species are of interest and quadrats contain more than one boundary, the quadrats are too large and the scale of sampling does not match the scale of the pattern.

In a transect of contiguous quadrats on an environmental gradient, provided that the length of the transect is sufficient, some quadrats will contain the upper boundaries and some will contain the lower boundaries of the species of plants in the area. Several questions can be asked about such data, either collectively (e.g. Are the upper boundaries clumped?) or at the level of the individual species (e.g. Is the range of Species A along the gradient greater in the absence of competition?). No single measure or test statistic will tell us everything we want to know, and we should be prepared to use several complementary approaches, even if they are not independent. The choice and effectiveness of methods may depend on the relationship between the number of species and the number of quadrats. The interpretation of a given number of quadrats containing no boundaries will be different if there are 20 quadrats and 200 species than if there are 200 quadrats and 20 species.

The analysis of quadrat data will depend in part on the null model under consideration, and there are a variety of ways of constructing null models for transects of contiguous quadrats along an environmental gradient. The usefulness of any null model as the basis for statistical tests will depend on the number of replicated transects, the number of quadrats, the number of species, the strength of the gradient and so on. Dale and Fortin (2014) have provided a detailed (if incomplete) discussion of possible approaches to analysis for these questions.

For presence–absence data, recorded continuously, we may represent the species' ranges as solid lines, although a more realistic representation of species occurrence would be an irregularly broken line with many small gaps. In the same way, an accurate continuous record of *density* (not just presence) might be extremely irregular and would then have to be smoothed to produce a curve. The use of such density data (original or smoothed), in the broad area of multivariate analysis, includes direct gradient analysis, niche overlap measurement and so on. Those topics are covered with varying degrees of detail and sophistication in other places (Legendre & Legendre 2012) and we will not repeat that coverage. Many of the kinds of analysis already presented in this chapter can be used with such data.

Gradients can be spatially continuous or spatially discontinuous. In the first case, an analysis may be straightforward, but, in the second, the gradient will have to be (re)constructed by the researcher. Perhaps the most important feature of density data is that it allows us a quantitative evaluation of a species' response to the environmental gradient within its range. The usual model of a species' response is the Gaussian or bell-shaped curve but, in general, few species are likely to have this idealised response, instead being either asymmetric, bimodal or irregular (Austin 1987; Legendre & Legendre 2012).

Density data also allow us to look at among-species patterns along a gradient, by examining the relative positions of the density modes of the species. For instance, Minchin (1989) examined some of the propositions of Gauch and Whittaker (1972) concerning the organization of species' responses by testing whether the modes of the major species were evenly distributed along the gradient, perhaps because of competition and resource partitioning, and whether the modes of the minor species were randomly distributed. For data from Tasmanian montane vegetation, they found that the modes of the major species were randomly distributed.

Spatial structure on environmental gradients is an important aspect of the spatial organization in ecological systems. This kind of spatial pattern includes characteristics such as the upper and lower boundaries of species, the ranges of their presence and how the densities of individual species respond to the gradient. All these characteristics can be used to help ecologists to generate ideas and to test hypotheses about how the systems are assembled and how they function. There is more research needed to understand this kind of spatial pattern and the processes that give rise to it, particularly as related to the detection of boundaries in multivariate data and the delimitation of multivariate (multispecies) patches. One-dimensional transect data allow only the detection of the sharpness and width of boundaries, but the determination of other properties requires boundary detection in two-dimensional area data, which will be addressed in Section 5.2.4.

5.2.4 Boundary Detection based on Two-Dimensional Data

In two dimensions, boundaries can be detected in either quantitative or categorical data. For quantitative data, a boundary can be defined as the location in space where the change of intensity in a set of variables is the greatest (Fortin 1994). For qualitative data, a boundary is a location in space where category turnover is the highest.

5.2.4.1 Lattice Data

Two-dimensional data can be derived from a complete regular lattice of sample locations with constant location spacing or an irregular sampling scheme. In a square lattice, each unit has four adjacent units, like a checkerboard; other regular lattices are possible, such as hexagonal. The main advantage of two-dimensional area data over one-dimensional transect data is that all boundary characteristics can be assessed, including sharpness, width, shape and sinuosity.

As in one dimension, most two-dimensional boundary detectors compute some difference among adjacent locations by a match–mismatch algorithm (Figure 5.8a), kernel filter (Figure 5.8b) or moving window (Figure 5.8c). The differences seem subtle, but they are not.

Kernel filters are usually squares of various sizes (3 × 3, 5 × 5, 7 × 7, etc.); each cell of the kernel contains a weight that is multiplied by the value at the correspondent location of the data lattice and the sum of these products is assigned to the centre cell. This is called a *convolution* in GIS and remote sensing applications. As for moving windows, the kernel filter slides over the entire area but the kernel filters produce a new value for each cell of the lattice, as many scores as the original number of locations: $n_{rows} \times m_{columns}$ (see Figure 5.8b). There are several filters available, with different weighting schemes and different sensitivities to edges (Shih 2009).

Moving windows compute a metric from a square of adjacent sampling locations (e.g. a 2 × 2 window) to quantify the difference among the four values (see Section 5.2.4.2). After sliding the window over all possible positions on the lattice, the number of metrics computed is less than the number of locations (Figure 5.8c). A lattice data of n rows and m columns will produce $n - 1 \times m - 1$ rates of change. The shape of the moving window can vary and, for example, the difference can be based on three adjacent sampling locations forming a triangle instead of a square (triangulation–wombling, Section 5.2.4.3). We will present only one of these algorithms: the first partial derivative algorithm (Womble 1951).

5.2.4.2 Lattice-Wombling

With quantitative lattice data, the differences in values among the four locations forming a 2 × 2 square can be estimated by the variable's first partial derivative (Womble 1951):

$$m = \sqrt{\left[\frac{\partial f(x, y)}{\partial x}\right]^2 + \left[\frac{\partial f(x, y)}{\partial y}\right]^2}, \qquad (5.3)$$

where the function $f(x, y)$ is linear in both x and y directions, with values z_i at the four sampling locations ($i = 1, 2, 3$ and 4):

$$f(x, y) = z_1(1 - x)(1 - y) + z_2 x(1 - y) + z_3 xy + z_4(1 - x)y. \qquad (5.4)$$

This formulation assumes that distances among the four locations are small and, for convenience, the x and y coordinates are scaled to range from 0 to 1. The rate of

Spatial Partitioning: Clusters and Boundaries

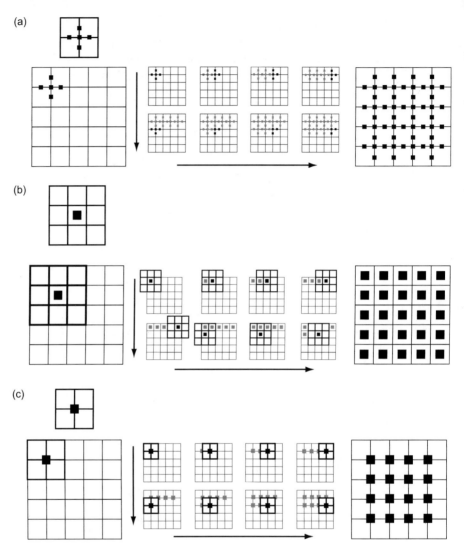

Figure 5.8 Difference measures computed by window (*a* and *c*) and kernel filters (*b*) to detect boundaries. In (*a*), the differences are based on the match–mismatch between two sampling units. This is comparable to the join count statistics presented in Chapter 4. The kernel (*b*) or the moving window (*c*) is slid, one sampling unit at a time, in the *x* and then in the *y* direction over the entire study area. The difference is computed at the central cell of the kernel (the square in (*b*)) or the centroid location of the window (the square in (*c*)). There are fewer difference values computed with the window approach than in the original data ($n_{rows} - 1 \times m_{columns} - 1$) but not in the kernel approach where there is the same number of difference values as the number of sampling locations.

change, m, is computed for the centroid of a square window and the square window slides over the entire lattice one location at a time. In a multivariate context, the difference among the four adjacent sampling locations is the average of the absolute values of the first derivatives of each variable. When there is only one variable, the detected boundaries reflect high differences in its values; with several variables, both the amount of species turnover as well as their difference in values affect the location of the boundaries.

The orientation angle of the change, θ, can be calculated as:

$$\theta = \tan^{-1}\left[\frac{\left(\frac{\partial f}{\partial x}\right)}{\left(\frac{\partial f}{\partial y}\right)}\right] + \Delta, \qquad (5.5)$$

(Barbujani *et al.* 1989) where

$$\Delta = \begin{cases} 0°, & \text{if } \left(\frac{\partial f}{\partial x}\right) \geq 0, \\ 180°, & \text{otherwise.} \end{cases} \qquad (5.6)$$

Such that the orientation is calculated by first doubling the angles, to avoid slopes of opposite direction cancelling, and then averaging and halving the result. When the highest rate of change in one variable occurs north-to-south and in another variable south-to-north, the two directions do not cancel but reinforce.

In essence, the magnitude of the rate of change, m, is the slope of the plane that can be fitted to the values at the four adjacent locations (Figure 5.9). Boundaries, as locations with high rates of change, correspond to steep gradients among the variable's four values. Weak differences among adjacent values will result in low rates of change. Adjacent locations of low rates of change can be considered spatially homogeneous: a patch. The major problem is to decide the threshold value of the rate of change for boundary detection (Figures 5.9 and 5.10a). When an arbitrary threshold is used, say the highest 10th percentile (Fortin 1994, 1997, 1999b), the rates of change

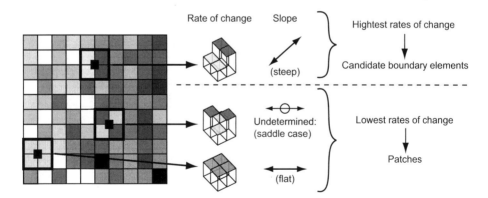

Figure 5.9 Lattice-wombling algorithm. Rates of change based on lattice-wombling are in essence the slope of the plane that fits the four values of the variable at the sampling locations. The orientation of the slope could also be useful in some studies. When the rates of change are ranked in decreasing order of magnitude (highest values of the slope), the candidate boundary elements are determined using an arbitrary threshold. When the rates of change are very low, close to zero, the sampling locations are most likely part of a patch.

identify what can be called *potential* or *candidate* boundary elements (Figure 5.10*a*). Subsequently, boundary properties (length, width, shape, etc.) can be measured in terms of spatially connected candidate boundary elements using the boundary statistics (Section 5.3). The threshold selection depends on the context (strength of the boundaries and their number) and the number of sampling locations (Fortin 1999*b*).

As an example, rates of change based on lattice-wombling (Figure 5.10*a*) were classified into four categories of 25% each where the highest rate of change is 1 and the lowest is 4. We chose an arbitrary threshold (here category 1) to determine candidate boundary elements. The candidate boundary elements that are adjacent to one another are linked in Figure 5.10*a*. Looking at the next rates of change (e.g. category 2), these locations coincide with the demarcation among the spatial clusters. Spatial clustering and boundary detection can be used as complementary methods to highlight different characteristics: complete spatial partitioning (the spatial clusters), and the spatial location and properties of boundaries (their width, shape, sinuosity) between patches (Fortin & Drapeau 1995).

These rates of change measure the slope of the gradient among adjacent samples, indicating boundary elements where it is steepest (Figure 5.10*b*). For a better indication of a boundary's width and its limits, second derivatives can identify the inflection point location corresponding to the limit of a boundary (Figure 5.11). In studies dealing with large-scale processes, as in Barbujani *et al.* (1989), studying the migration path of human populations in Europe, the orientation of the rates of change may provide interesting insights.

Complete randomization is not an appropriate test of the significance of candidate boundary elements because they are not independent of each other. A complete randomization procedure cannot be applied as the spatial structure of patches and gaps can create higher rates of change than in the observed data (Oden *et al.* 1993). Instead, a restricted randomization test, maintaining some spatial structure, is

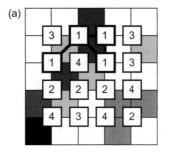 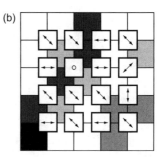

Figure 5.10 (*a*) Lattice-wombling of abundance data based on 25 sampling units with a potential of rates of change ($5_{rows} - 1 \times 5_{columns} - 1 = 16$) classified into 10 classes of 10 percentiles each: the highest rates of change are indicated by 1 and the lowest by 10. The bold lines link spatially adjacent candidate boundary elements, which can be connected into boundaries. (*b*) The orientation associated with each rate of change is classified into directions (N–S, north–south; E–W, east–west; NE–SW, northeast–southwest; SE–NW, southeast–northwest). Note that one orientation cannot be determined as it is a saddle case, indicated by an open circle.

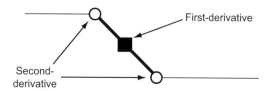

Figure 5.11 Lattice-wombling first derivatives identify the boundary as the magnitude of the slope of a plane (square), whereas the second derivates identify the inflection points where the boundary ends (circle).

recommended (Fortin et al. 2012a). When several variables are analysed, one can determine whether the candidate boundary elements are significant using a binomial test (Barbujani & Sokal 1991; Fortin 1994). This tests the significance of each candidate boundary element separately from all the other candidate boundary elements. For example, using the arbitrary threshold of the 10th percentile, each rate of change has a probability $p = 0.1$ of being classified as a candidate boundary element. Suppose a of the b variables analysed are candidate boundary elements at a given locality. In that case, the probability that this location has significant rates of change for all variables is the binomial:

$$\Pr(a \mid b) = \binom{b}{a} 0.1^a 0.9^{b-a}, \tag{5.7}$$

based on the number of ways to choose a elements out of b. The location is said to be significant when the binomial probability of the actual count a, given the maximum possible number b, is below some threshold, say 5%. To test whether connected candidate boundary elements form cohesive boundaries, boundary statistics are used (Oden et al. 1993; Fortin & Drapeau 1995; Polakowska et al. 2012; Dale & Fortin 2014).

Lastly, there are cases when the bilinear algorithm for lattice-wombling will not provide accurate estimates of the gradient, for example a saddle point with two diagonal values that are high and two that are low (Figure 5.9). Depending on the spatial autocorrelation within patches, some boundaries may fall inside patches, making the detection of cohesive boundaries more difficult. To minimize within-patch boundaries and facilitate the detection of between-patch boundaries we can apply boundary detection at several spatial resolutions of sampling units. Such scaling allows the assessment of the degree of boundary persistence across scales (Fortin 1999b; Philibert et al. 2008).

5.2.4.3 Non-lattice Data: Triangulation-Wombling

In the field, ecological data are rarely completely surveyed on a lattice but rather data are sampled where the sampling locations are irregularly spaced. With such a data set, lattice-wombling cannot be carried out unless the data are initially interpolated onto a regular lattice. This is not recommended, because most interpolation techniques smooth the data, reducing the strength of boundaries. Fortin (1994) proposed a triangular window instead, with triplets of locations determined by the Delaunay triangulation algorithm (Chapter 2). Then a plane can be fitted to the values at the vertices of the triangle. The magnitude of the rate of change, m, for that plane is computed by Equation (5.3) but with $f(x, y)$ now

$$f(x, y) = ax + by + c, \tag{5.8}$$

and

$$\begin{bmatrix} a \\ b \\ c \end{bmatrix} = \begin{bmatrix} x_1 & y_1 & 1 \\ x_2 & y_2 & 1 \\ x_3 & y_3 & 1 \end{bmatrix}^{-1} \begin{bmatrix} z_1 \\ z_2 \\ z_3 \end{bmatrix}. \tag{5.9}$$

The position of the centroid is $(x, y) =$

$$\left(\frac{x_1 + x_2 + x_3}{3}\right), \quad \left(\frac{y_1 + y_2 + y_3}{3}\right). \tag{5.10}$$

As for lattice-wombling, we can compute the average rate of change and the orientations of the rates of change (Equation 5.9). With a triangle window, the saddle problem cannot occur (Figure 5.12). On the other hand, when the four nearest locations form a perfect square, there are two possible pairs of triangles and a selection of two triangle windows is required, arbitrarily or not. When the four values are similar, that selection will not have a big impact but if they are very different the choice may affect the detection of boundaries.

In triangulation-wombling, the number of rates of change is smaller than the number of sampling locations, but this number depends on the spatial arrangement of locations. Examples show that lattice- and triangulation-wombling often agree on

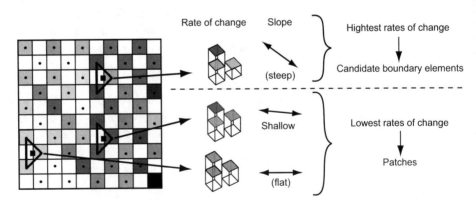

Figure 5.12 Triangulation-wombling algorithm. Rates of change based on triangulation-wombling are the slope of the plane that fits the three values of the variable at the sampling locations (indicated by a dot within a sampling unit): the three locations are based on the Delaunay triangulation algorithm. The orientation of the slope could also be useful in some studies. When the rates of change are ranked by decreasing magnitude, the candidate boundary elements are determined using an arbitrary threshold, starting with the highest values. Low rates of change, near zero, indicate sampling locations that are most likely part of the same patch.

the boundaries they detect (Fortin & Drapeau 1995). A *Julia* package is available to implement wombling algorithms for either regularly or randomly sampled landscape data (Strydom & Poisot 2023).

5.2.4.4 Categorical-Wombling

It is common to have only presence–absence data for species over a two-dimensional area. In such cases, boundaries are located where there is high species turnover. These boundaries can be established using either spatial clustering or by computing a *match–mismatch* measure between adjacent sampling locations (Oden *et al.* 1993; Figures 5.8*a* and 5.13). This last method is categorical-wombling (Fortin & Drapeau 1995), where mismatches between adjacent locations (i.e. not the same species in adjacent locations or one species present in one location and absent in another) are summed over all the categorical variables (here species). Sampling location adjacency can be determined by any connectivity algorithm (Chapter 2). The number of mismatches can be ranked as in the wombling algorithms and the highest values are represented at the midpoint between the linked sampling locations.

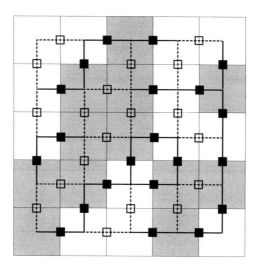

Figure 5.13 Categorical-wombling of presence–absence data for 25 sampling units for a total of 40 rates of change based on the rook connectivity algorithm. The solid lines indicate candidate boundary elements (21 mismatches) that are spatially adjacent. The dashed lines indicate potential patches (19 matches).

When we want to find boundaries using only presence–absence data of one species (e.g. species geographical range limit), the categorical-wombling method is not appropriate and, instead, home-range delimitation methods based on kernel approaches could be used.

5.2.4.5 Other Boundary Detection Methods

Several other boundary detection methods have been developed, including Bayesian point wombling (Liang *et al.* 2009) for spatial point processes, Bayesian areal wombling (Lu & Carlin 2005; Fitzpatrick et al. 2010; Gelfand & Barnejee 2015) for polygons instead of points and a hierarchical Bayesian approach for multiscale data (Ku et al. 2019). As for many spatial analysis methods, wombling has been extended to treat spatio-temporal data from spatio-temporal processes (Halder 2020). Another approach is the simultaneous local boundary detection and global spatial clustering method proposed by Jacquez et al. (2008) based on a probabilistic framework.

5.3 Boundary Overlap Statistics

Once cohesive boundaries have been delineated by any boundary algorithms and image segmentation techniques (Figure 5.2) and their significance tested using boundary statistics, interesting ecological questions can be investigated using overlap statistics (Jacquez 1995; Fortin et al. 1996; 2005; St-Louis et al. 2004; Hall 2008; Polakowska et al. 2012) that quantify the degree of spatial relationship between the locations of boundaries. Do boundaries directly overlap? Are boundaries spatially associated positively or negatively? For animals' responses to forest boundaries, overlap statistics can identify and test which spatial relationships prevail. There are four overlap statistics; one measures the perfect spatial overlap between boundaries while the three others account for small spatial lags between the two boundaries due to sampling measurement errors:

- The *direct overlap statistic*, O_s, is the number of the candidate boundary elements that are at the same location. In Figure 5.14, $O_s = 7$.
- The *mean minimum nearest distance statistic*, O_1, is an asymmetric measure of the distance from boundary 1 to boundary 2:

$$O_1 = \frac{\sum_{i=1}^{n_1} \min(d_{i.})}{n_1}, \tag{5.9}$$

where n_1 is the number of candidate boundary elements in boundary 1 and $\min(d_{i.})$ is the minimum Euclidean distance between the ith candidate boundary element of boundary 1 to a candidate boundary element of boundary 2. In Figure 5.14, the minimum distance between vegetation boundaries and animal boundaries is 15.5 units.

- The *mean minimum nearest distance statistic*, O_2, is an asymmetric measure of the distance from boundary 2 to boundary 1:

$$O_2 = \frac{\sum_{j=1}^{n_2} \min(d_j)}{n_2}, \tag{5.10}$$

where n_2 is the number of candidate boundary elements in boundary 2 and $\min(d_j)$ is the minimum Euclidean distance between the jth candidate boundary element of boundary 2 to a candidate boundary element of boundary 1.

In Figure 5.14, the minimum distance between animal boundaries and vegetation boundaries is 6.5 units.

5.3 Boundary Overlap Statistics

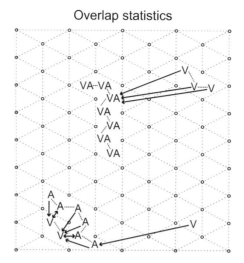

Figure 5.14 Overlap statistics between boundaries based on forest vegetation and animal abundance data for 81 sampling locations (open circles). The Delaunay links are the dashed lines from which 128 triangles can be formed. Using the triangulation-wombling algorithm with a 10% threshold, there are 13 candidate boundary elements for a total of four vegetation boundaries (indicated by V) and a total of two animal boundaries (indicated by A). The lines with an arrow at one end indicate the minimum nearest distances between the two types of boundaries (from one type to the other), while the lines with double arrows indicate the cases where the minimum nearest distances are symmetric in both directions. The overlap statistics are: O_s (direct overlap) is 7 candidate boundary elements (as indicated by the VA); the mean minimum nearest distance statistic, O_V, is 15.5 units; the mean minimum nearest distance statistic, O_A, is 6.5 units; and the overall mean minimum nearest distance statistic, O_{VA}, is 11.0 units. This example illustrates well that the animal boundaries are closer to the vegetation ones than the reverse; suggesting that animal boundaries are spatially associated with vegetation boundaries but that vegetation boundaries are not spatially associated with animal boundaries.

- The *overall mean minimum nearest distance statistic*, O_{12}, between boundaries 1 and 2:

$$O_{12} = \frac{\sum_{i=1}^{n_1} \min(d_{i \cdot}) + \sum_{j=1}^{n_2} \min(d_{\cdot j})}{n_1 + n_2}. \tag{5.11}$$

- In Figure 5.14, the overall mean minimum distance between the two boundaries is 11.0 units.

The statistic O_s allows us to test whether boundaries spatially coincide with one another, whereas the three other statistics, O_1, O_2 and O_{12}, can help discriminate between boundaries that are spatially associated (small significant values) or repulsing one another (large significant values). To test the significant spatial relationship between boundaries, it is recommended to randomize the rates of change, rather than the raw data, because the rates of change already include the inherent spatial structure of each variable and that is what is of interest.

The overlap statistics have been used to investigate the spatial relationship between forest edges and soil discontinuities (Fortin *et al.* 1996), as well as to test the relationship between bird and forest boundaries (Hall & Maruca 2001; St-Louis *et al.* 2004; Hall 2008) or landscape cover types (Polakowska *et al.* 2012). These overlap statistics offer a new means of investigating forest edge effects on other wildlife and environmental variables.

5.4 Hierarchical Spatial Partitioning

With only one quantitative variable (e.g. vegetation productivity based on the normalized difference vegetation index, NDVI), the lattice-wombling algorithm, being a local boundary detector, may not accurately detect a boundary, especially when there is local noise and spatial autocorrelation in the data. In such circumstances, hierarchical global edge detectors and kernel filters are more appropriate.

A hierarchical global boundary detector, such as the wavelet transform analysis, can identify boundaries from quantitative data in a transect (Redding *et al.* 2004; James *et al.* 2010) or in an area (see Csillag & Kabos 2002 for mathematical details; James *et al.* 2011). In Chapter 3, wavelet variance was introduced as a method to characterize and determine the scales of spatial patterns. Wavelet analysis is also used to compress an image to use less storage (Daubechies 1992) by using only a few wavelet transformation coefficients. This feature can be used to detect boundaries by partitioning and characterizing an image into relatively homogeneous areas using as many waveforms as are needed to model local patterns. Relatively homogeneous areas require few coefficients of low values, whereas contrasting locations, such as edges, require more coefficients with larger values. Wavelet analysis allows local multiscale analysis of the data by partitioning the data into relatively homogeneous spatial subareas. The determination of these spatially homogeneous areas is obtained using a hierarchical procedure based on *quadtree* decomposition (Csillag & Kabos 1996). Quadtree is a recursive algorithm that partitions an area initially into four quadrants and continues to divide each quadrant into four smaller quadrants in a hierarchical way until relatively homogeneous subareas are obtained. Depending on the spatial structure of the data, a single partition may be sufficient to create homogeneous subareas where a few wavelet transform coefficients adequately describe the structure; in other cases, more partitions may be necessary. The resulting subareas are the 'leaves' of the quadtree partition and the smallest possible leaf is a single sampling unit. At each partition, wavelet transform coefficients are added to describe the structure and to indicate the level of partitions needed. For each leaf, there is an equation. To obtain a higher degree of fit with the spatial pattern of data, more wavelet transform coefficients are required at each level. For image compression purposes, however, there is a trade-off between the amount of resolution retained and the storage of wavelet transformation coefficients

5.4 Hierarchical Spatial Partitioning

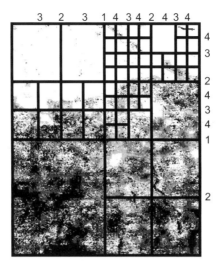

Figure 5.15 Spatial partition using the wavelet transforms at four hierarchical levels based on percentage coverage of black spruce in Quebec boreal forest (high values in white and low values in black). Most of the spatial partitions are in the upper right quadrant where the juxtaposition of high (white) and low (black) values necessitates more partition, that is coefficients, to describe the data. Indeed, trend patterns such as those seen in the upper left quadrat are easier to characterize than several small patches of low percentage coverage in forests of high percentage coverage.

at each scale and leaf. Usually more precision implies more leaves, thus more coefficients. For small data sets (e.g. 65 values), partitions cannot be performed for more than two hierarchical levels, whereas, for a larger region (e.g. 425 rows × 350 columns = 148,750 in Figure 5.15), the partition can be extended to the four hierarchical levels.

5.4.1 Edge Enhancement with Kernel Filters

The kernel filters are local edge detectors that enhance the contrast of adjacent pixels of an image to delimit objects. There are several algorithms, called operators (Xuan & Hong 2018), that are available in GIS and remotely sensed packages. Lastly, deep learning algorithms, like convolutional neural networks, can segment images (Abrams *et al.* 2019; Niedballa *et al.* 2022). The first operators developed measured the gradient on adjacent pixels using first- and second-order derivatives, such as the Laplacian filter. The 3 × 3 discrete approximation kernel version of the second-derivative Laplacian is:

$$\begin{array}{|c|c|c|} \hline 1 & 1 & 1 \\ \hline 1 & -8 & 1 \\ \hline 1 & 1 & 1 \\ \hline \end{array}, \quad (5.12)$$

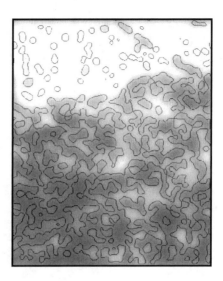

Figure 5.16 Spatial partition using the scale-space technique of the percentage coverage of black spruce in Quebec boreal forest, as in Figure 5.15. First, the data are smoothed using a Gaussian filter with a scaling factor of 40 cells (high values in white and low values in black) and then edges are delineated using a Laplacian filter (solid lines). All the edges are mapped without knowing which ones indicate the sharpest difference among cells.

with sum zero, and identical to the template for 9TLQV (Figure 3.14b). This filter produces zeros everywhere except at edge locations, thus detecting boundaries, but it is sensitive to noise, making prior smoothing necessary. This can be achieved either by aggregating adjacent cells, obtaining fewer larger cells (Fortin 1999b), or by using a Gaussian filter that preserves the number of cells. The most efficient kernel filters both reduce noise and detect edges, such as the Canny adaptive filter (Canny 1986; Richardson et al. 2009; Xuan & Hong 2018) or the scale-space techniques using the Laplacian or Gaussian algorithm (Faghih & Smith 2002). The scale-space techniques perform a series of smoothing using a Gaussian kernel of increasing size, allowing detection of the persistence of boundaries across scales. Figure 5.16 combines the smoothed data based on a Gaussian filter using a scaling factor of 40 cells and the delineated edges based on the Laplacian filter. Unlike the wavelet analysis that finds partitions based on the entire region (Figure 5.15), the scale-space approach identifies many more local boundaries because all the edges are mapped (Figure 5.16). The Laplacian algorithm identifies an edge where the kernel sum is different from zero so that strong or weak boundaries are treated similarly. An overly large kernel size can distort the spatial partitioning by isotropic smoothing of the data. The scale-space approach has been used in forestry to identify individual trees in a forest from high-resolution aerial imagery (Brandtberg 1999).

Other kernels are available for use and some are non-linear, such as those based on polynomials or global threshold kernels. To learn more about kernel filters, we recommend the literature on computer vision and image recognition for remote sensing data and medical imagery (Zhan et al. 2011).

5.5 Concluding Remarks

Spatial partitioning methods (spatial clustering and boundary detection) presented in this chapter are only the tip of the iceberg because we have focused on methods that are commonly used by ecologists (Table 5.2). There are many other methods, mostly developed for image recognition, that enhance edges by smoothing or thresholding noise. These sophisticated kernels are useful for univariate data but seeking to smooth out noise with larger kernels may deform the spatial pattern by imposing an isotropic form on the analysis.

For the multivariate data that is common in ecological research studying ecotones or cohesive ecological boundaries, several conceptual and methodological aspects need to be addressed. In this context, all variables and species may not have the same weight in the detection of boundaries. Hence, rare or omnipresent species may be ineffective as indicator species or species responding to specific environments. From a methodology perspective, novel ways to measure the spatial overlap between boundaries are needed so that we can compute the distance between a line boundary (vector mode) and a difference boundary (raster mode). This follows a theme that arises in several contexts for spatial analysis: the current and future relationship between methods based on geometry and vector representations and those that arise from raster approaches.

Table 5.2 Summary of the spatial analysis methods presented in Chapter 5

Spatial analysis method	Template
Spatial clustering	
Moving split-window	
Lattice-wombling	
Triangulation-wombling	
Categorical-wombling	
Enhancement filter	
Wavelets	
Scale-space	

Once boundaries have been detected, many ecological questions can be addressed regarding the relationship between the location of boundaries and the underlying processes that generate them. Then, one can also investigate how the width and shape of boundaries or edges influence ecological processes like in the distance to edge influence (Redding *et al.* 2004; Harper *et al.* 2005).

The greatest challenge is to integrate ecological concepts with statistical theory. What are the most appropriate ways to generate restricted randomization procedures that can test whether boundaries are cohesive? Indeed, complete randomization is not an appropriate procedure to test the significance of rates of change, boundary statistics and overlap statistics; however, restricted randomization that retains the basic spatial structure of the data is recommended. With ecological data, restricted randomizations need to reflect our ecological understanding of the processes. In the case of ecological boundaries, several processes are acting: the processes that create the patches and those that generate the boundaries. Usually, they are all different, or at least the processes that generated the patches on either side of their boundary are different. Consequently, patches and boundaries indicate non-stationarity such that global spatial statistics and global randomization are inappropriate. Hence, we are in a 'chicken and egg' situation where the proper way to restrict the randomization by area is first to identify subareas with stationarity but this is exactly what we seek through spatial partitioning. James *et al.* (2011) provided a first attempt to test boundaries detected by wavelets with restricted randomization based on the data's spatial structure using autoregressive and Gaussian random field models. This approach is a promising avenue for testing the significance of boundaries using empirical data but other developments will require modelling of the processes that generated the boundaries.

6 Spatial Autocorrelation and Inferential Tests

Introduction

Parametric statistics assume independence of the errors in the model to be fitted, as in linear regression (Legendre & Legendre 2012; Kim 2021). This assumption is often violated when analysing ecological data by increased Type I error and biased estimation of regression parameters. This lack of independence can arise because objects (samples, observations, etc.) near each other may tend to be more similar than those farther apart due to spatial, temporal and phylogenetic processes (Ives & Zhu 2006; Peres-Neto & Legendre 2010). The spatial dependence may be causally within the variable itself; as in the abundance of a single species which has constraints on mobility and dispersal. This form is sometimes called 'true autocorrelation' or 'inherent autocorrelation'; it is distinguished from 'induced spatial dependence' where the observed variable, such as abundance, depends on an underlying covariable, such as soil moisture, which is itself autocorrelated. The two may not be distinguishable in ecological data, and both may occur together in the same variable (Wagner & Fortin 2005). Autocorrelation in the dependent response variable, y, in the independent covariates, x's, or in both, will affect estimation accuracy and the power of any significance test. While the presence of spatial autocorrelation in the response variable can be accounted for by the relationship with the covariate variables, autocorrelated covariates still influence the detection of the predictor effects (Peres-Neto & Legendre 2010).

In ecological data, autocorrelation is scale- or distance-dependent, changing with spatial distance, and it can appear related to patterns like patchiness or gradients over a wide range of scales. Hence, determining its degree and scale is the first step in relating the ecological response to covariates. Spatial autocorrelation is often considered a statistical nuisance because it causes each observation to provide less than a full degree of freedom.

This chapter presents ways to deal with significant and strong spatial autocorrelation in inferential tests. We first describe how models from time series analysis can be modified to model spatial data. Then, we present a series of solutions to account for spatial autocorrelation in parametric inferential models using spatial regression and restricted randomization procedures. Lastly, we discuss how to accommodate spatial autocorrelation in sampling and experimental designs.

6.1 Models of Autocorrelation in One Dimension

We start with the simplest models for one-dimensional data before describing more general models. These first models of serial correlation were developed for time series and then modified for spatial data. Consider observations of a variable y in a series of n locations $(y_1, y_2, \ldots, y_i, \ldots, y_n)$ under the following conditions:

- Complete independence, Model CI:

$$y_i = \varepsilon_i;$$
$$\varepsilon_i \approx N(0, \sigma_\varepsilon^2). \tag{6.1a}$$

Here, ε_i is an independent 'error' term, usually following a Normal distribution with a mean of zero and variance σ_ε^2. Figure 6.1a illustrates this model with the expected autocorrelation between observations being 0. The term ε_i usually follows a Normal distribution, but only independence is required and other distributions can be considered.

An alternative model of independence retains spatial independence but includes functional dependence on a covariate:

- Spatial independence with functional dependence, Model SI(x):

$$\begin{cases} y_i = \beta x_i + \varepsilon_i, \\ x_i = \xi_i. \end{cases} \tag{6.1b}$$

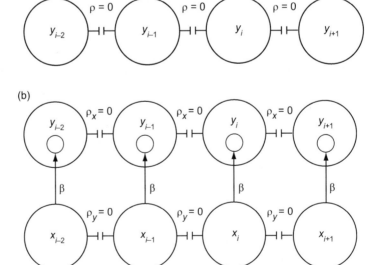

Figure 6.1 (a) Model CI: The series of values is independent of each other and so the expected correlation between adjacent values is 0 (see Equation 6.1a). (b) Model SI: The series of y values are dependent on the x-series, which are independent of each other. The y values remain spatially independent so the expected correlation between adjacent values is 0 (see Equation 6.1b).

6.1 Models of Autocorrelation in One Dimension

Both ε_i and ξ_i are independent 'error' terms following some statistical distribution, and β is the linear regression parameter. The expected correlation between adjacent values of y and between adjacent values of x is 0, as illustrated in Figure 6.1b.

Consider a response variable that affects its own spatial pattern through serial dependence:

- Inherent autoregression, Model IH:

$$y_i = \rho y_{i-1} + \varepsilon_i \text{ with } -1 \leq \rho \leq +1. \tag{6.2}$$

Here, ε_i is as before and ρ is the autocorrelation parameter. Model IH determines the strength of the autocorrelation. Figure 6.2 illustrates the model; the expected correlation between adjacent values is ρ.

The spatial structure of y can be affected by the serial dependence in the covariate x:

- Induced autoregression, Model ID:

$$\begin{cases} y_i = \beta x_i + \varepsilon_i \\ x_i = \rho x_{i-1} + \xi_i \end{cases} \quad \xi_i \approx N\left(0, \sigma_\xi^2\right). \tag{6.3}$$

Both ε_i and ξ_i are error terms, β is the usual regression parameter and ρ is the autocorrelation parameter. The correlation between adjacent values of y has an expected value that is a non-zero function of β and ρ (Figure 6.3). The second error term, ξ_i, need not be normally distributed, but that is sometimes assumed.

Lastly, the structure of y can be due to dependence in both variables:

- Double autoregression, Model IH–ID:

$$\begin{cases} y_i = \beta x_i + \rho_y y_{i-1} + \varepsilon_i, \\ x_i = \rho_x x_{i-1} + \xi_i. \end{cases} \tag{6.4}$$

Two autocorrelation parameters are required, one for y and one for x. This model is doubly autoregressive because it includes both inherent and induced dependence, as illustrated in Figure 6.4. The expected value of the serial correlation in y is a function of the three parameters ρ_y, β and ρ_x.

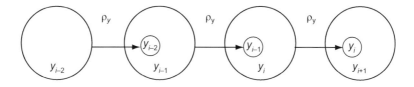

Figure 6.2 Model IH: The values in the series are not independent, with each being directly dependent on its predecessor and thus indirectly dependent on all preceding values. The expected correlation between adjacent values cannot be 0, but ρ (see Equation 6.2).

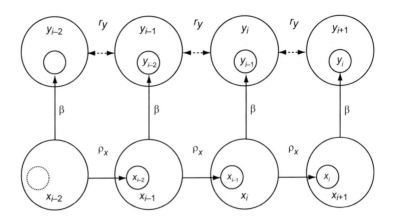

Figure 6.3 Model ID: The values in the y-series are not independent, but their observed autocorrelation, r_y, is induced by their linear dependence on the x-series, which is governed by a Model IH process (see Equation 6.3).

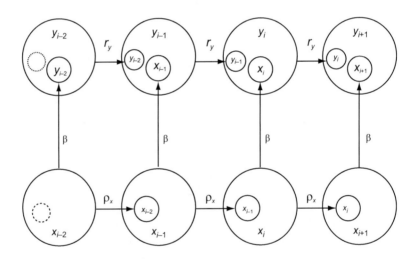

Figure 6.4 Model IH–ID: Correlation of y values comes from two sources, inherent in y itself and induced by dependence on x, giving a doubly autoregressive model.

Model IH produces a first-order autoregressive structure that is the most frequently studied model of autocorrelation; the correlation of any two variates, y_i and y_j, depends only on the number of intervening steps:

$$\mathrm{Cor}(y_i, y_j) = \rho^{|i-j|}. \tag{6.5}$$

Substituting for y_{i-1} in Equation (6.2), we get:

$$y_i = \rho(\rho y_{i-2} + \varepsilon_{i-1}) + \varepsilon_i = \rho^2 y_{i-2} + \rho \varepsilon_{i-1} + \varepsilon_i, \tag{6.6}$$

giving $\mathrm{Cor}(y_i, y_{i-2}) = \rho^2$.

6.1 Models of Autocorrelation in One Dimension

This first-order version in Equation (6.2) is the simplest of the general kth-order autoregressive model (Cressie 1996):

$$y_i = \sum_{j=1}^{k} \rho_k y_{i-k}. \tag{6.7}$$

In Model ID, the dependence in y is induced by dependence on autocorrelated x's, resulting in the correlation:

$$\mathrm{Cor}(y_i, y_j) \propto \rho^{|i-j|} \quad \text{for } i \neq j. \tag{6.8}$$

The values differ between the two models, being proportional to σ_ε for Model IH and to σ_ξ for Model ID, but the response to distance is the same.

These serial models seem to have been formulated for only temporal series because of the explicit dependence of y_i on y_{i-1} (Model IH) and thus only on preceding values of y. This may seem inappropriate for spatial data, where we would expect the dependence to be equal in both directions, but it is appropriate because the correlation between variates depends only on the distance between them, not on the direction (Equation 6.5).

We can examine the covariance structure of our data by calculating the covariance for all pairs separated by distance 1, by distance 2, and so on. These are estimates of the underlying covariances, and that is what a covariogram or correlogram does (see Chapter 4). The estimate of covariance for lag d is:

$$C_y(d) = \frac{\sum_{i=1}^{n-d} y_i y_{i+d} - \sum_{i=1}^{n-d} y_i \sum_{j=d+1}^{n} y_j / n}{n - d - 1} = \frac{\sum_{i=1}^{n-d}\left(y_i - \overline{y}_{(1,n-d)}\right)\left(y_{i+d} - \overline{y}_{(d+1,n)}\right)}{n - d - 1}, \tag{6.9}$$

where the sample means are for subsets of the data based on the first and the last $n - d$ values in the series (the 'regional' means). In Figure 6.5, for $n = 12$ and a lag of 3, the sample means of y_4 to y_{12} and of y_1 to y_9 are used. To calculate the correlation coefficient for lag d, using the 'regional' sample variances (in Figure 6.5 that is of y_4 to y_{12} and of y_1 to y_9), leads to poor estimates (Legendre & Legendre 2012). A better estimate is:

$$r_y(d) = \frac{C_y(d)}{s^2_{y(1,n)}} = \frac{C_y(d)}{C_y(0)}. \tag{6.10}$$

Equation (6.10) will be used for estimating autocorrelation in what follows.

Figure 6.5 The calculation of autocovariance or autocorrelation for $n = 12$ and lag $d = 3$: the 'regional' sample means are calculated from y_4 to y_{12} and from y_1 to y_9. Using the regional values of sample variance results in poor estimates of the autocovariance or autocorrelation.

So far, we focused on spatial data along one dimension, but many of the comments apply equally well to two or more. In two dimensions, we usually estimate the covariance using distance classes, as described (Equation 6.9), but in some cases, where anisotropy is a concern (see Chapters 2 and 4), direction classes may be used as well. In two or more dimensions, the concepts are the same as those in one dimension, but the technical aspects of setting up models tend to be more complicated. Whittle (1954) showed that, in one dimension, bilateral dependence (where y_i depends on both y_{i-1} and y_{i+1}) can be reduced to unilateral dependence, which is more tractable but a similar reduction is not easily achieved in two dimensions.

To investigate the characteristics of these models in one dimension, we need to generate a large number of comparable realizations, say using $n = 500$ and $\rho = 0.4$. For each realization, we calculate the sample mean, $\bar{y}$, and sample variance, s^2, and the t statistic to test the null hypothesis that the true mean of y is zero: $H_0 \equiv (\mu_y = 0)$:

$$t = \frac{\bar{y} - 0}{s/n}. \tag{6.11}$$

For 10,000 realizations of each model, we count the number of times the test statistic is less than the 0.05, 0.5, 2.5 and 5% critical values of t_{n-1} and the number of times it is greater than the 95, 97.5, 99.5 and 99.95% critical values. For Models ID and IH–ID, which include induced autocorrelation, we examine the results for the original variable y and for its residual after the linear dependence on x is removed: $y' = y - (a + bx)$, with a and b estimated from the data using standard techniques. Table 6.1 shows some typical results. For the independence models, Model CI or SI(x), the rates are close to the expected counts (as they should be), but for Models IH and ID the rates are much higher. This demonstrates the effect of positive autocorrelation on statistical tests: they produce more apparently significant results than the data justify (Cliff & Ord 1981). This effect is one of the main topics of this chapter and the subject of much of the discussion that follows, but we should comment on the last line of Table 6.1. In Model ID, the autocorrelation is caused by the linear dependence of y on x, which has inherent autocorrelation, and so removing that dependence might just

Table 6.1 Results of 10,000 simulations for Models CI to IH–ID ($\rho = 0.4$; $\beta = 0.6$)

Critical value (%)[a] Expected count	0.05 5	0.5 50	2.5 250	5 500	95 500	97.5 250	99.5 50	99.95 5
Model CI	4	43	227	472	531	262	66	3
Model SI	6	54	272	628	442	198	42	6
Model IH	116	419	913	1,317	1,575	1,086	490	126
Model ID (y)	32	181	497	857	617	369	124	24
Model IH–ID (y)	309	648	1,252	1,600	1,635	1,210	624	237
Model ID or IH–ID (y') (residuals)	0	0	0	0	0	0	0	0

[a] The values tabulated are the number of trials in which the test statistic was more extreme than the critical value associated with the probability given by the column heading. The null hypothesis (of mean 0) is true and, therefore, these rates represent Type I errors.

remove the effect of the induced autocorrelation from y, giving results like CI or SI(x). The zeros arise because, by using $y' = y - (a + bx)$, with a and b estimated from the data, the observed mean of y' is forced to be zero, causing the test statistic to also be zero in every case and, thus, never outside the critical values.

The main message of Table 6.1 is that positive autocorrelation, whether inherent or induced, produces 'too many' significant results, and a lot too many. For a two-sided test with $\alpha = 0.05$, as often in ecological studies, using the 2.5 and 97.5% critical values, Model IH gives almost 2,000 apparently significant results, four times the nominal rate. Clearly, the magnitude of the effect could lead to serious errors in decision-making based on the test results. An intuitive understanding of this effect can be based on the fact that, because the n observations are not completely independent of each other, we do not get the full n units of information, but something less. The 'effective sample size', n', is the equivalent number of independent observations that would provide the same amount of information as n non-independent observations. Here, n' is less than n and so, when we use s/n in the divisor of the t statistic, rather than s/n', we are dividing s by a number that is larger than it deserves to be, thus underestimating the variance of the mean and producing a test statistic too large. Section 6.2 discusses how to approach the fact that spatial autocorrelation changes the rates at which statistical tests produce significant results.

6.2 Dealing with Spatial Autocorrelation in Inferential Models

6.2.1 Simple Adjustments

The simplest approach might be to acknowledge the existence of spatial autocorrelation and to adjust the Type I error rate, α, to a more conservative value using a Bonferroni correction that accounts for the number of tests performed: for $k = 5$, try $\alpha' = \alpha/k = \alpha/5$. Depending on the autocorrelation structure underlying the data, this adjustment can be too liberal or too conservative.

The commonly invoked first-order autoregressive structure (Model IH) has autocorrelation that declines exponentially with distance. This suggests that, for distances greater than some particular value, the autocorrelation is effectively zero and observations further apart can be treated as independent. If this is true, and with abundant data, it should be possible to use a widely spaced subset of the data to ensure independence. In an analysis of landscape patterns in two dimensions, Ostendorf and Reynolds (1998) determined that autocorrelation did not extend beyond 20 pixels in their data and, therefore, used what they considered to be a non-autocorrelated subset of pixels 20 units apart, representing 1/400 of the pixels available. This approach has two major drawbacks: first, it seems very wasteful of data (Legendre & Legendre 2012) and, second, the concept of a 'distance to independence' may be mistaken for real spatial data, where the values having declined to zero then rebound and where non-zero autocorrelation may have an effect even when not significantly different from zero. We will elaborate in Section 6.2.2.

6.2.2 Adjusting the Effective Sample Size

As described, autocorrelation produces an effective sample size that is not the number of samples, $n' \neq n$. Positive autocorrelation reduces the effective sample size and the estimated degree of spatial autocorrelation determines the ratio of n' to n (Clifford et al. 1989; Dutilleul 1993). Consider tests of a mean (Section 6.1); its variance is estimated by the sample variance divided by n:

$$\text{var}(\bar{y}) = s^2/n. \quad (6.12)$$

With autocorrelation, n is replaced by n', which we wish to determine. In general, the variance of the mean of $y_1, y_2, \ldots, y_n$ can be adjusted using the covariances of the y's, $\text{Cov}(y_i, y_j)$ (Cressie 1996):

$$\text{var}(\bar{y}) = n^{-2} \sum_{i=1}^{n} \sum_{j=1}^{n} \text{cov}(y_i, y_j). \quad (6.13)$$

We can estimate n' by re-arranging:

$$n' = n^2 s^2 \div \sum_{i=1}^{n} \sum_{j=1}^{n} \text{cov}(y_i, y_j) = n^2 \div \sum_{i=1}^{n} \sum_{j=1}^{n} \text{cor}(y_i, y_j). \quad (6.14)$$

For a first-order autoregression, $\text{cor}(y_i, y_j) = \rho^{|i-j|}$, and the effective sample size for large n is approximately:

$$n' \simeq n \frac{1-\rho}{1+\rho} = n\Theta, \quad (6.15)$$

where Θ is $(1-\rho)/(1+\rho)$. If $n = 500$ and $\rho = 0.4$, then $\Theta = 0.429$ and $n' = 214$. Numerical simulations of artificial data with this autoregressive structure confirm the correctness of using $n' = n\Theta$ for one- and two-sample t-tests and for ANOVA F-tests comparing means. Simulations also confirm this correction for paired sample t-tests (Dale & Fortin 2002). Of course, for this correction to be useful for any test, the autoregressive model must be a good description of the data's structure or the approach must be robust to departures from that model.

The next step is to investigate estimating the effective sample size, not from model parameters but from the correlations calculated for each lag distance, $r(d)$, giving the correlation matrix, $\mathbf{R}$, with elements r_{ij}. Because we cannot estimate the correlation of individual pairs of variates, we use the correlation for each lag, d, from all $(n-d)$ such pairs. The correlation matrix of pairs of variates (Table 6.2) contains n 1s on the main diagonal, and two of each of the other entries, symmetric about the diagonal. Therefore, the estimate of n', based on these correlations is:

$$n'(\mathbf{R}) = \frac{n^2}{\sum_{i=1}^{n}\sum_{j=1}^{n} r_{ij}} = \frac{n^2}{n + 2\sum_{d=1}^{n-1}(n-d)r(d)}. \quad (6.16)$$

To evaluate this estimate of the effective sample size, $n'(\mathbf{R})$, we can examine other autocorrelation structures.

6.2 Spatial Autocorrelation in Inferential Models

Table 6.2 First-order serial autocorrelation structure. The main diagonal gives the correlation at distance zero, the first off-diagonals those at distance 1, and so on.

$$\begin{vmatrix} 1 & \rho & 0 & 0 & \cdots & 0 \\ \rho & 1 & \rho & 0 & \cdots & 0 \\ 0 & \rho & 1 & \rho & & 0 \\ 0 & 0 & \rho & 1 & & 0 \\ \vdots & & & & \ddots & \vdots \\ 0 & 0 & 0 & 0 & \cdots & 1 \end{vmatrix}$$

For the case of first-order serial autocorrelation, with correlation ρ between adjacent observations but 0 for all other pairs (Table 6.2), $n' = n^2/(n + 2(n-1)\rho)$. For $n = 500$ and $\rho = 0.4$, this gives $n' = 278$. As a second case, consider a first-order moving average model (Model MA_1) determined by the sum of adjacent independent error terms:

$$y_i = \varepsilon_i + \varepsilon_{i-1}. \tag{6.17}$$

This gives $\rho = 0.5$ and for large n, $n' = n/2$. Figure 6.6 gives an intuitive illustration of why this is so. If half the information in y_i is contained in y_{i-1} and the other half is contained in y_{i+1}, then only every second y is required to recover all the information in the series, and $n' = n/2$.

Model MA_1 is the simplest member of a class of autocorrelation models called 'moving average' models, which, with k being the model's order, have the general form (Chatfield 1975):

$$y_i = \varepsilon_i + \sum_{j=1}^{k} a_j \varepsilon_{i-j}. \tag{6.18}$$

Figure 6.7 illustrates the name, with a moving window averaging the εs.

In the second model of this moving average series, the correlation between adjacent observations is ρ_1; it is ρ_2 for pairs at one remove and 0 for all other pairs. The effective sample size is:

$$n' = n^2/[n + 2(n-1)\rho_1 + 2(n-2)\rho_2]. \tag{6.19}$$

The values $\rho_1 = 0.67$, and $\rho_2 = 0.33$ can be generated by this model (Model MA_2; Figure 6.8):

$$y_i = \varepsilon_i + \varepsilon_{i-1} + \varepsilon_{i-2}. \tag{6.20}$$

For these values and large n, $n' = n/3$.

Table 6.3 shows the results for a range of autocorrelation structures, generated in the same fashion as Models MA_1 and MA_2 (see Dale & Fortin 2002). With the exception of the last two lines, computer experiments show that the effective sample

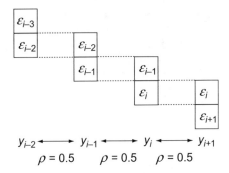

Figure 6.6 Model MA_1: $y_i = \varepsilon i + \varepsilon_{i-1}$, a first-order moving average model. Half the information in y_i is contained in y_{i-1} and the other half in y_{i+1} (see Equation 6.17). Unlike Model IH, there is no dependence beyond lag 1. The correlations are: $\text{cor}(y_i, y_{i-1}) = 0.5$ and $\text{cor}(y_i, y_j | j \neq i; j \neq i \pm 1) = 0$.

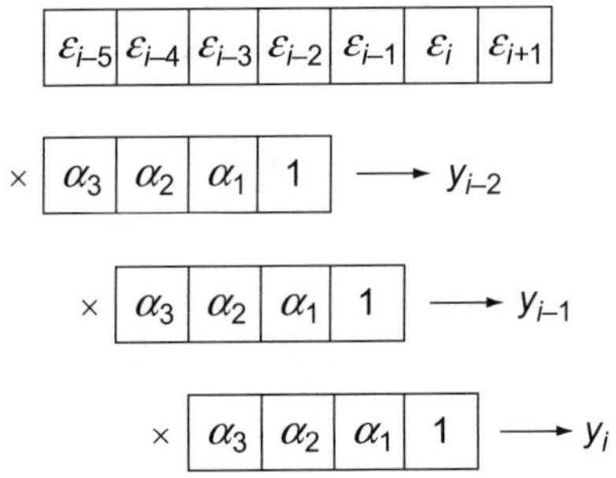

Figure 6.7 Illustration of a general moving average (MA) model, here of order 4. A template with weights moves along the data series and calculates a weighted average at each position to create the value of y.

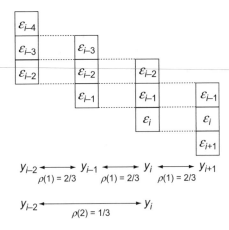

Figure 6.8 A particular second-order moving average model: $y_i = \varepsilon_i + \varepsilon_{i-1} + \varepsilon_{i-2}$, $\text{cor}(y_i, y_{i-1}) = 2/3$ and $\text{cor}(y_i, y_{i-2}) = 1/3$.

6.2 Spatial Autocorrelation in Inferential Models

Table 6.3 Effective sample sizes based on autocorrelation values, $n'(\rho)$, for artificial data from 'moving average' (MA) models of autocorrelation.

$\rho(d), d = 0, 1, 2, \ldots$	$\Sigma\Sigma\rho_{ij}/n^a$	$n'(\rho)$
1, 0.5, **0**[b]	2	250
1, 0.67, 0.33, **0**	3	167
1, 0.8, 0.6, 0.4, 0.2, **0**	5	100
1, 0.25, 0, −0.25, **0**	1	500
1, 0.5, 0, −0.17, −0.33, −0.17, **0**	2/3	750
1, 0.4, −0.2, −0.4, −0.2, **0**	1/5	2,500
1, −0.83, 0.67, −0.5, 0.33, −0.17, **0**	≈0	?
1, −0.75, 0.5, −0.25, **0**	≈0	?

[a] The actual sample size is $n = 500$.
[b] **0** means that correlations at this lag and all larger lags are zero.

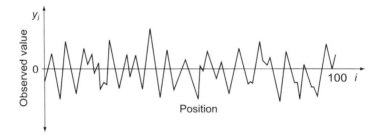

Figure 6.9 Cyclic behaviour of the variable y induced by the moving average model $y_i = \varepsilon_i + \varepsilon_{i-1} - \varepsilon_{i-2} - \varepsilon_{i-3} - \varepsilon_{i-4}$. The behaviour of y is, however, aperiodic, with autocorrelation expected to be 0 at lags of 5 and beyond.

sizes in the last column are 'correct', with 10,000 realizations producing rejection rates close to the nominal values. The most interesting autocorrelation structures are those which have some negative autocorrelation added. Consider the model $y_i = \varepsilon_i + \varepsilon_{i-1} - \varepsilon_{i-2} - \varepsilon_{i-3} - \varepsilon_{i-4}$, which produces apparently cyclic behaviour in the data, as shown in Figure 6.9. Sufficient negative autocorrelation actually increases the effective sample size and it can be greater than n. This means that strong positive autocorrelation at small scales does not necessarily compromise statistical tests if cyclic behaviour produces negative autocorrelation at larger scales. In fact, the test statistic may require inflation rather than deflation to achieve significance at the nominal rate, but rarely.

The second related point is that autocorrelation at all lags must be included in the calculation of effective sample size, even if the individual value does not seem to be statistically significant itself. Many small negative values at greater distances may counteract the effects of larger positive values at short distances. This situation arises often in ecological data with patchiness, which can give rise to cycles of positive and negative autocorrelation.

Based on the correctness of the effective sample sizes for a range of models given in Table 6.3, it is tempting to suggest that the solution is to calculate the autocorrelation matrix, **R**, from the data and then use its values to find the correct effective sample size. Alas, the computer runs that gave rise to Table 6.3 demonstrated the real problem for this approach. That problem is that the realizations of a very simple structure, such as that generated by $y_i = \varepsilon_i + \varepsilon_{i-1}$, can have very different estimates of n'. In a set of 10,000 runs with $n = 500$ of that simple model (the first line of Table 6.3), although the average effective sample size calculated from the data was 250 (good!), individual realizations ranged from 48 to 492. Clearly, if this approach cannot be used for artificial data with a simple underlying structure, it will be more dangerous for real data with an unknown and possibly complex structure.

Given these concerns about estimating the effective sample size from the data, it may seem that we should abandon this approach, but no... Before we return to general solutions, we consider other models that produce spatial autocorrelation, as background information, and then we present some corrections available in the literature.

6.2.3 More on Induced Autocorrelation and the Relationships between Variables

So far, we have used the terms autocorrelation and autoregression without drawing a clear distinction between correlation and regression. In general, correlation refers to the positive or negative relationship between two quantitative variables, both possibly measured with error, where it is not known that one has a direct causal effect on the other (Sokal & Rohlf 1995). In the model of induced autocorrelation (Model ID), adjacent x's have non-zero correlation although they have no direct causal effect on each other. Correlation is a measure of the covariance of the two variables, relative to their variances:

$$r_{xy} = \frac{\text{Cov}(x,y)}{\sqrt{\text{Var}(x)\text{Var}(y)}}. \tag{6.21}$$

Given three quantitative variables, x, y and z, the partial correlation of x and y with z held constant is:

$$r_{xy \cdot z} = \frac{r_{xy} - r_{xz}r_{yz}}{\sqrt{\left(1 - r_{xz}^2\right)\left(1 - r_{yz}^2\right)}}. \tag{6.22}$$

Figure 6.10 illustrates an artificial example, in which both x and y are positively correlated with z (and apparently with each other: $r_{xy} = 0.76$), but they are negatively correlated with each other when the relationships with z are removed: using the formula given in Equation (6.22), $r_{xy \cdot z} = -0.91$.

Linear regression, by contrast, evaluates the strength of the linear dependence of a dependent variable on an independent variable, based on a model such as:

$$y_i = \alpha + \beta x_i + \varepsilon_i. \tag{6.23}$$

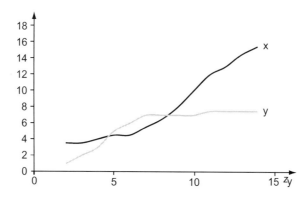

Figure 6.10 Both x and y are correlated with z ($r_{xz} = 0.96$; $r_{yz} = 0.90$), as illustrated in part and positively correlated with each other ($r_{xy} = 0.76$); but they are negatively correlated with each other when their dependence on z is controlled for ($r_{xy \cdot z} = -0.91$).

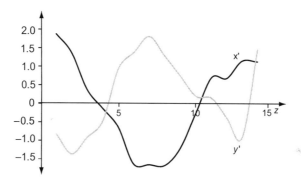

Figure 6.11 The residuals of x and y when the linear dependence on z is accounted for.

When the linear dependence of y on x is controlled for or removed, what is left is the residual:

$$y'_i = y_i - (\hat{\alpha} + \hat{\beta} x_i). \tag{6.24}$$

For Figure 6.10, when the linear dependence of x and y on z is removed, negative correlation of the residuals is obvious (Figure 6.11) and the correlation coefficient of x' and y' is -0.91. It is tempting to speculate that if autocorrelation in x and y could be attributed to their dependence on z, its effects could be similarly controlled by accounting for their dependence on z.

With real data, we observe spatial dependence in y, but we do not know its origins. It may be inherent, induced, or both. We may not be able to distinguish among the possibilities. For Models IH, ID and IH–ID, the autocorrelation declines exponentially with increasing distance, although that may not be true of all models of autocorrelation. Given the exponential decline, we may not be able to identify the model as IH, ID or IH–ID. In addition, given a single data set, it may not be possible to identify the underlying structure as an autoregressive versus a moving average model.

To introduce 'the trace method' for correctly testing the correlation of two variables with autocorrelation, we begin with models in which the relationship between x and y arises from their dependence on a third variable, z. Begin with the simple model:

$$\begin{cases} y_i = \beta_x z_i + \varepsilon_i, \\ x_i = \beta_y z_i + \eta_i, \\ z_i = \xi_i. \end{cases} \tag{6.25}$$

We then add autocorrelation to x and y, or to z, in turn, either in the moving average (MA) or autoregressive (AR) form. After their linear dependence on z is controlled, we examine tests of correlation between x and y for non-zero correlation or inflation of significance rates in all nine combinations. The results are given in Table 6.3, providing several clear messages:

(1) Controlling for the dependence on z does not remove the effects of spatial autocorrelation in x and y, in that the rates are inflated (MA) or the correlation is greater than zero (AR).
(2) The autocorrelation in z is not an important factor in this context.
(3) The moving average and autoregressive models produce *qualitatively different* behaviours. (Models that combine AR and MA terms will also have a non-zero correlation.)
(4) Interpretation of a significant correlation will be especially difficult if it is not obvious whether the AR or MA model is the better description of the data.

These features are especially important in the ecological context because some form of inherent autocorrelation can be expected in most biological variables *and* some form of induced autocorrelation can be expected due to autocorrelation inherent in abiotic factors in the environment. Usually, we will not be able to determine the relative strength of these two sources in our biological variables, and it is probable that both will exhibit double autocorrelation.

6.2.4 Correlation and Related Measures

The correlation coefficient measures the strength of the linear relationship between two variables. For the correlation coefficient between two independent variables, each with autoregressive correlation and parameters ρ_1 and ρ_2, Bartlett (1935) showed that its variance is approximately:

$$s_r^2 = \frac{1 + \rho_1 \rho_2}{n(1 - \rho_1 \rho_2)}. \tag{6.26}$$

This result suggests that:

$$n' = n\Psi = n \frac{1 - \rho_1 \rho_2}{1 + \rho_1 \rho_2}. \tag{6.27}$$

For $n = 500$ and $\rho_1 = \rho_2 = 0.4$, $n' = 362$. If only one of the series has autocorrelation, then no correction is required; one of the parameters is 0 and $n' = n$ (Bivand 1980).

Clifford et al. (1989) suggested a method for using the t-test to assess the significance of the correlation coefficient in the presence of spatial autocorrelation, with the sample size, n, replaced with an effective sample size. In this version, $n'_r = 1 + s_r^{-2}$, using an estimate of the variance of the correlation coefficient based on the autocorrelations of x and y. Using the autocorrelations at lag k, $r_x(k)$ and $r_y(k)$, with $n(k)$ being the number of pairs in distance class k, here $n - k$, the effective sample size n' is

$$n'_r = 1 + \frac{n^2}{n + 2\sum_{k=1}^{n-1}(n-k)r_x(k)r_y(k)} = 1 + n\frac{1}{1 + \frac{2}{n}\sum_{k=1}^{n-1}n(k)r_x(k)r_y(k)}. \quad (6.28)$$

This resembles the estimates of the effective sample sizes for x and y considered separately; it can also be written as:

$$n'_r = 1 + \frac{n^2}{\text{tr}(\mathbf{R}_x\mathbf{R}_y)}, \quad (6.29)$$

where $\mathbf{R}_x$ and $\mathbf{R}_y$ are the autocorrelation matrices estimated from the data and 'tr' is the matrix trace, the sum of the major diagonal.

Dutilleul (1993) refined this method and provided a generalized and exact form of it, thus correcting problems that can occur with small sample sizes. This approach calculates the variance of the sample covariance from the (auto)covariance matrices and provides this adjustment: $n'_r = 1 + s_r^{-2}$.

The procedure is to calculate the covariance (not correlation!) matrices for x and y as estimates based on the distance classes $d = 0, 1, 2, \ldots$; call the matrices $\mathbf{S}_x$ and $\mathbf{S}_y$. Let $\mathbf{B}$ be the matrix with $b_{ii} = 1/n - 1/n^2$ on the main diagonal and $b_{ij} = -1/n^2$ elsewhere. Then:

$$n' = 1 + \frac{\text{tr}(\mathbf{BS}_x)\text{tr}(\mathbf{BS}_y)}{\text{tr}(\mathbf{BS}_x\mathbf{BS}_y)}, \quad (6.30)$$

where 'tr' refers to the trace, the sum of the major diagonal. With both x and y following first-order simultaneous autoregressive processes on a lattice, Dutilleul (1993) provided some examples of this correction. For a 10×10 lattice ($n = 100$) and $\rho_x = \rho_y = 0.1$, the effective sample size is 80; if the autocorrelation parameters are of opposite signs, $\rho_x = -\rho_y = 0.1$, the effective sample size is 119, illustrating again the effects of negative spatial autocorrelation. Investigations by Legendre et al. (2002) confirm the effectiveness and robustness of Dutilleul's correction.

6.2.4.1 Correlation

We investigated the robustness of spatial autocorrelation corrections based on the 'trace' approach for several bivariate tests, including those for the correlation coefficient, based on a range of artificial data generated by AR and MA models (Dale & Fortin 2009). We included the potentially difficult cases of artificial data with waves, from fairly strict to very irregular. The effective sample size for the bivariate test, n'_r is much less variable than those of the single variables n'_x or n'_y. In fact, it is possible for n'_r to be much less than n, even while n'_x and n'_y are greater because of negative autocorrelation. The effective sample size n'_r cannot be derived from the values of n'_x and n'_y alone. It is the combination of the autocorrelation structures that determine the effective sample size; and so, we cannot characterize spatial autocorrelation simply by its 'strength' or 'intensity' in the individual data series.

The 'trace' correction is robust for most MA or simple AR models. More complex AR models can have incorrect rates of rejection of the null hypothesis, but only when the models produce reinforced wave structures and when the two data sets have the same period or periods that are integer multiples of each other. These results confirm that this correction for spatial autocorrelation in this particular bivariate test is broadly successful but not infallible.

6.2.4.2 Partial Correlation

Partial correlation evaluates the linear relationship between two variables when a third variable is considered. Given three variables, x, y, and z, the partial correlation of x and y, controlling for z is:

$$r_{xy.z} = \frac{r_{xy} - r_{xz}r_{yz}}{\sqrt{(1 - r_{xz}^2)(1 - r_{yz}^2)}}. \tag{6.31}$$

This coefficient measures the relationship of x and y as if z were held constant and the effective sample size should be the same as for the correlation coefficient and the 'trace' correction should work. Our computer simulations confirmed this result as presented in Alpargu and Dutilleul (2006).

6.2.4.3 Linear Regression

The calculations for linear regression are also similar, and so the same effective sample size correction should apply. This was shown by Alpargu and Dutilleul (2003) and our own computer runs confirmed that the robustness conditions for this 'trace' correction for the F-test is the same as for the t-test for the correlation coefficient. We have also speculated that the same correction will work for logistic regression (see Dale & Fortin 2009).

6.2.4.4 Tests for Proportions

Tests for proportions are different from these correlation methods; they are based on contingency tables of counts and goodness-of-fit statistics. A familiar example is testing for positive or negative association of two species based on the counts of presence or absence in sample units. These counts are summarized in a 2×2 contingency table, and a goodness-of-fit statistic is compared to the χ_1^2 distribution. If the sample units are in a contiguous series, the data will have spatial autocorrelation and

6.2 Spatial Autocorrelation in Inferential Models

we should correct for this lack of independence. Consider two species recorded in a string of contiguous sampling units as 0 for absence and 1 for presence. A reversible Markov model of these sequences would be based on the underlying transition probabilities, which can be estimated from the data. For 2×2 contingency tables derived from data with such serial correlation, Tavaré (1983) provided a correction based on the non-unit eigenvalues, λ, of the transition matrices (details in Tavaré 1983). The deflation factor is:

$$\Phi = \frac{1 + \lambda_1 \lambda_2}{1 - \lambda_1 \lambda_2}, \qquad (6.32)$$

(Tavaré 1983; Tavaré & Altham 1983). The structural similarity to the Bartlett correction factor is striking. The test statistic is calculated in the usual way, but it is divided by the deflation factor before being compared to the reference distribution. For the goodness-of-fit statistic, dividing the statistic by Φ is equivalent to an effective sample size of $n' = n\Phi^{-1}$.

A more general correction for spatial autocorrelation in $2 \cdot 2$ contingency tables can be based on the correlation structure of the data, rather than properties of a model (Cerioli 1997). It is similar to the approach of Clifford et al. (1989) and Dutilleul (1993) and seems to be robust. Harmonizing notations, Cerioli's effective sample size for a $2 \cdot 2$ table is:

$$n' = n\left(\frac{1}{1 + \varphi}\right) \qquad (6.33)$$

with

$$\varphi = \frac{2}{n} \sum_{k=1}^{D} r_x(k) r_y(k) n(k). \qquad (6.34)$$

Here, D is the number of distance classes and $r_x(k)$ and $r_y(k)$ are the autocorrelations of x and y in the kth distance class, with $n(k)$ pairs in the class. This gives:

$$n' = \frac{n^2}{n + 2\sum_{k=1}^{D} r_x(k) r_y(k) n(k)}. \qquad (6.35)$$

Apart from the addition of 1, for a one-dimensional data series this is identical to the correction of Equation (6.33).

This same approach has since been extended to $R \times C$ tables (Cerioli 2002). For R rows and C columns, the proportions of observed values in the ith row and jth column are designated as $p_{i\cdot}$ and $p_{\cdot j}$. The correction also requires $r_{I(x,i)}(k)$, the autocorrelation coefficient for x belonging to category i and distance class k, as indicated by the index function $I(x,i)$, and so on:

$$\varphi = \frac{2}{n} \sum_{i=1}^{R} \sum_{j=1}^{C} \left\{ (1 - p_{i\cdot})(1 - p_{\cdot j}) \sum_{k=1}^{D} r_{I(x,i)}(k) r_{I(y,j)}(k) n(k) \right\} \qquad (6.36)$$

and then

$$n' = n\left(\frac{(R-1)(C-1)}{(R-1)(C-1)+\varphi}\right) = n\left(\frac{1}{1+\varphi/(R-1)(C-1)}\right). \qquad (6.37)$$

We evaluated this correction for goodness-of-fit tests using a range of spatial models of binary data. That evaluation showed that the correction for goodness-of-fit tests is mainly robust, failing only for models in which reinforced waves in the data created strongly cyclic autocorrelation. We conclude that, except in unusual circumstances, this correction based on the structure of the data can be used to correct autocorrelation for $R \times C$ goodness-of-fit tests.

The main problem with this general approach in ecological studies is that its applicability depends on the appropriateness of a Markov model. Often, that will not be a good description of the data and the test is not particularly robust to departures from the underlying assumption. The concerns about Markov models aside, for bivariate and multivariate testing there is cause for optimism. For bivariate data, if the autocorrelation is 'well-behaved' with any cycles being irregular or decaying, the 'trace' corrections are reliable, but ecologists should be concerned that the autocorrelation may cycle persistently because of patchiness in the data. Our advice is to check the autocorrelation structure, particularly the product of the autocorrelations: if the product is regularly cyclic and decreases only slowly, the corrections may not work. If any autocorrelation cycles are irregular or their amplitude decreases with distance, the correction procedures usually can be applied with confidence.

Corrections for bivariate tests are more robust than for univariate tests because they use the product of the autocorrelations for the correction; these have magnitudes less than one, so the products are small. Note that if only one variable has autocorrelation, no correction is needed because the product is always zero (cf. Tavaré 1983; Cerioli 1997). Corrections for univariate tests remain elusive.

6.3 Randomization Procedures

6.3.1 Restricted Randomization and Bootstrap

Randomization can be a general approach to testing significance by rearranging the original data to generate a reference distribution (Legendre & Legendre 2012; Manly 2018). A test statistic calculated from the data is compared with the distribution of the same statistic after the data have been randomized (see Chapter 2). For ecological applications, restricted randomizations retain the structure of the data (spatial or temporal) as much as possible, whereas complete randomization would erase it (Fortin & Jacquez 2000; Fortin et al. 2012a). Figure 6.12 illustrates the difference between complete and restricted randomization. A 20×20 grid contains 400 cells in four density classes, which we will reduce to two for analysis. For the 180 black (high density) cells (Figure 6.12a) and counting 'rook's move' contiguous neighbours, the observed number of black–black joins ($J_{BB} = 220$) is greater than the expected

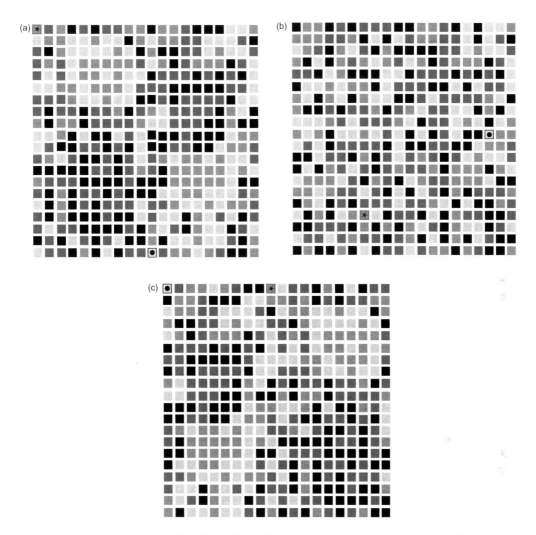

Figure 6.12 A comparison of complete and restricted randomization. In the original data (*a*) there is significant autocorrelation. With complete randomization, shown in (*b*), that structure is destroyed, but randomization by toroidal shift (shown in (*c*)) retains most of the structure. The dot and the asterisk identify individual squares where shifts in locations from (*a*) to (*c*) occurred.

number, 153.4. This reflects the autocorrelation of neighbours in the data with obvious patches of low and high density.

When the data are completely randomized by exchanging randomly chosen pairs of grid cells (Figure 6.12*b*) the black–black join count is close to the expected value ($J_{BB} = 154$). In Figure 6.12*c*, the data have been randomized with the restriction of a toroidal shift of 10 on the *x*-axis and 1 on the *y*-axis, preserving much of the spatial structure ($J_{BB} = 212$). The toroidal shift and related methods are the most common restricted randomization techniques. For transect data of tree-canopy density and understorey cover, we can test the significance of their correlation, preserving the

Figure 6.13 The 'caterpillar' randomization of one-dimensional data: the original data series (upper) is broken at a randomly chosen point (indicated by the arrowhead) and then recombined (lower). If there are two data sets to be related, either set can be shifted in this way to provide new relative positions.

spatial structures, by iteratively shifting their relative positions and recalculating the correlation (Figure 6.13). This is the 'toroidal shift' in one dimension, the 'caterpillar randomization'. Longer transects with more quadrats will have more 'shifted' relative positions and greater sensitivity. Very short transects may not have enough relative positions to allow this test.

Another application of randomization is to derive variance estimators using jack-knifing estimation, as described by Lele (1991) and Heagerty and Lumley (2000). In particular, it derives the reference distribution by resampling the data, leaving out one observation in each iteration. This jackknife approach enables confidence intervals to be derived for parameters of interest, such as those of regression models, which can then be tested for significant difference from 0. Cohn (1999) recommended a bootstrap procedure for comparisons of multivariate structures in the presence of serial correlation and we suggest (without rigorous proof) that it should apply well to other forms of autocorrelation in two dimensions. Bootstrapping is another randomization technique that uses resamples of the data, allowing each datum to be used more than once. Bjørnstad and Falck (2001) proposed a bootstrap algorithm to create a confidence envelope for a non-parametric estimate of a spatial covariance function. There have been several advances in the use of bootstrapping in spatial situations (Lin et al. 2011; Garcia-Soidan et al. 2014; Castillo-Paez et al. 2019), including the use of block bootstrapping which attempts to preserve the dependence structure within the data (see Lahiri 1999), but there have been few applications of this in ecological studies.

6.3.2 Monte Carlo Markov Chain

Another approach to autocorrelation is to develop a relatively simple model of the autocorrelation structure and then use a Monte Carlo simulation (Monte Carlo Markov Chain or MCMC methods) to generate artificial data sets to compare with the observed. Begin by finding a parametric model of the spatial dependence in the data by standard model selection procedures. The model is then used in a Monte Carlo approach to finding good confidence intervals for the test statistics (Manly 2018). This last approach is the one we advocate in the absence of a robust 'analytic' solution for a particular set of circumstances. It may seem somewhat indirect, but it does allow for tests of significance. It is also the procedure recommended by Mizon (1995) in the context of economic analysis. He suggested that, starting with a very general model which well describes the data structure, test for valid reductions by determining which explanatory variables can be omitted. Over-specified models, which have more variables than they need, still produce valid inferences; they are merely inefficient (Mizon 1995). We could start, therefore, with a very general autoregressive model of the data, considering observations as far as 20 steps apart, such as:

$$y_i = \sum_{j=1}^{20} \beta_j y_{i-j} + \varepsilon_i, \qquad (6.38)$$

where ε_i is $N(0, \sigma^2)$.

By eliminating many of the variables, we might end up with a model in which only two or three of the βs were significantly non-zero. It is appropriate to omit the non-significant terms. As an illustration of this idea, we generated 10 realizations of the model $y_i = 0.4y_i - 1 - 0.2y_i - 4 + \varepsilon_i$, with $n = 100$. We then fit the model given in Equation (6.38) and determine the best-fitting sub-model using a maximum likelihood backward selection procedure. Table 6.4 gives examples of the results.

Table 6.4 shows a range of possible outcomes for $n = 100$; many are 'close' to the original underlying model, particularly at lag 1, but the effective sample size ranges from one-third to almost double the 'true' value. Clearly, however, these results show that the best-fit model is not always the model that generated the data. A larger sample size, $n = 500$, will produce best-fit models that are more like the original, showing the advantage of larger sample sizes, but they still exhibit considerable variability. The 'model and Monte Carlo' (MCMC) procedure is an attractive approach in the absence of more direct solutions.

6.4 Considerations for Sampling and Experimental Design

In the statistical literature, discussions of spatial autocorrelation are often predicated on the simplest autoregressive model, giving an exponential decline in autocorrelation with distance (Figure 6.14a) or the 'typical' variogram depicted in Figure 4.6a. One concern is that these usual models are not good descriptions of the structure of ecological data: ecological variograms tend to fluctuate non-randomly, because of patchiness in the data. ('The natural world is a patchy place'! Dale 1999). That means that corrections based on the first-order autoregressive model may be misleading, although the matrix trace correction avoids the weakness of model-based corrections. We must consider how our understanding of spatial autocorrelation and its effects can inform our decisions in designing sampling or the layout of experiments.

Table 6.4 Coefficients of models fit to realizations of $y_i = 0.4y_{i-1} - 0.2y_{i-4} + \varepsilon_i$.

β_1	β_2	β_3	β_4	β_5	β_6	β_7	n'(approx.)[a]
0.47	−0.04	−0.04	−0.18	0.33	•[b]	•	22
0.56	−0.22	•	•	•	•	•	44
0.30	0.19	−0.04	−0.21	•	•	•	59
0.30	•	•	•	•	−0.26	0.08	77
0.21	0.05	−0.08	−0.32	•	•	•	128

[a] The effective sample size, based on the underlying model, is $n' = 65$ (approximately).
[b] Symbol '•' indicates a non-significant term.

6.4.1 Sampling

The assessment of spatial autocorrelation is an important step in understanding ecological data and one that must be considered in analysis and interpretation (Chapter 2). It is also apparent that, if the underlying autocorrelation structure were known in advance, the design for sampling or for laying out an experiment could be adjusted to minimize its effect, before any analysis was conducted. If the autocorrelation declines rapidly with distance and becomes effectively zero beyond some distance, R (Figure 6.14a), samples spaced at R or greater might be treated as independent (a big 'if'!). This is the concept of 'distance to independence', which seldom applies in ecological studies (see Chapter 12). Ecological variables are typically patchy, so their autocorrelation fluctuates between positive and negative with increasing distance (Figure 6.14b). Under those circumstances and with the locations of low values and high values known, the design could be adjusted appropriately. On the other hand, if autocorrelation declines slowly with distance, the distance between samples or experimental units may not be that important. In the intermediate case, where autocorrelation declines appreciably at the scale of the extent of the study, we may choose a design that provides a balanced set of distances (van Es & van Es 1993; Legendre et al. 2004).

The important first step is to assess the spatial autocorrelation, and thus a *pilot study* is required (Legendre et al. 2002). A pilot study involves taking many samples designed to evaluate how autocorrelation changes with distance and direction. Consider point samples, measuring altitude, estimating population density or detecting a particular substrate. These point samples can be arranged in many ways, each with advantages and disadvantages. Random placement has the advantage of simplicity but the disadvantage of a lack of control over the spacing of the points, the coverage of the study plot and the range of lags available in each direction. A regular grid of equally spaced points has an inherent direction and scale of its own, which may interact with patterns in the variable being studied. The 'wagon wheel' design of radiating lines of sample points will produce a trend in sampling intensity and circumferential distances from the centre of the wheel to the edge. A design based on the Fibonacci spiral avoids trend and directionality problems, but the full spiral requires a high intensity of sampling and has a trade-off between the number of samples and the distances it

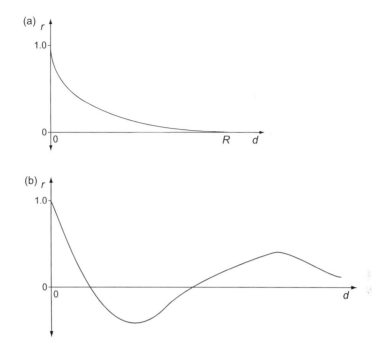

Figure 6.14 Two models of how autocorrelation may vary with distance: (*a*) Autocorrelation falls to zero at some distance R, known in geostatistics as the range. This is equivalent to the 'exponential' model in geostatistical analysis. (*b*) Autocorrelation falls below zero and then fluctuates above and below zero.

can evaluate. Partial Fibonacci designs in which only a systematic or randomly chosen subset of the full spiral is sampled seem to combine most of the advantages with a few disadvantages. Whichever subset is chosen, such a design offers the advantages of no directional bias, a good range of inter-sample distances and a lack of trend in sampling intensity.

We conclude this section with a summary of the findings of two studies on the effects of spatial structure on the design and analysis of field surveys and field experiments by Legendre *et al.* (2002, 2004). Both studies used simulations of an environment with one of several different kinds of spatial structure (gradient, waves, etc.) in the underlying environmental variable as well as autocorrelation in it and in the variable of interest to address questions about the effect of these structures on design and analysis.

In the study on surveys (Legendre *et al.* 2002), the simulation structure can be described using E_{ij} as the environmental variable, $\rho_{E,ij}(R_E)$ as its autocorrelation component with range R_E and ε_{ij} as a standard normal error term:

$$E_{ij} = S_{ij} + \rho_{E,ij}(R_E) + \varepsilon_{ij}, \qquad (6.39)$$

and

$$V_{ij} = \beta\, E_{ij} + \rho_{V,ij}(R_V) + \eta_{ij}. \qquad (6.40)$$

V_{ij} is the variable of interest, $\rho V_{,ij}(R_V)$ is its autocorrelation component with range R_V, β is a measure of its linear dependence on the environmental variable and η_{ij} is a standard normal error term. The questions asked included:

(1) Can the effect of spatial autocorrelation on tests be reduced by the survey design?
(2) Which designs provide the greatest power for given combinations of spatial structure and autocorrelation range?

One finding was that spatial autocorrelation in both the environmental and response variables affects the standard tests, but spatial autocorrelation in only one does not. A broad-scale spatial structure in the underlying environmental variable, however, combined with spatial autocorrelation in the response variable, inflated Type I error, just as spatial autocorrelation in both variables would. The major advice from this exercise was to use a pilot study to identify the underlying structure. If a gradient is present, its effect can be accounted for by using it as a covariate. For different zones, representing non-stationarity, a covariable that distinguishes the zones should be used. This paper provided the reassurance that Dutilleul's method for correcting the t-test of the correlation coefficient was robust in the various spatial structures tested.

6.4.2 Experimental Design

The study of experiments (Legendre et al. 2004) had questions like those for the survey study, and a similar approach was used. The only modification was that the variable of interest includes a treatment effect for experimental units, τ, with effect sizes of 0.0, 0.18 and 0.5:

$$V_{ij} = \beta E_{ij} + \rho_{V,ij}(R_V) + \eta_{ij} + \tau_{ij}. \tag{6.41}$$

This study provides several general lessons:

(1) If either spatial autocorrelation or repetitive structures like waves occur in the environmental variable, randomly positioned units should not be used; blocking is recommended, with the different treatments applied in neighbouring units.
(2) With spatial autocorrelation present, for a given number of units, more and smaller blocks spread throughout the study provide greater statistical power.
(3) Short-range spatial autocorrelation (related to the size of the experimental units and of the blocks) affects ANOVA tests more strongly than long-range spatial autocorrelation. For example, where the blocks were 3, 6 or more units in linear extent, autocorrelation with a range of 4 units (as opposed to 16 or 40 units) caused the greatest decrease in statistical power.

The approach used in these two studies is clearly helpful and we can find out more about mitigating the effects of spatial autocorrelation on statistical tests by pursuing it further. The 'bottom line', however, is that it is ESSENTIAL to have a good assessment of the nature and range of spatial autocorrelation before designing or carrying out a survey or experiment (see Figure 6.15).

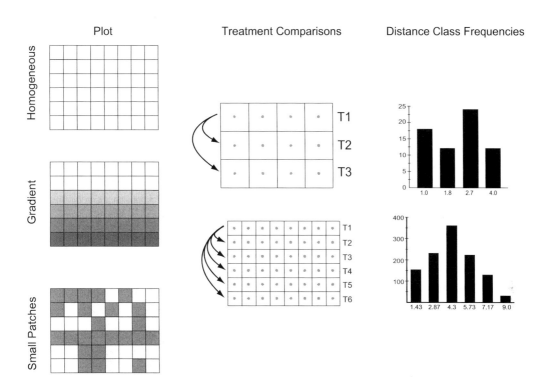

Figure 6.15 Illustration of how the inherent spatial pattern of the data (e.g., homogeneous, gradient, small patches) prior to the experiment can influence its outcome depending on the experimental design: the number of treatments and the geographical distance between the experimental units (Figure modified from van Es & van Es 1993).

6.5 Concluding Remarks

Spatial autocorrelation is an important feature of ecological data and it often has several sources that are not distinguishable. Autocorrelation affects the outcomes of statistical tests and it cannot be ignored. Even apparently non-significant amounts of autocorrelation can have a serious cumulative effect.

Furthermore, because patchiness is also common, autocorrelation may fluctuate between positive and negative with increasing distance, which poses a challenge for correction and interpretation. In addition, we might question whether the similarity of adjacent samples (such as contiguous quadrats in a transect) really represents redundancy in the data, the way it would in a simple model of autocorrelation. Having two samples in the same patch tells us something more than sampling the same individual twice.

In considering solutions to the effects of this phenomenon on statistical tests, we can offer some observations. The first and most important is that the problem seems to be mainly solved or solvable for bivariate and multivariate tests using the properties of the data themselves, thanks to the work of researchers like Dutilleul and Cerioli.

It remains a major difficulty for univariate tests, although several solutions have been proposed. Thinning the data is not a good idea because it is wasteful of information and because it is based on the concept of distance to independence, which may well be wrong. Adjusting the effective sample size is a possible solution *if* the chosen model is a good description of the data's structure; we cannot usually calculate the effective sample size from the data. Randomization methods may work, but they must be applied carefully and with an awareness of the potential problems. Complete randomization of the data cannot and does not control the effects of spatial autocorrelation on statistical tests of significance. Restricted randomizations that maintain much of the spatial structure of the data are the most appropriate, as are related approaches such as block bootstrapping. The 'model and Monte Carlo' method has several features that recommend it, but it is not always appropriate. As we showed for artificial data, the best-fitting model may not be the model that generated the data; for field data, we will not know the relationship between the model fit and the underlying structure. Larger sample sizes should improve the accuracy of the modelling exercise, but they increase the risk of encountering non-stationarity if the extent of the sampling is also increased.

Studying the relationship between autocorrelated variables requires careful consideration of the structure of the variation, particularly in ecology, where autocorrelation may be both inherent and induced in a single variable, and where it may not be well described by one of the standard models. This is why more and more ecologists are using spatial regression models. Yet, we need to understand well the source of the spatial structure to account for it appropriately. Of particular importance is the fact that removing the variables' dependence on an underlying factor does not avoid the general problems associated with analysis in the presence of spatial autocorrelation.

In considering the design of sampling or experimentation, the important first step is to determine the characteristics of the spatial autocorrelation before doing anything else. Knowing that structure will enable the researcher to develop a design that avoids, or at least reduces, the effects of autocorrelation on subsequent analysis. Pilot studies are necessary and may be combined with prior knowledge to produce effective designs.

In 1993, Legendre posed the question 'Spatial autocorrelation: trouble or a new paradigm?' as the title of a paper that provided a wide-ranging discussion of this topic. The answer, 30 years later, is that it is both. It is troublesome, not in the sense of being just a nuisance, but because it is not easy to deal with, and it is certainly part of the current approaches to ecology, which include spatial structure (they must) or provide spatially explicit results. We must remember, however, that it is this lack of independence through space and time that makes prediction possible. We would be in trouble, indeed, if ecological phenomena were spatially and temporally independent (if that were possible). Clearly, the characteristics, effects and corrections for spatial autocorrelation and other forms of spatial dependence require and are worth more effort and thought, before we can suggest that we understand them truly and thoroughly.

7 Spatial Regression and Multiscale Analysis

Introduction

Both correlative and inferential methods (this chapter and Chapters 6) can help to determine the factors and processes affecting the spatial patterns that are expressed in ecological data. The data may be univariate (e.g. species abundance, plant height) or multivariate (e.g. species composition, environmental variables). For either, the first step is to evaluate the degree of spatial autocorrelation and its significance for each variable. When data contain a significant degree of spatial autocorrelation, the parametric inferential statistical tests cannot be used because the standard significance evaluation is compromised (Chapter 6). The problem is that each observation does not provide a full degree of freedom, but only a portion determined by the amount of spatial autocorrelation. In such circumstances, the significance of the regression models needs to be assessed using a modified effective sample size (Chapter 6) or restricted randomization procedures (Chapter 2 and Sections 7.3 and 7.6). Another common feature of current ecological studies is that they can be conducted over large study areas (extents) increasing the risk of combining several non-stationary processes operating at different spatial and temporal scales. The sources of non-stationarity can be amplified when several data types with different grains and extents are combined without accounting for sampling efforts (Fletcher et al. 2019). Researchers should, therefore, perform first an *exploratory spatial data analysis* (ESDA) to determine the key spatial scale(s) at which statistical analyses should be performed (Fortin et al. 2012b).

This chapter present methods that quantify the spatial relationship between spatially autocorrelated variables (both univariate and multivariate) using correlation and regression in the spatial context, and shows how multiscale analysis can depict the scales relevant to the spatial structure. Then, in Chapter 8, we will show how these methods can be extended to spatio-temporal data.

7.1 Spatial Causal Inference

We begin with causal inference in the spatial context, a fascinating aspect of spatial analysis in ecological research. Note that causal *inference* is not the same as causal *proof*. The basic argument for the outcome process, Y, and the covariate process, X, goes like this: If knowledge about X does not significantly improve our ability to predict Y, then we conclude that X is not the cause (or *a* cause) of Y. Note that this is different from the converse. If knowledge about X significantly improves our ability to predict Y, we do not conclude that X is the cause of Y. Other possibilities include: (1) both X and Y are caused by some other process; (2) X is acting through a third process, M; or (3) we are seeing the effects of an unobserved factor, U.

The discussion of causal inference should start in the non-spatial context (based on Pearl 2009; Pearl & MacKenzie 2018; Brumback 2022) and then proceed to the challenges and possible solutions for spatial applications, based mainly on Thaden and Knieb (2018), Reich et al. (2021), Gao et al. (2022) and Akbari et al. (2023).

The general case is best understood from a simple graph depicting a hypothesized causal network, with the nodes being processes or variables and the directional edges showing conditional dependence or conditional independence. Figure 7.1 (also presented later as Figure 12.14) presents a simple causal network with Y as the outcome variable and X as a possible cause. Other potential variants are (a) including U as an unobserved factor affecting both X and Y; (b) M as a mediator of X's affects on Y (indirect effects), but with the spatial confounder measured as S; or (c) with Z as a third observed variable, an instrumental variable that affects only X, so that Y and Z are independent.

Such a causal diagram has no cycles (in the graph theory sense) and is a *Directed Acyclic Graph* (DAG). Causal inference analysis has arisen in many different areas of science and from many sources, hence several frameworks have developed for understanding cause in a hypothesized causal network. For most ecologists, Structural Equation Models may be the most familiar, related to path analysis, and in our recent book on network analysis (Dale & Fortin 2021) we described this approach to causal analysis in some detail (see also Grace 2020; Kimmel et al. 2021). Pearl's approach to structural causal models is a refinement of this method (Pearl 2009; Pearl & MacKenzie 2018) but there are other approaches available such as the technique of Instrumental Variables (Pearl 2009). We are not going to give details about these various approaches, or describe their strengths and weaknesses, but will focus on the peculiarities of causal inference within a spatial structure.

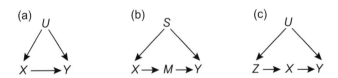

Figure 7.1 Hypothesized causal networks, with the nodes being processes or variables and the directional edges showing conditional dependence or conditional independence. Y is the outcome process or variable, with X as a possible cause. (*a*) U is an unobserved factor that may affect both X and Y; (*b*) M is a mediator for the effect of X on Y (indirect effect), but the spatial confounder is measured as S; (*c*) Z is a third observed variable (instrumental variable) that only affects X, so Y and Z are independent.

What is the problem? The problem is that many aspects of spatial data are inconsistent with the assumptions on which non-spatial causal analysis is founded (Akbari *et al.* 2023). You will have already guessed that spatial autocorrelation, called 'spatial spillover' for treatments, is an important one, with spatial heterogeneity and spatial scale effects (like the Modifiable Areal Unit Problem, MAUP) also on the list (again not a surprise). The question is how to deal with these problems in pursuing causal inference in spatial data and spatial statistics. In considering spatial statistics, Gao *et al.* (2022) suggested combining these with spatial regression to develop techniques for spatial causal inference. They conclude that spatial heterogeneity and autocorrelation are the main problems for this endeavour but the relationship between spatial patterns and spatial abnormalities deserve to be further investigated.

Reich *et al.* (2021) provided much detail on models of the spatial structure of interference (none, general, partial and spatial network). Of these models, ecologists will be most interested in pursuing an understanding of spatial network interference (Reich *et al.* 2021, figure 5d) which affects only nodes joined to each other by network edges, the network neighbours. This version has clear implications for the spatial graph or spatial network methods we advocate and describe in Chapter 10 and can allow a focus on local effects by looking at individual nodes and their immediate neighbours. Spatial heterogeneity can also be addressed by evaluating local average treatment effects instead of the usual global average (Imbens & Angris 1994). For localized inference, Delgado and Florax (2015) advocated the use of a 'difference of differences' approach (Ashenfelter 1978) with a spatial estimator to assess local autocorrelation and spatial interactions. As with many of the methods discussed here, care must be taken to ensure that the assumptions of the approach are valid.

In many discussions, the spatial structure under consideration takes the form of the values of a factor associated with subregions of the study area, such as air pollution associated with zip-code districts (Reich *et al.* 2021) or household disposable income in economic planning districts (Thaden & Kneib 2018). These values can 'spill over' into neighbouring units as a form of spatial autocorrelation. Thaden and Kneib (2018) examined structural equation models for dealing with spatial confounding, using the technique of neighbourhood 'penalties' to adjust the estimates. They illustrated the recommended method with artificial data with known spatial structure (gradient, central cluster, and random unstructured) and the results for likelihood-based and Bayesian estimation turned out to be very similar. They also provided an example of the relationship between average age and disposable household income in German economic districts in which the unemployment rate may be the spatial confounder, as a measured rather than unknown factor, as in Figure 7.1*b*.

In observational studies through space and time, confounding can be temporal and spatial. Dee *et al.* (2023) described the importance of such confounding effects in assessing the relationship between biodiversity and productivity in natural grassland ecosystems. The contrast between their finding that increased plot-level richness caused a decline in productivity is attributed to differences between dominant native species and rare non-native species, as well as the incomplete control of a range of confounding factors (Dee et al. 2023, figure 1b). Their study supports the contention that understanding species richness requires careful use of spatio-temporal approaches to determine the processes acting on the system (White et al. 2010).

Considering spatio-temporal situations, we note that spatial effects have no inherent directionality and this symmetry itself may be a problem for spatial causal inference (Akbari et al. 2023). However, for the greater challenges of spatio-temporal analysis, the directionality of time may allow the use of the Granger causality test, at least for the time component (Reich et al. 2021). That test tells us that if predictions about Y are better when *past* information about X is included than when it is omitted, we can conclude that X is a cause of Y. Reich et al. (2021) suggested that extending the method to spatio-temporal context is straightforward with the spatial dependence following a SAR or CAR structure (see Chapter 8). On the other hand, Herrera et al. (2016) produced a spatial analogue of this approach, including blocking the data in a bootstrap procedure; that also deserves further investigation.

Reich et al. (2021) discussed methods for 'point-referenced' data, such as those of a marked point process (Chapter 3). The use of the potential outcomes framework for causal inference allows simplification when no interference can be assumed, but without that assumption the analysis becomes complicated with weighting functions and a spatial integral that may be intractable because the treatment locations are not infinitely many. Some workarounds may be available, but also complicated by requiring assumptions; see Reich et al. (2021) for details.

For the purely spatial situation, it seems that most solutions for causal inference still have assumptions that will not hold for the messy world of ecological spatial data. In the absence of more definitive solutions from the experts, we continue with what we are already familiar with: local statistics and restricted randomization procedures to understand the effects of heterogeneity, as well as Model-and-Monte-Carlo procedures to investigate interactions.

7.2 Correlation between Spatially Autocorrelated Variables

To determine how environmental factors affect ecological data, it is usual to compute the correlation among the variables, but some of these will contain various degrees of spatial autocorrelation. As already mentioned, positive autocorrelation can reduce the information provided by n observations to something less than would be provided by n independent observations (Chapter 6). Therefore, we estimate the 'effective sample size' (n') of the n dependent observations. The effective sample size for bivariate tests, such as correlation, can be estimated from the degree of autocorrelation in the two variables (Clifford et al. 1989; Dutilleul 1993). Chapter 6 provided a detailed account of how to compute the effective sample size, but here is a summary.

The adjustment requires computing the degree of spatial autocorrelation for each variable, x and y, in k distance classes, producing a Moran's I spatial correlogram.

These are then used to estimate the covariances in the k distance classes, recorded in the matrices $\mathbf{S}_x$ and $\mathbf{S}_y$. The weighted connectivity matrix, $\mathbf{B}$, contains $b_{ii} = 1/n - 1/n^2$ on the main diagonal and $b_{ij} = -1/n^2$ elsewhere. The effective sample size is based on the products of $\mathbf{B}$ with $\mathbf{S}_x$ and $\mathbf{S}_y$:

$$n' = 1 + \frac{\text{tr}(\mathbf{BS}_x)\text{tr}(\mathbf{BS}_y)}{\text{tr}(\mathbf{BS}_x\mathbf{BS}_y)}, \qquad (7.1)$$

where 'tr' is the sum of the elements in the main diagonal (Dutilleul 1993).

Testing the significance of the linear correlation between x and y is based on $n' - 2$ degrees of freedom, instead of $n - 2$. The estimated effective sample size is inversely related to the estimated strength of spatial autocorrelation. It will vary depending on the number of distance classes used: fewer distance classes usually result in weaker spatial autocorrelation such that the estimated value of n' will be closer to n. In comparison, more distance classes can result in stronger spatial autocorrelation, hence the estimated value of n' will be much smaller than n (Fortin & Payette 2002).

As an illustration, consider the correlation between the species 'Spp.1' and the environmental variable 'Moisture', as depicted in Figure 7.2. The Pearson correlation between the two variables is 0.838, which is significant using $n - 2$ degrees of freedom (here $10 - 2 = 8$; $p < 0.002$). However, correcting for the degree of spatial autocorrelation using two distance classes to estimate the linear spatial pattern of the species 'Spp.1' reduces the effective sample size to 2.273 and its significance to $p = 0.131$ (computed using the function *modified.ttest* of the R package *SpatialPack*). This example demonstrates how statistical inference could change when data are spatially autocorrelated (see also Fortin & Payette 2002).

7.3 Mantel Test

Correlation between two sets of multivariate data can be computed as the cross-product between two distance matrices as proposed by Mantel (1967). The null hypothesis for a Mantel test is that the distances in some matrix $\mathbf{X}$ (e.g. species abundances at a set of sites) are independent of the distances in another matrix $\mathbf{Y}$ (e.g. environmental variables sampled at the same locations). The Mantel statistic computes a linear relationship between two symmetric matrices:

$$Z_\text{M} = \sum_{\substack{i=1 \\ i \neq j}}^{n} \sum_{\substack{j=1 \\ j \neq i}}^{n} x_{ij} y_{ij}. \qquad (7.2)$$

Z_M is the Mantel statistic and x_{ij} and y_{ij} are similarity, dissimilarity, Euclidean distance or connectivity matrix ($\mathbf{X}$, $\mathbf{Y}$) elements from the sampling locations. The advantage of this matrix approach is that any dissimilarity or distance coefficient can be used (see Legendre & Legendre 2012 for a full list of coefficients), that reflect the

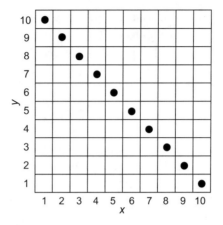

Figure 7.2 Two sets of variables (three species and two environmental variables) sample at 10 sampling locations (x, y). These data will be used to create symmetrical distance matrices (one for the x–y coordinates, one for the species and one for the environment) of $n(n-1)/2$ elements (here $10 \times 9/2 = 45$ elements per matrix) to compute the standardized Mantel statistic, r_{SE}, $r_{S\text{-}xy}$ or $r_{E\text{-}xy}$.

community structure of the variables of interest. Equation (7.2) can be rewritten in the form

$$Z_{XY} = \mathbf{X} \bullet \mathbf{Y}. \qquad (7.3)$$

Z_{XY} is the sum of all element-by-element products as indicated by '•' which is the Hadamard product of the distance matrices $\mathbf{X}$ and $\mathbf{Y}$ (Figure 7.3).

The measure Z_{XY} is unbounded and, to facilitate its interpretation, each matrix can be standardized to obtain a bounded Mantel statistic, r_M, ranging from -1 to 1:

$$r_M = \frac{1}{d-1} \sum_{i=1}^{n-1} \sum_{j=i+1}^{n} \left(\frac{x_{ij} - \bar{x}}{s_x}\right) \left(\frac{y_{ij} - \bar{y}}{s_y}\right), \qquad (7.4)$$

where d is $n(n-1)/2$, the number of elements in the lower (or upper) triangular matrices. From Equations (7.3) and (7.4), it is evident that r_M computes a Pearson linear correlation coefficient between the elements of two matrices and so the Mantel test and Pearson correlation values are closely related. Legendre and Fortin (2010)

7.3 Mantel Test

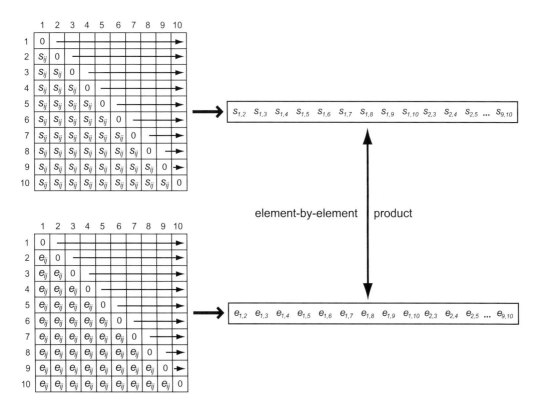

Figure 7.3 Illustration of how the element-by-element matrix multiplication (Hadamard product) is performed to compute the Mantel statistic. The square distance matrices on the left can be unfolded into vectors from which the Hadamard product is computed between the elements of the $D_{Species}(s_{ij})$ and the $D_{Environment}(e_{ij})$ matrices.

showed that the Mantel test is approximately the square root of the coefficient of determination R_M^2 of the linear regression between the elements of the matrices. The equivalence is not perfect because the sample size is n for correlation and $n(n-1)/2$ for the Mantel test.

Mantel statistics can be tested either by an approximate t-test when the sample size is large or by a randomization test when the sample size is small. The randomization is a restricted form so that it retains the relationships between pairwise distances measured between sampling locations; this is accomplished by randomly shuffling the rows and columns of one of the matrices (Fortin & Gurevitch 2001; Legendre & Legendre 2012). When the null hypothesis is true, the observed Mantel's statistic is expected to have a value near the middle of the reference distribution. With a significant relationship between the two matrices, whether positive or negative, the observed Mantel statistic is expected to be more extreme, either higher or lower, compared to the reference distribution. The precision of the probability value is directly related to the number of randomization iterations. We recommend at least 10,000 randomizations (Fortin et al. 2012a; Manly 2018), with the observed statistic included in the reference distribution.

As the Mantel computation is performed using distance values rather than the raw data, r_M values are not directly comparable to those of Pearson's correlation coefficient (Legendre & Fortin 2010; Legendre et al. 2015; Franckowiak et al. 2017) but are considerably weaker. Therefore, it is important to test the significance of the Mantel statistic, as even quite small values could be significant given the large sample sizes of the matrices: $n(n-1)/2$ instead of n. For example, the Mantel statistic, r_M, based on Euclidean distances of the three species and two environmental data, as depicted in Figure 7.2, is 0.372 ($p = 0.024$; computed using the function *mantel* of the R package *vegan*).

The Mantel test was developed to compute a linear correlation between the distances, even if the relationship is not linear. Hence, one should check the linearity between the two distance matrices by plotting them as a scattergram (Legendre et al. 2015). If the relationship between the two matrices is monotonic, then the Mantel test can be computed using a Spearman correlation; however, the magnitude of the effect is often weaker than with the raw data, as illustrated with the data from Figure 7.2. Usually, the relationship computed is between the distance measures, such that sampling pairs can have the same degree of dissimilarity because they both have high or low values. Consequently, the magnitude of r_M should not be used as the equivalent of the Pearson or Spearman correlation of the raw data.

The significance of the observed r_M is tested against the restricted randomization procedure of the rows and columns of distance matrices (Mantel 1967), and it can be significant even if the value is very small (close to zero) because the sample size for distances is greater $(n(n-1)/2))$ than the original number of sampled locations, n. Yet, when there is any strong spatial structure in the original data or the distance matrices, the estimates of the Mantel statistic will be biased, also affecting the Type I error rate (Guillot & Rousset 2013; Crabot et al. 2019; Somers & Jackson 2022). To minimize these effects, Somers and Jackson (2022) proposed computing the Mantel test from the simple difference between the elements of each matrix rather than using distance matrices like those of Euclidean or squared Euclidean distance.

Mantel statistics can be used to compute a *Mantel correlogram* that is a multivariable version of a spatial correlogram such as a Moran's *I* correlogram (Oden & Sokal 1986; Legendre & Fortin 1989; Borcard & Legendre 2012). In essence, a Mantel correlogram can be obtained by computing a Mantel statistic using a distance matrix from species data and a distance class matrix. The Mantel statistic values are then plotted against lag distance classes. In a simulation study, Borcard and Legendre (2012) showed that the Mantel correlogram's significance level is valid (i.e. the Type I error is correct). Hence, the Mantel correlogram can evaluate the spatial structure at increasing spatial lags. Using the tree assemblage data from Fortin (1992), a trend is identified (Figure 7.4) by the Mantel correlogram with a spatial range at about 60 m and where the r_M at the first distance class is 0.09 ($p < 0.001$; computed using the function *mantel.correlog* of the R package *vegan*).

There is a long list of 'pros and cons' of using the Mantel test. It is popular because it is one of the few tests that computes the relationship between two sets of multivariate data, based on a single synthetic distance measure. However, this is a key concern

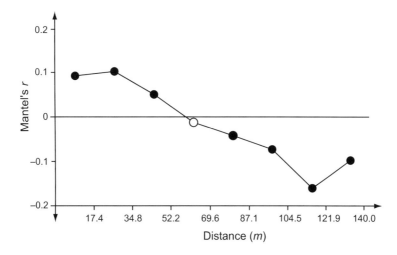

Figure 7.4 Mantel correlogram based on the abundance of 14 tree species where the standardized Mantel statistic, r_M, is plotted against the distance in metres. The overall spatial structure of the tree community is a trend with a spatial range of around 60 m. Data from Fortin (1992).

if multivariate data are synthesized into a single coefficient, not allowing the determination of the variable(s) responsible for the overall correlation between the two matrices. To avoid this limitation, Legendre and Fortin (2010) suggested using redundancy analysis (RDA or dbRDA; see Legendre & Legendre 2012; Capblancq & Forester 2021).

In addition, a misconception has developed over the years that the Mantel test can correct for the presence of spatial autocorrelation in the data. This is not true *per se*; what the Mantel test can do is account for the relative distance between the sampled locations. Furthermore, the Mantel test assumes independence among the distances and that will not be the case in the presence of a spatial structure. We will return to this issue in Section 7.3.1.

7.3.1 Partial Mantel Tests and Multiple Regression on Distance Matrices

Correlation between the distances of two matrices can be attributable to covariates or unmeasured data. To determine these effects and potential spurious correlations, a partial correlation approach can be used by extending the Mantel two-matrix test to a partial Mantel three-matrix test (Smouse *et al.* 1986). The partial Mantel test quantifies the relationship between two matrices, **X** and **Y**, while controlling for the effects of a third, **Z**. The partial Mantel statistic, $r_{XY.Z}$, is computed by detrending (or controlling or removing) the linear effects of matrix **Z** on matrix **X** and matrix **Y**, using a linear regression (Figure 7.5) for **X** on **Z** and **Y** on **Z**. Then, the residuals from both regressions, $Res_{X|Z}$ and $Res_{Y|Z}$, are the basis for a Mantel test, as in Equation (7.3). The partial Mantel can also be computed from correlations using all three matrices (Legendre & Legendre 2012):

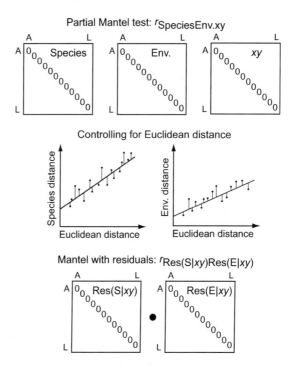

Figure 7.5 Partial Mantel test. This figure illustrates how the correlation between two variables (species and environment) sampled at the same 12 sampling locations can be computed by first factoring out the effect of a third matrix (here the x–y coordinates of the sampling locations), using linear regression and then computing a standardized Mantel statistic, $r_{Res(S|xy)\,Res(E|xy)}$, with the residuals of these two linear regressions.

$$r_{XY.Z} = \frac{r_{XY} - r_{XZ}r_{YZ}}{\sqrt{1 - r_{XZ}^2}\sqrt{1 - r_{YZ}^2}}. \tag{7.5}$$

By comparing r_M values computed with two and three matrices (partial Mantel test, $r_{XY.Z}$), we can test alternative causal relationships among the three matrices as we do in path analysis (Legendre & Fortin 1989; Legendre & Legendre 2012). Furthermore, the third matrix, **Z**, can be a *design matrix* with dummy variables corresponding to treatments and control locations (see Fortin & Gurevitch 2001) or a set of covariables, allowing the test of specific hypotheses by coding as in an ANOVA contrast matrix (Sokal et al. 1993) or by using geographic locations as surrogates for unsampled variables (Fortin & Payette 2002).

Considering the data analysed to compute a Mantel test (Figure 7.2), the partial Mantel test of the relationship between the three species (S) and the two environmental (E) data controlling for the Euclidean distances between the sampling locations (xy), $r_{SE.xy}$, is now -0.120 ($p = 0.777$; from function *mantel.partial* in R package *vegan*). Hence, controlling for spatial distances between the sampling locations reduces the value of the statistics and changes the sign so that the relationship is not significant. Yet, when we control for the effects of a third matrix, **Z**, such as the Euclidean distance matrix based on the sampling locations, we were not controlling for the degree of spatial autocorrelation of the variables but only for the relative distance among locations.

When the data are strongly spatially autocorrelated, the restricted randomization (by rows and columns of the matrices) no longer provides unbiased outcomes, so the significance of the partial Mantel test is not adequately evaluated (Oden & Sokal 1992; Guillot & Rousset 2013). This problem has been acknowledged often and different restrictions in the randomization procedure have been proposed (Legendre & Fortin 2010; Crabot et al. 2019). The problem stems from the lack of independence in the data so that the complete randomization procedure will produce inflated levels of Type I error (Cliff & Ord 1981). The inflated level of Type I error is not unique to the Mantel and the partial Mantel tests and it is inherent to most inferential tests; we will expand on this topic in Chapter 8. Yet, Oden and Sokal (1992) showed that, when the spatial autocorrelation between the distance values of the matrix is weak, partial Mantel tests, as proposed by Smouse et al. (1986), can still be used because they are unlikely to reject the null hypothesis incorrectly.

The partial Mantel test provides a way to identify potential causality among the matrices using a path analysis framework (Legendre & Legendre 2012). The partial Mantel, however, accounts for only one matrix and there are several cases where more matrix sets could explain the variability of species assemblage. An alternative is to use a *multiple regression on distance matrices* (MRDM) approach (Lichstein 2007; Franckowiak et al. 2017). Here, there is a clear 'response' distance matrix ($\mathbf{D}_Y$) and a series of predictor distance matrices ($\mathbf{D}_{X1}, \mathbf{D}_{X2}, \ldots, \mathbf{D}_{Xn}$) that are unfolded into their corresponding vector form on which a multiple regression is then performed (Figure 7.6). The significance of the regression is tested by a randomization procedure. Legendre et al. (1994) suggested using a stepwise selection procedure to determine which predictor vectors are the most parsimonious to be retained in the regression. Nowadays, ecologists and evolutionary biologists mostly choose the Akaike information criterion (AIC) as a model selection tool (Akaike 1974; Burnham & Anderson 2002; Johnson & Omland 2004). AIC is a log-likelihood model that assumes a least-squares estimation with normally distributed errors from the residual sums of squares (RSS):

$$\text{AIC} = n + n \times \log(2\pi) + n \times \log(\text{RSS}/n) + 2(p+2), \tag{7.6}$$

where n is the sample size, p is the number of predictors in the model with the addition of two to account for the intercept and the error term. There are some statistical issues (Franckowiak et al. 2017) when using AIC to select predictor vectors: (1) it should not be used with linear mixed models; and (2) regression with vector dependence between the distances can occur, implying that the sample size in an MRDM is inflated to $n(n-1)/2$, and a correction should be used to account for the strength of dependence present in the data (Van Strien et al. 2012). For further comments on AIC and alternatives, see Section 7.4.4.

$\mathbf{y}$	~	$\mathbf{x}_1$	$\mathbf{x}_2$	$\mathbf{x}_3$
$s_{1,2}$		$e_{1,2}$	$d_{1,2}$	$g_{1,2}$
$s_{1,3}$		$e_{1,3}$	$d_{1,3}$	$g_{1,3}$
$s_{1,4}$		$e_{1,4}$	$d_{1,4}$	$g_{1,4}$
$s_{1,5}$		$e_{1,5}$	$d_{1,5}$	$g_{1,5}$
$s_{1,6}$		$e_{1,6}$	$d_{1,6}$	$g_{1,6}$
$s_{1,7}$		$e_{1,7}$	$d_{1,7}$	$g_{1,7}$
$s_{1,8}$		$e_{1,8}$	$d_{1,8}$	$g_{1,8}$
$s_{1,9}$		$e_{1,9}$	$d_{1,9}$	$g_{1,9}$
$s_{1,10}$		$e_{1,10}$	$d_{1,10}$	$g_{1,10}$
$s_{2,3}$		$e_{2,3}$	$d_{2,3}$	$g_{2,3}$
$s_{2,4}$		$e_{2,4}$	$d_{2,4}$	$g_{2,4}$
$s_{2,5}$		$e_{2,5}$	$d_{2,5}$	$g_{2,5}$
...		...	...	...
$s_{9,10}$		$e_{9,10}$	$d_{9,10}$	$g_{9,10}$

Figure 7.6 Multiple regression on distance matrices. From the unfolded distance matrices (Figure 7.2), compute the multiple regression of the response vector $\mathbf{y}$ of the distances of the species assemblage data against the predictor vectors ($\mathbf{x}_1$: distances of environmental data, $\mathbf{x}_2$: Euclidean distances, $\mathbf{x}_3$: distances of geomorphological data).

7.4 Spatial Regressions

An alternative to any sort of correlation analysis for these data is to consider the environmental variable(s) as independent predictors of the species abundance and use *spatial regression* to examine that relationship, acknowledging the potential effects of the spatial context. Let us first model a response variable, y, as a function of independent predictors in vector $\mathbf{X}$ using linear regression (ordinary least squares, OLS; Figure 7.7*a*) in the absence of spatial structure:

$$y \sim \boldsymbol{\beta}\mathbf{X} + \boldsymbol{\varepsilon}, \tag{7.7}$$

where $\boldsymbol{\beta}$ is a vector of regression coefficients, and $\boldsymbol{\varepsilon}$ gives the random errors; these are assumed to be independent so that $\text{var}(\boldsymbol{\varepsilon}) = \sigma^2\mathbf{I}$, where $\mathbf{I}$ is the $n \times n$ identity matrix. The off-diagonal elements in the covariance matrix are zeros.

To illustrate how spatial regression methods behave, we model the relationship between ovenbird relative abundance (Figure 7.8*a*) and forest relative cover (Figure 7.8*b*) in southern Ontario ($n = 909$ sampled locations). Using the function *lm* in the R package *stats*, the adjusted R^2 is 0.307 and the linear regression is significant ($p < 0.001$).

When either the response or the predictor variables are spatially autocorrelated, the errors may not be independent, precluding linear regression. To determine whether the

7.4 Spatial regressions

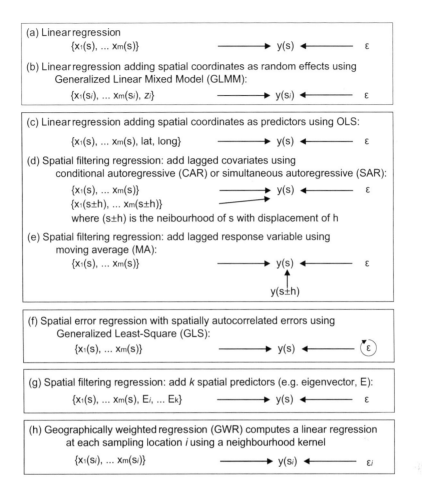

Figure 7.7 Regression approaches to account for spatial autocorrelated data within a regression framework: (*a*) linear regression, (*b*) adding spatial predictor variables or random effect (GLMM), (*c*) adding spatial coordinates as predictors, (*d*) spatial lagged models (CAR, SAR), (*e*) spatial lagged models (MA), (*f*) spatial error models (GLS), (*g*) spatial filtering models (e.g. Moran's Eigenvector Map) and (*h*) geographical weighted regression (GWR).

errors of OLS are autocorrelated we can plot the residuals and compute a Moran's *I* correlogram with these residuals. Significant spatially autocorrelated residuals may be due to (1) important predictors missing from the model, (2) incorrect model specification (e.g. using a linear model when the relationships are non-linear), (3) spatial mismatches between the spatial patterns of the response and predictors due to the grain and/or extent used or (4) spatial confounding that can emerge because multiple predictors affect the data at multiple scales.

While we presented some of these reasons in Chapter 6, here we focus of the general concept of *spatial confounding*. Spatial confounding indicates that the relationship between a covariate, *x*, and an outcome variable, *y*, results when unmeasured variables with their own spatial structures may be affecting either *x* or *y* or both

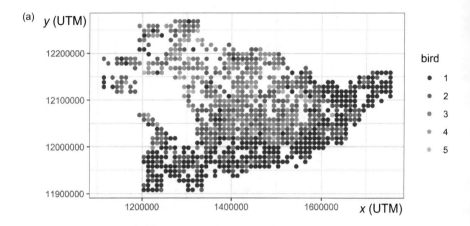

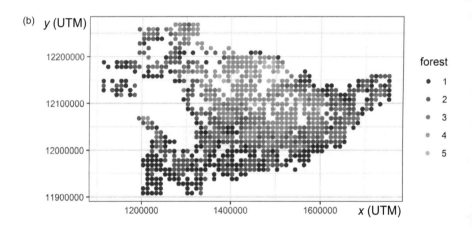

Figure 7.8 Spatial distribution of the ovenbird relative abundance (*a*) and relative forest cover (*b*) in southern Ontario ($n = 909$ sampling locations of 10×10 km each).

(Gilbert et al. 2022). That unmeasured variable is a spatial confounder, as mentioned in the last example. Several methods have been proposed to deal with this concern both in regression and related models (Urdangarin et al. 2023) and in geostatistics (Gilbert et al. 2022). We will not go into detail on the range of methods, but Urdangarin et al. (2023) recommended the 'Spatial+' method developed by Dupont et al. (2022). Gilbert et al. (2022) focused on a double machine learning approach which uses flexible regression models for both x and y on the confounders to derive a causal estimator (see Gilbert et al. 2022). Spatial confounding is obviously a concern for spatial causal inference, especially in the ecological context which is both multivariate and potentially with complex relationships among many variables.

One solution is to add environmental and ecological predictors to the OLS regression. Indeed, in some cases adding exogenous predictors could account for the spatial structure of the data. However, in other situations, depending on the spatial scales of the endogenous processes it may not be sufficient and spatial patterns persist in the residuals (Melles *et al.* 2011; Kim 2021). Spatial coordinates can be added as predictor variables in Ordinary Least Squares (OLS; Figure 7.7c). Using the *lm* function, the adjusted R^2 increases to 0.473 when the northing UTM and easting UTM were added as predictors compared to the R^2 of 0.307 with forest as the unique predictor.

Another option is to use a Generalized Linear Mixed Model (GLMM; Bolker *et al.* 2009; Zuur *et al.* 2009; Harrison *et al.* 2018) in which a random effect is added to account for the effects of location. In a GLMM the added random effect, zi, accounts for the variation in y at location i, given the effects of the other sampling locations that are not explained by predictor variables (Figure 7.7b):

$$y_i \sim \beta Xi + zi + \varepsilon. \tag{7.8}$$

When analysing the relationship between ovenbird and forest data with two spatial zones as a random effect in a GLMM (divided at UTM northing 12,000,000), the R^2 of the fixed effect (i.e. forest) is 0.207 (function *lmer* of the R package *lme4*) and of the random effect (zones) is 0.083 (function *r.squaredGLMM* of the R package *MuMIn*; Nakagawa & Schielzeth 2013).

GLMM may not be the best approach when there are mismatches between the spatial scales in the response and the predictors (Hefley *et al.* 2017; Teng *et al.* 2018; Fletcher *et al.* 2019). In such cases, a spatial regression technique can be used. *Spatial regressions* are models that include 'space' explicitly and are extensions of the familiar statistical technique with modifications to accommodate spatial dependence (Figure 7.7). Anselin (2009) provided an especially clear introduction to spatial regressions, but also see Kim (2021).

Figure 7.7 outlines several forms of spatial regressions. The basic idea is that physical position affects the variables: the value of the dependent variable y is affected by its position, **s**, on 1, 2, or 3 spatial axes, which can be written as $\mathbf{s} \rightarrow y$. More formally, $y = f(\mathbf{s}) + \varepsilon$ (meaning 'y is some function of **s** plus an independent error term'). The position **s** can be determined by a frame of reference created by spatial axes, the 'absolute position', or relative to a local neighbourhood. Position may combine with other variables, the x_is, to affect y:

$$\{x_1(\mathbf{s}), \ldots, x_m(\mathbf{s}); \mathbf{s}\} \rightarrow y \text{ or } y = f(\mathbf{s}) + g(\{x_i(\mathbf{s})\}) + \varepsilon. \tag{7.9}$$

In this spatial context, y can be affected by spatial autocorrelation in y, in the predictor variables (the xs), or both, for example:

- Spatial error models account for the spatial structure in the errors, such as generalized least-squares regression (GLS; Beguería & Pueyo 2009; Figure 7.7f) and regression Kriging (Karl 2010).
- Geographically weighted regression (GWR; Fotheringham *et al.* 2002) that creates a regression model for each sampled location, based on the values in its neighbourhood (Figure 7.7h).

7.4.1 Spatial Filtering Using Autoregressive Models

So far, our presentation of autoregressive models has dealt with continuous variables for which the calculation of correlation was a logical approach, but we also need to consider how autocorrelation applies to discrete variable data, the simplest being sequences of 0s and 1s. Autocorrelation in such data can be created when the value at a particular location is dependent on the values at preceding locations. Spatial filtering models, in which the spatial signal in the response or covariate variables, is accounted for using kernel approaches, including conditional autoregressive (CAR), simultaneous autoregressive (SAR) and moving average (MA) models ('spatial lag regression', Figure 7.7c). These structures are Markov models, as introduced in Section 6.2. The implicit assumption for spatial filtering is that the spatial structure is mostly autocorrelation which decays with distance (Figure 6.14a). SAR models can be computed using the function *lagsarlm* of the R package *spatialreg*.

An autoregressive model in which y_i is expressed as a function of y_{i-1} appears to have directionality, which is logical in time series but not for spatial data, particularly in two dimensions. Unidirectionality seems obvious in the first-order autoregressive model (Model IH, Equation 6.2): $y_i = \rho y_{i-1} + \varepsilon_i$. Bidirectionality is present, just implicitly, and the forward and backward neighbours could be included explicitly:

$$y_i = \rho y_{i-1} + \varepsilon_i + \varphi y_{i+1}. \tag{7.10}$$

Implementing this structure requires the simultaneous solution of n equations for the n values of the ys. In one dimension, each location has two neighbours but, in two dimensions, more neighbours adjoin and their effects may have different weights depending on position and distance. Weights are given in a 'proximity matrix', **W**, with w_{ij} being greater than zero if the value at location i depends on the value at location j. In a regular square lattice, all eight 'queen's move' neighbours might receive equal weighting (say $1/8$), all others being 0. Autocorrelation can occur in many forms, and we will describe two: simultaneous autoregressive models (SAR) and conditional autoregressive models (CAR), both most easily explained using matrix notation (Ver Hoef *et al.* 2018).

7.4.1.1 SAR Models

Simultaneous autoregressive models (SAR) are based on the concept that the equation defining y_i contains y_{i-1} and y_{i+1}, which are defined by equations containing other ys, resulting in a system of *simultaneous* equations. Consider the model in which variable y, given as vector **y**, is linearly dependent on some independent variables, $x_1, x_2, x_3, \ldots$, given in matrix **X**:

$$\mathbf{y} = \beta \mathbf{X} + \mathbf{u}, \tag{7.11}$$

where **u** is a vector of possibly non-independent errors with a mean of zero and variance–covariance matrix **C**. With the matrix **W** of neighbour weights (standardized to row totals of 1), the variance–covariance matrix (associated with **u**) is:

7.4 Spatial regressions

$$C = \sigma^2 \left[(I - \rho W)^T (I - \rho W)\right]^{-1}. \tag{7.12}$$

The weight matrix, **W**, is not necessarily symmetric, allowing the inclusion of effects like water currents, prevailing winds or any factors that might impose directionality.

7.4.1.2 CAR Models

Although the SAR model is used extensively, many statisticians emphasize the CAR model. The conditional autoregressive model is based on the probability that a location takes a particular value that is *conditional* upon the neighbour values:

$$P(y_i = y) = P(y_i = y | \{y_j; w_{ij} > 0\}). \tag{7.13}$$

It is not much different from the SAR model (Equation 7.12) but it requires that the weight matrix, **V**, must be symmetric. The model is:

$$y = X\beta + u, \tag{7.14}$$

where **u** is a vector of errors with a mean of zero and a variance–covariance matrix that includes the autocorrelation parameter φ:

$$C = \sigma^2 (I - \varphi V)^{-1}. \tag{7.15}$$

CAR and SAR models are closely related and any SAR process is a CAR process with

$$V = W + W^T - W^T W, \tag{7.16}$$

but the converse is not true (Ripley 1988).

7.4.2 Other Spatial Filtering Models

A key concept is spatial filtering, a process to separate the spatially structured random component (the focus of interest!) from trend and random noise (Tiefelsdorf & Griffith 2007; Griffith & Chun 2014); and it forms the basis of several methods.

Haining (2003) advocated the advantages of moving average (MA) models. For lattice or grid data, it seems natural to consider a moving average based on the four 'rook's move' neighbours:

$$y_{ij} = \varepsilon_{ij} + \rho(\varepsilon_{i-1,j} + \varepsilon_{i+1,j} + \varepsilon_{i,j-1} + \varepsilon_{i,j+1})/4. \tag{7.17}$$

MA models have the advantage that autocorrelation can decline sharply with distance and to zero at close range, whereas in autoregressive models it tends to persist over greater distances. In general form, the MA model is

$$y = X\beta + \rho W\varepsilon + \varepsilon, \tag{7.18}$$

and the variance–covariance matrix is:

$$C = \sigma^2 \left[(I - \rho W)^T + (I - \rho W)\right]. \tag{7.19}$$

This looks to have some structural similarities to the SAR model (Equation 6.45) but, for the same symmetric proximity matrix, **W**, the variance–covariance matrices for the three different models will all be different.

These models should aid understanding; estimated model parameters that give good agreement with the data may not give much information about the underlying process. For real data, we do not even know the proximity matrix, **W**. Such models are helpful because we can use them to generate artificial data of known structure, with which to compare the data we are trying to analyse. The implementation and comparison of models seem to be a good approach to studying this phenomenon.

Other spatial filtering models add orthogonal spatial predictors like Moran's eigenvectors (MEMs, to be described in detail in Section 7.6.1; Figure 7.7g). Then spatial cross-regressive models (SARMA; Anselin 2001) can also be used.

7.4.3 Spatial Error Regression

If the residuals of a linear regression are spatially autocorrelated, then $\text{var}(\varepsilon) = \sigma^2 \Sigma$, where Σ is a spatially structured $n \times n$ matrix. Its values are proportional to the distance between the sampled data based on autocovariance functions and so the off-diagonal elements are not zero. In this circumstance, we should reanalyse the data using a spatial error regression method such as generalized least-squares regression (GLS; Zuur *et al.* 2010) or regression Kriging (Hengl *et al.* 2004, 2007). GLS can be computed using the function *gls* of the R package *nlme*.

Generalized least-squares regression can parameterize this spatially structured covariance matrix based on an autocorrelation function, which can be an inverse distance function, an autoregressive function or a variogram model. Hengl *et al.* (2007) proposed using variogram models for the covariance matrix and named their model a regression Kriging model. In essence, GLS and regression Kriging are the same, but the advantage of using variogram models is that anisotropic spatial structures can be modelled.

Comparing several non-spatial and spatial regression models, Beale *et al.* (2010) concluded that overall GLS performs well as evaluated by absolute bias and root mean square error (RMSE). Hence, using GLS to account for spatially autocorrelated data in a regression framework is a good way to obtain reliable parameter estimates.

7.4.4 Geographically Weighted Regression

When the spatial structure of the data includes both spatial autocorrelation and spatial heterogeneity (several different regions) then other kinds of spatial regression are needed and one such is geographically weighted regression (GWR; Fotheringham *et al.* 2002; Fotheringham 2009). This computes a regression model for each sampling location using its neighbouring locations to parameterize the model. Hence, the usual regression equation expressing the dependence of y on a set of xs,

7.4 Spatial regressions

$$y = \alpha + \sum_k \beta_k x_k + \varepsilon, \qquad (7.20)$$

is modified so that the parameters are not stationary, but vary with location, s:

$$y(\mathbf{s}) = \alpha(\mathbf{s}) + \sum_k \beta_k(\mathbf{s}) x_k + \varepsilon. \qquad (7.21)$$

Thus, each observation of y has a separate model with its own parameter estimates, based on a weighting function (also referred to as a kernel) that allows closer samples to have a greater effect on the estimates. The shape and the size of the kernel can vary. Usually, a bell-shaped Gaussian kernel is used, so that the closest observations have the most weight. The size of the kernel can be constant throughout the study area (fixed kernel) or it can vary according to location to include a constant number of observations (adaptive kernel). Because the estimates are local, they can be mapped to produce a spatially explicit result.

This technique illustrates the interplay of independent predictor variables (the xs) and location in explanatory modelling. One version is to combine the effects of the two kinds of variables, the spatial coordinates, $\mathbf{s} = (s_1, s_2)$, with potentially non-linear effects, and the environmental variables, the xs, with possibly non-stationary effects, in a model to explain the ith observed value of y:

$$y_i = \sum_{p=0}\sum_{q=0} \alpha_{pq} s_{1i}^p s_{2i}^q + \sum_{k=1} \beta_k(\mathbf{s}_i) x_{ki} + \varepsilon_i. \qquad (7.22)$$

If the spatial response is simple and linear and the response to the xs is stationary, omitting the subscript i gives:

$$y = \alpha_{00} + \alpha_{10} s_1 + \alpha_{01} s_2 + \alpha_{11} s_1 s_2 + \sum_{k=1} \beta_k x_k + \varepsilon. \qquad (7.23)$$

Because the xs may depend on location, spatial effects and environmental effects may be confounded, even if the parameters are stationary so that the βs do not depend on $\mathbf{s}$. GWR can be computed using the function *gwr* of the R package *spgwr*. When computing the relationship between the ovenbird and forest data, the pseudo-R^2 ranges from 0.015 to 0.785 with a mean of 0.313.

Explanation and the effort to distinguish space effects from 'environment' effects originated in ecological studies, but the 'environment' can be any set of factors other than spatial location. The total variation in the data can be partitioned into four by identified sources of the variation:

(a) explained by environmental variables alone;
(b) explained by confounding environmental and spatial variables together;
(c) explained by spatial variables alone; and
(d) explained by neither.

GWR has been used with good results (Osborne *et al.* 2007; Windle *et al.* 2010) but, because there are as many models as sampling locations, GWR should be used as an Exploratory Spatial Data Analysis (ESDA) to identify subregions within the study area where the relationship gives comparable parameter values. GWR cannot be used as a predictive method and Griffith (2008) stressed that GWR overfits, thus requiring caution.

7.4.4.1 A Note on Model Selection

Legendre et al. (1994) suggested a stepwise selection procedure to determine which predictor vectors are the most parsimonious and to be retained in the regression. In the past, ecologists and evolutionary biologists tended to choose the Akaike Information Criterion (AIC) as a model selection tool (Akaike 1974). AIC is a log-likelihood approach that assumes a least-squares estimation with normally distributed errors from the residual sums of squares.

AIC is often used to compare alternative models given that some models and/or outputs of R packages do not provide adjusted R^2. In the case of the relationship between the ovenbird and forest, the best AIC was with the GLS model using a spherical variogram and the worst was with the OLS model (see Table 7.1). Overall, the AIC improved using spatial regression models.

Reliance on AIC has been replaced by more general approximations such as the Deviance Information Criterion (DIC) and Watanabe's Widely Applicable Information Criterion (WAIC); for a good discussion of these criteria (and, yes, the Bayes Information Criterion), we recommend the commentary available in McElreath (2020). The same source provides advice on these and alternatives, as well as R code examples with WAIC.

7.4.5 How to Remove Spatial Autocorrelation from the Residuals

Removing spatial autocorrelation from model residuals can be accomplished by several methods, based on Bayesian methods or wavelets. Aing et al. (2011) described the use of a Bayesian hierarchical occupancy model for assessing the occurrence of river otter (*Lontra canadensis*) based on snow-track data from helicopter surveys. The data included autocorrelation from several sources and their model included three levels: occupancy by otters; availability of tracks for detection, conditional upon occupancy; and track presence or absence, conditional upon availability. Spatial dependence was included using an intrinsic conditional autoregressive (CAR) model, with parameters evaluated by a Markov Chain Monte Carlo (MCMC) procedure. The

Table 7.1 AIC values used to compare non-spatial regressions with spatial regressions.

Model	Degrees of Freedom	AIC values
Linear model (OSL)	3	−427.180
GLMM	4	−471.483
OLS + x–y coordinates	5	−606.305
GLS + linear variogram	4	−514.273
GLS + spherical variogram	5	−727.200
SAR	4	−683.517

7.4 Spatial regressions

Table 7.2 Examples of relationships among spatial scales, the data type and key ecological processes

Local scale	Regional scale	Broad scale
Data model: vector	Data model: vector, raster	Data model: raster
Data type: qualitative and quantitative	Data type: qualitative and quantitative	Data type: presence and absence
Foraging	Resource use	Abiotic and climate factors
Biotic factors/intraspecific	Biotic factors/interspecific	Speciation/extinction
Population dynamics	Metapopulation	Species range shift
Species interactions	Metacommunity	Species assemblage shift
Daily movement	Natal dispersal	Migration

spatial model gave more accurate estimates and better credibility intervals for this spatially autocorrelated data.

Ver Hoef *et al.* (2006) proposed anisotropic autocovariance to model dendritic networks such as stream systems, by using directional functions to capture the anisotropy:

$$\mathbf{y} = \mathbf{X}\boldsymbol{\beta} + z_{TU} + z_{TD} + z_E + \varepsilon. \tag{7.24}$$

The functions are as follows: z_{TU} is the tail-up vector based on a moving-average function (tail upstream), z_{TD} is the vector based on a moving-average function (tail downstream), and z_E is a correlation structure based on Euclidean distance between the sampling locations. This model has much potential for the study of stream data (Peterson *et al.* 2013).

Beguin *et al.* (2012) introduced an alternative to MCMC procedure, integrated nested Laplace approximations (INLA), for fitting Bayesian hierarchical spatial models with general covariance structures. The class of models considered is that of latent Gaussian models, with a latent Gaussian field, x, at least partially observed through the data variable, y. The field x can include non-spatial or spatial random effects and y can follow any of several distributions in the exponential family such as Poisson. This allows much latitude for application and Beguin *et al.* (2012) provided details and an example investigating woodland caribou in Eastern Canada's boreal forest. Using both the conditional autoregressive model (CAR) and the Matérn model (see Minasny & McBratney 2005), they found that the INLA method and the Matérn model had advantages compared to MCMC, being accurate and rapid, and effectively removing autocorrelation from the model residuals. This approach also allowed a good evaluation of the parameter estimates' uncertainty.

Haas *et al.* (2011) described an application of INLA to ecological data, using the intrinsic conditional autoregressive error model in a study of the effects of forest diversity on disease risk during the invasion of a generalist plant pathogen. Their results suggest that disease risk is less where species diversity is higher, once the effects of host density and landscape heterogeneity are accounted for. Clearly, the most appropriate treatment of spatial autocorrelation in such data is essential for the correct interpretation of the disease data and the insight into the process they provide.

Another solution for autocorrelation in linear model residuals is based on wavelets. In a series of papers, Carl and colleagues (Dormann et al. 2007; Carl et al. 2008; Carl & Kühn 2008, 2010) have developed a technique called the 'wavelet revised model'. The wavelet of choice, like the Haar wavelet, removes a localized mean component from the data (the 'smooth' component), which is subtracted to leave the 'detail' component. For a range of scales of wavelet, say 2×2 blocks of the original data, then 4×4, 8×8 and so on, the 'detail' components are added together. They are the basis for subsequent analysis with the autocorrelation substantially removed. In concept, this approach resembles the method proposed by Bartlett (1948) for the localized analysis of periodograms, removing differences due to the local means and summing the results. The wavelet approach has several advantages, including the fact that it requires no *a priori* knowledge or assumptions about the spatial structure and it deals well with non-stationarity because the wavelets make the 'corrections' locally.

Exciting developments continue in this area, offering an ongoing challenge to keep up and to make the right choices. The main question is to what extent the good performance of any one approach is robust or sensitive to departures from the models and assumptions. That seems to be a strength of methods where corrections are derived from the characteristics of the data, rather than from a priori structural details. It provides the ability to respond to characteristics of the data, providing greater confidence in the outcomes of complex methods.

7.5 Canonical (Constrained) Ordination

The spatial regression techniques discussed in Section 7.4 assess the dependence of a single variable, y, on predictor variables, the xs, and on location in space, s, in 1, 2 or 3 dimensions. Consider more dependent variables. Figure 7.8 illustrates a simple example of moving to the multivariate case of assessing the dependence of a set of species abundances, in matrix Y, on environmental variables, in matrix X, and on spatial location, s (data from Figure 7.2). To determine the contribution of each variable to the relationship among multivariate data sets, we use canonical ordination techniques, including canonical correspondence analysis, CCA, or redundancy analysis, RDA (Legendre & Legendre 2012). Here 'canonical' refers to the relationship between two sets of variables. Redundancy analysis is an ordination technique which corresponds to a multiple regression for multivariate data, maximizing the variance explained by the linear relationship between the canonical axis of response matrix Y and the explanatory variables in matrix X. RDA is a direct ordination technique because the relationship between two matrices is computed using the eigenvalue–eigenvector properties of the multivariate matrices. In the eigenvalue notation, RDA is:

$$\left(S_{YX} S_{XX}^{-1} S'_{YX} - \lambda_k I\right) u_k = 0, \tag{7.25}$$

where S_{YX} is the matrix of covariance between Y and X and S'_{YX} is its transpose, S_{XX}^{-1} is the inverse of the variance–covariance matrix of X and k is an index of the dimensions analysed. The matrix I is the identity with 1s on the main diagonal and

0s elsewhere, λ_k is the kth eigenvalue and $\mathbf{u}_k$ is its eigenvector. Any RDA ordination of the $\mathbf{X}$ axes is linearly constrained by the variables in $\mathbf{Y}$. The best linear fit between the species and environmental ordination axes is obtained by an iterative procedure. RDA is, in essence, an extension of a principal component analysis (PCA) where the correlation between the two matrices is linear, so that RDA is a Euclidean representation of the objects in multidimensional space.

Often the relationship between species and environmental variables is induced by other underlying factors such as climate, topography or historical events. One way to account for or to control for the effects of these other variables is to use partial canonical ordination techniques such as partial CCA or partial RDA (Borcard et al. 1992; Legendre & Legendre 2012). The advantage of these partial ordination techniques over partial Mantel tests is that the relationship is computed on the raw data values rather than on derived distance measures. Also, these techniques allow an assessment of the relative contributions of each variable to the overall relationship between the two matrices.

The partial RDA of the multivariate data from Figure 7.2, relating the species assemblage to environmental variables, while constrained by location, is shown in Figure 7.9. The triplot indicates that only species, Spp1, is related to the environmental variable of moisture. Based on randomization procedure, this partial RDA seems to be significant ($F_{2,6} = 4.59$, $p = 0.059$, from 9,999 randomizations). By next partitioning the variance into four components ([a] to [d] in Figure 7.10) and adjusting for the number of parameters (see Peres-Neto 2006), we find the following: (1) overall 75.17% of the variance is explained by the environmental data and location, components [a + b + c]; (2) 22.72% is explained solely by the environmental variables; (3) 55.09% is explained by the spatially structured environmental variables; (4) −2.19% due to the other space components [c]; and (5) 24.83% is not explained, the residual component [d]. The main reason for the negative contribution in the space fraction [c] is that there is high collinearity between the response and predictor variables, all showing spatially autocorrelated values based on Moran's I computed at the distance of 1.41 units: Spp1 = 0.187; Spp2 = 0.499; Spp3 = 0.428; soil = 0.784; and moisture = 0.631. In some circumstances, it is recommended to detrend the data and analyse the detrended data (Legendre & Legendre 2012). Unlike the Mantel test, which cannot distinguish the contributions of individual variables to the overall result, the RDA could determine that the relation between the species assemblage is mostly due to the 'Spp1' relationship with 'moisture'. We used RDA in this example because the species response is approximately linear due to the relatively narrow range of the environmental gradient. Where the environmental gradient is wide and the species response is unimodal, CCA should be used instead (Legendre & Legendre 2012).

7.6 Multiscale Analysis

Many ecological studies aim to understand the processes acting on ecological response variables to generate patterns, whether the patterns are spatial, temporal or

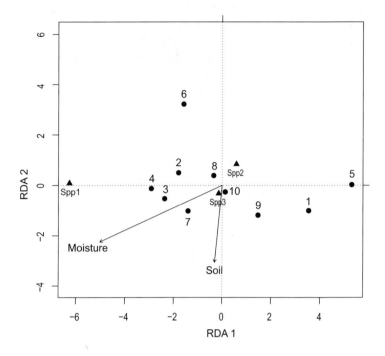

Figure 7.9 Partial RDA based on the 10 sampled data points in Figure 7.2: the relationship between species assemblage and the environmental variables are constrained by the x–y coordinates of the sampling locations. The first axis explained 60.44% of the variance and the second 0.026%. The plot depicts clearly that only Spp1 is related to the environmental variable Moisture. Dots with number labels indicate the positions of the sampling locations in the RDA plot.

phylogenetic. Chapters 1 and 2 described how several factors and processes can produce more than one spatial pattern at more than one scale (e.g. large trends and local patchiness) embedded in data. At each spatial scale (local, regional and broad), different environmental factors and processes are responsible for shaping the spatial patterns we observe (Table 7.2; Figure 7.10; Fortin et al. 2012b). As two examples of scale-dependency, Desrochers et al. (2010) showed that each increase in the study's window radius corresponds to different ecological processes for forest birds and Weaver et al. (2012) compared selected radii to determine which scales are the most important for each of three key environmental factors for an invasive non-native bird species. At the various spatial scales, both the spatial resolution (data model) and the ecological data type vary, including vector (point, line, polygon) and raster (pixel, cell, quadrat) types. It is, therefore, imperative to use the appropriate analytical methods to determine the key spatial scales according to the data model and the data type.

The effects of space on ecological data are not limited merely to the physical locations of the samples but include spatial legacy effects, in that the current spatial pattern is affected by past spatial patterns, and 'spatial contingency', in that the current spatial pattern at location i is influenced by the nearby spatial pattern at locations surrounding location i (Chapters 1 and 2; Peres-Neto & Legendre 2010; Fortin et al.

7.6 Multiscale Analysis

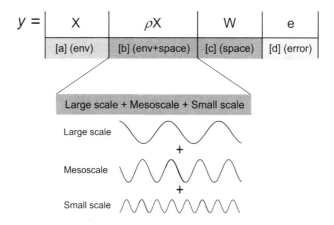

Figure 7.10 Variance partitioning analysis where the variability of the response variable y is explained by: (1) solely the environmental variables, X, and is known as the fraction [a]; (2) the spatially autocorrelated portion, ρ, of the predictor, environmental, variables [b]; and (3) [c] 'space' which includes may be due to a mismatch between the spatial sampling design and the spatial scales that act on the response and predictor variables or non-linear relationships. The residual errors form the fraction [d]. Here we explicitly depict that the spatial autocorrelated structure can be due to the addition of spatial patterns (here for illustration, a series of stationary sine–cosine patterns according to the spatial scales along a transect) occurring at three spatial scales (small scale, mesoscale and large scale).

2012b). These various aspects of space are often used as surrogate predictors in multiple regressions and constrained ordinations in the absence of factor variables. Attempts have been made to add space through predictor variables by (1) simply using the spatial coordinates of the samples (e.g. in a multiple regression); (2) combining spatial coordinates using a polynomial regression as in trend surface analysis (Borcard *et al.* 1992; Chapter 6) modelling only larger spatial scales; (3) defining a neighbourhood matrix based on local adjacency or connectivity among the sampling locations (Pelletier *et al.* 1999); (4) defining a Euclidean distance matrix based on spatial coordinates, as with the Mantel and partial Mantel tests (Legendre & Legendre 2012; Guillot & Rousset 2013); (5) defining a least-cost or quantitative resistance matrix based on landcover type values between the samples, as it is done with the Mantel and partial Mantel tests (Spear *et al.* 2010); and (6) applying eigenfunction-based methods using the spatial coordinates of the locations (Griffith 1996; Diniz-Filho & Bini 2005; Dray *et al.* 2006; 2012). The advantage of the eigenfunction methods is their ability to provide predictors for more than one spatial scale, reflecting a broader spectrum of potential factors influencing the spatial patterns observed.

If the processes can be assumed to be stationary, a multiscale analysis can be performed by multiscale ordination (MSO) based on multivariate variograms (Wagner 2003). This MSO analysis of multivariate species data produces variance–covariance matrices that match the various spatial scales as determined by an empirical variogram matrix (Chapter 6), allowing us to determine the spatial scale of individual species and species composition. A generalized MSO method can also be used with canonical ordination axes (Wagner 2004; Couteron & Ollier 2005) and can be used to analyse beta-diversity in a way comparable to the PCoA-based approach of Legendre and Condit (2019).

We present two types of decomposition analysis that can be used when the assumption of stationary is not met: (1) the generalized Moran's eigenvector maps (MEM; Dray 2011; Dray et al. 2012) which provide spectral decomposition based on an eigenfunction method for sampled data (qualitative or quantitative vector data); and (2) multiresolution analysis (MRA) based on hierarchical wavelet transform for continuous population data (quantitative raster data).

7.6.1 Generalized Moran's Eigenvector Maps

To begin, the spatial structure of sampled data can be coded into a spatial weighted matrix, $\mathbf{D_W}$, in several ways: (1) a connectivity matrix (1 = connected; 0 = not connected), (2) a Euclidean distance matrix or (3) a weighted matrix based on either cost or resistance values or on probabilities. For any $\mathbf{D_W}$ matrix, we have a choice of spectral decomposition methods based on eigenfunctions to produce sets of spatial eigenvectors that are orthogonal and thus uncorrelated with each other.

As the $\mathbf{D_W}$ matrix for spectral decomposition based on a principal coordinate analysis (PCoA), Borcard and Legendre (2002) proposed a matrix of truncated Euclidean distances, $\mathbf{D}_{trunc}$, calling the approach the Principal Coordinates of Neighbour Matrices (PCNM). A truncated distance matrix allows the separation of the effects at small distances from those operating at large distances, and so determining the truncation distance threshold (i.e. a Euclidean distance) is a crucial step in producing these spatial eigenvectors. Without other knowledge, a threshold value can be the length of the longest edge in a Minimum Spanning Tree constructed on the spatial coordinates of the sampling locations. For computational purposes, four times this threshold value is designated as the numerical equivalent of 'far away' and replaces all the distances larger than this value; it is also inserted on the main diagonal of the $\mathbf{D}_{trunc}$ matrix, so that each sampling location is not considered to be connected to itself. The PCoA of the $\mathbf{D}_{trunc}$ matrix, which is the next step, produces $n - 1$ non-zero eigenvalues and their $n - 1$ non-zero eigenvectors. Perhaps surprisingly, some of the eigenvalues are negative, but that is an effect of truncation (see Dray et al. 2006). Each of the $n - 1$ spatial eigenvectors is related to a spatial scale. When the sampling locations are regularly spaced, the corresponding spatial eigenvectors give sine waves. The spatial eigenvectors are produced with decreasing periods so that both large and local spatial scales can be determined from the set of Euclidean distances between the sampling locations. The orthogonal spatial eigenvectors thus produced can then be used as surrogates for the effects of 'space', forming a matrix of explanatory variables in an RDA, partial RDA or multiple regression.

Since the Borcard and Legendre (2002) paper, there have been many developments leading to mathematical refinements of spectral decomposition based on eigenfunctions (Griffith & Peres-Neto 2006; Dray et al. 2012). The original PCNM is a specific case of a family of spectral decomposition methods based on eigenfunctions and it is now seen as a distance-based version of Moran's eigenvector maps (see Dray 2011; Dray *et al.* 2012; Legendre & Legendre 2012). The generalized MEM differs from the distance-based version (dbMEM) in its use of a $\mathbf{D_W}$ matrix that is the Hadamard product of two matrices, a connectivity matrix and a similarity matrix, the product of which is then centred by removing the column- and row-means. The eigenanalysis of this centred $\mathbf{D_W}$ symmetric matrix produces both eigenvalues and eigenvectors where the eigenvalues are Moran's *I* coefficient values. The first eigenvalue, λ_1, is the largest eigenvalue and largest value of Moran's *I* and the last eigenvalue, λ_n, is the smallest eigenvalue and smallest value of Moran's *I*. The sign of Moran's *I* coefficient of spatial autocorrelation depends on the sign of the eigenvector and its associated eigenvalue.

The method has several advantages, including the potential to compare the effects of scale in different connectivity networks. McLeish *et al.* (2021) used four of the familiar neighbour networks (Chapter 2) to study spatial structure in agricultural landscapes: Delaunay, Gabriel and two versions of the relative neighbour graph, customized to account for the linear connections of major river valleys. While the global spatial autocorrelation was found to be positive by Moran's *I*, there were patterns of positive and negative autocorrelation at 'fine' and 'broad' scales, as determined by the MEMs, for four *a priori* habitat types (Crop, Oak, Edge and Wasteland) in the Tagus River Valley (Spain). They suggest that spatially-optimized variables with the right definitions of connectivity can improve the understanding of functional differences among communities.

Wagner and Dray (2015) used the MEMs approach to generate restricted randomization data to test the significance of linear regression against a null reference distribution of data with comparable spatial patterns. Such restricted randomization procedure using MEMs has also been proposed by Crabot *et al.* (2019) to generate null distributions to assess the significance of the Mantel tests.

Given that the MEM is based on an undirected connectivity matrix, it is assumed that the spatial patterns are isotropic. There could be circumstances, however, where directionality between the sampling locations affects the degree of similarity of the data. Such anisotropic cases can occur when the data are sampled along streams, for example. The MEM can therefore be modified to account for directionality using a directional connectivity matrix (Blanchet *et al.* 2008, 2011). This directional method is called the asymmetric eigenvector maps (AEM).

One drawback of these eigenfunction decomposition methods is that they produce $(n-1)$ eigenvectors, which may be too many! When the sample size is small this is not so much an issue but, with a large sample size, one needs to select a subset of these eigenvectors. Jombart *et al.* (2009) proposed a multiscale pattern analysis based on a PCA of a matrix of the coefficients of determination, the multiple regressions among the ecological variables, and the spatial eigenvectors, with an alternative being some form of forward selection procedure (Legendre & Legendre 2012).

To illustrate the MEM multiscale decomposition analysis, we simulated data on a lattice of 128×128 units, following the model of Legendre *et al.* (2002) (Figure 7.11). This provides data with several scales of pattern: (1) a large north–south trend; (2) large

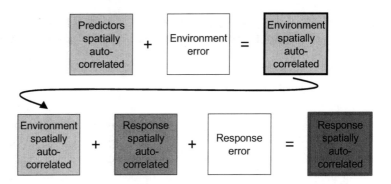

Figure 7.11 Flow chart on how predictor (e.g. environmental variables) and ecological response variables can be simulated using a conditional simulation method based on variogram, as implemented by Legendre et al. (2002) to include spatial autocorrelation structure in the data. Environmental error is added to spatially autocorrelated predictors to produce the spatially autocorrelated environmental data. The spatially autocorrelated predictors can be the sum of several predictors having different spatial scales. Then spatially autocorrelated environmental data are the starting point of the spatially autocorrelated ecological response data where a new spatially autocorrelated variable is simulated. The final spatially autocorrelated ecological response is the sum of the spatially autocorrelated environmental response as well as the response error.

scale patchiness from a spherical variogram with a spatial range of 100 units; (3) small scale patchiness from a spherical variogram with a spatial range of 20 units; and (4) random noise with a mean of 0 and variance of 1 (Figure 7.12). Four subsamples were taken from this dyadic grid (a power of 2; here 2^7) using random and systematic sampling designs with 100 and 400 samples. Based on these subsamples and their respective Minimum Spanning Trees (Figure 7.13), the threshold distance values to truncate the Euclidean distance matrix into small versus 'far' distances are: 14.0 for the 100 random points; 10.0 for the 100 systematic points; 14.14 for the 400 random points; and 6.0 for the 400 systematic points. Using these thresholds, Moran's I eigenvectors were computed. When performing forward selection to determine key spatial scales, the number of significant eigenvectors drops drastically (the number of spatial scales) for both sample sizes, where the largest spatial scales (i.e. the lowest ranked eigenvectors) are those retained to explain the pattern of the simulated data (Figure 7.14). The variance partitioning of the effects of the spatial coordinates (i.e. fraction [b]) and the significant eigenvectors (i.e. fraction [c]) show that the MEMs explain 32–40% of the simulated spatial pattern while spatial coordinates explain only 7–15% (Table 7.3). Overall, the results are consistent, given the good spatial coverage of the samples over the entire simulated extent; but, if the sample size is small relative to the extent of the study, the results can be quite different, resulting in less of the total variance being explained by the MEM, as Gilbert and Bennett (2010) have demonstrated.

These results illustrate the ability of MEMs to identify the key spatial scales and to help explain the variability of the simulated data. Although MEMs were used here as surrogate predictor variables, they can help to determine the spatial scales at which environmental or other processes may affect ecological variables.

7.6 Multiscale Analysis 217

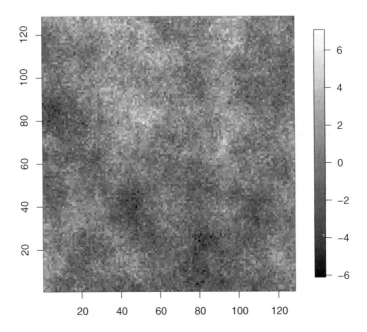

Figure 7.12 Simulated spatially autocorrelated response data (grid of 128 × 128 points) using the flowchart illustrated in Figure 7.11. The data are the sum of three spatially autocorrelated variables based on spherical variograms: (1) north–south gradient; (2) large scale patchiness with a range of 100 units; (3) small scale patchiness with a range of 20 units; and (4) random noise.

7.6.1.1 MEMs and Spatial Filtering

Looping back to spatial regressions and related techniques, we note that the spatial filtering models (SAR, CAR, MA) deal with spatial structure mostly on neighbourhood scales, but intermediate and large scales also affect the spatial pattern. To account for multiple spatial scales within a regression framework, Moran's eigenvector maps (MEMs) can act as spatial predictors. Because the number of orthogonal MEMs is n, we select a subset to explain as much as possible of the variation in the response variable y. This subset is $\mathbf{E}_k$, giving an $n \times k$ matrix of eigenvectors to be added as spatial predictors in a linear model:

$$\mathbf{y} = \boldsymbol{\beta}_x \mathbf{X} + \boldsymbol{\beta}_k \mathbf{E}_k + \varepsilon, \qquad (7.26)$$

where $\boldsymbol{\beta}_x$ gives the coefficients for $\mathbf{X}$ and $\boldsymbol{\beta}_k$ for the $\mathbf{E}_k$. The MEM eigenvectors can also be used as spatial predictors in a multiple regression or ordination analysis (e.g. RDA, CCA) or to partial out spatial effects in an ordination (partial RDA or partial CCA). The selected MEM eigenvectors can provide insights about spatial scales of responses and may help to identify processes that may act at these scales. The MEM eigenvectors are not process-based factors for prediction in different regions or at different time periods. Furthermore, Beale et al. (2010) found biased parameter estimates and inflated Type-I error rates when using MEM eigenvectors in spatial

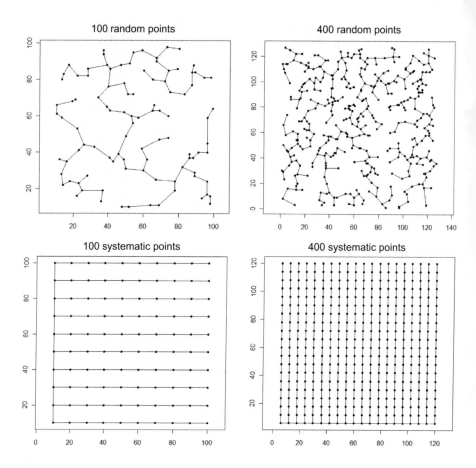

Figure 7.13 Four subsamples of the 128 × 128 points in Figure 7.12: random or systematic samples having 100 or 400 points. The edges shown are Minimum Spanning Trees for each sample. The threshold distances for the Moran's eigenvector maps are based on the Minimum Spanning Trees: 14 for the 100 random points; 10 for the 100 systematic points; 14.14 for the 400 random points; and 6 for the 400 systematic points.

filtering models. On the other hand, Wagner (2013) has proposed a modification called *spatial component regression* that successfully integrates the linear regression model with the Moran eigenvector maps and is worth exploring for this purpose.

7.6.2 Multiresolution Spectral Decomposition Analysis with Wavelets

Unlike the eigenfunction-based decomposition methods, multiresolution decomposition is based on the spectral signal of quantitative data observed at a set of contiguous locations. Such contiguous data should be in the form of a dyadic grid (the dimension is a power of 2), forming either a transect or a lattice. Spectral decomposition of this signal can be achieved by a Fourier transform analysis which assumes stationarity, requiring that the underlying process has the same

7.6 Multiscale Analysis

Table 7.3 Summary of the results of the MEM and variance partitioning of the four subsamples.

	100 random	100 systematic	400 random	400 systematic
Positive eigenvectors	33	44	120	189
Significant eigenvectors	5	6	25	40
Rank of the first four eigenvectors for comparison	1, 3, 4, 6	1, 3, 7, 12	1, 2, 4, 7	2, 4, 6, 8
% of variance explained by fraction [b]	13	7	7	15
% of variance explained by fraction [c]	34	33	32	40
% unexplained variance explained (fraction [d])	53	60	61	45

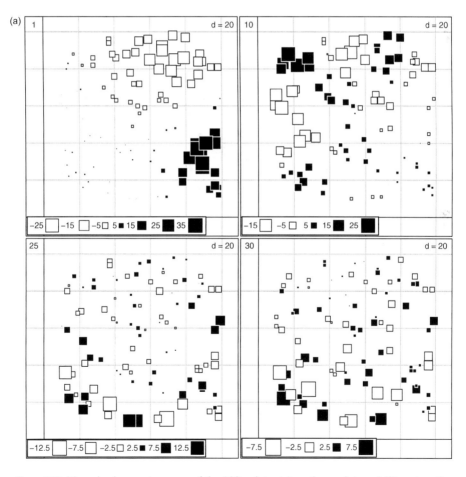

Figure 7.14 Moran's eigenvector maps of the 100 points: (*a*) random points and (*b*) systematic points. Here the 1st, 10th, 25th and 30th eigenvector maps are shown. The positive values are in black and the negative ones in white. The size of the square is proportional to their magnitude.

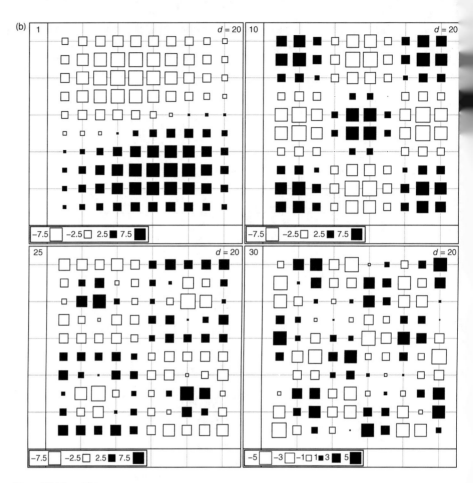

Figure 7.14 (cont.)

characteristics throughout. If this condition is obtained, then the resulting decomposition's spatial scales will have a sine–cosine wave pattern, but the likelihood of a process being stationary over an entire study area is low, especially for large areas. In such circumstances, a wavelet transform analysis is more appropriate because it does not require stationarity. Indeed, wavelet analysis can identify different types of spatial patterns at the local scale, as we describe in Chapter 3. Wavelet analysis can use templates of various shapes and at many spatial scales or resolutions (Daubechies 1992; Torrence & Compo 1998; Saunders *et al.* 2005) and so wavelets of multiple resolutions allow the identification of key spatial scales in the data (Keitt & Urban 2005; James & Fortin 2012).

There are different ways in which the multiresolution analysis can be performed, providing spatial scales that are either orthogonal or not. To obtain orthogonal spatial scales as powers of 2, a discrete decomposition wavelet transform (DWT) is applied (see Torrence & Compo 1998). The alternative, the continuous decomposition wavelet transform (CWT), can produce non-orthogonal results that will include redundant spatial signals throughout the range of decomposed spatial scales. The maximum overlap discrete wavelet transform (MODWT) method is a compromise between the DWT and CWT and it will result in fewer spatial scales while keeping the original spatial resolution of the data (James & Fortin 2012; see Roushangar et al. 2021 for an application). Here we will focus on the DWT method because it is more directly comparable to the MEM, resulting in orthogonal scales.

The discrete wavelet transform, DWT, for the data series f, indexed by the location variable s, is based on a wavelet function, $\Psi(s)$. The transform is computed as follows:

$$\text{DWT}_\psi f(m, n) \propto \sum_s f(s) \Psi^*_{m,n}(s), \tag{7.27}$$

where m is the scaling coefficient which determines the wavelet's width and thus the scale to which the wavelet is sensitive and n is the translating or shifting coefficient which determines the wavelet's position in the data grid. The wavelet function, now scaled and located, is $\Psi_{m,n}(s) = 2^{-m} \times \Psi(2^m s - n)$ for our spatial analysis using the wavelet defined by the function $\Psi(s)$. (As a technical detail, the asterisk indicates that this is a complex conjugate, which is one of a pair of complex numbers with a real part and an imaginary part of opposite signs). On a dyadic grid, this DWT decomposes the signal into discrete orthogonal spatial scales by dilating the wavelet defined by $\Psi(s)$. In general, the results of wavelet analysis are presented by a two-dimensional plot of the wavelet transform scores, coded by colour (from blue for strongly negative values to red for high positive values), with the dimensions of the plot being scale (m) on one axis and location (n) on the other. Averaging the squares of the scores over all locations gives a plot that is the equivalent of 'variance as a function of scale', as in quadrat variance methods (Chapter 3), called a *scalogram* for obvious reasons.

The number of spatial scales that can be computed by this discrete wavelet scaling approach is given by the dyadic grid size of the observed data. In the case of the simulated data illustrated in Figure 7.15, the grid size is 2^7, and so: the maximum number of scales we can examine is seven. Then, the multiresolution decomposition analysis provides the proportion of the total wavelet variance of the data that corresponds with each decomposed spatial scale in the DWT approach. Figure 7.15 shows the spatial pattern analysis of the simulated data at each orthogonal spatial scale. Remember that the DWT is applied directly to the quantitative grid data and not to the Euclidean distance between the sampling locations as was the case for the MEM. When a priori knowledge is available, we can combine the spatial pattern from two or more scales. In the absence of other knowledge about the key spatial scales, a wavelet variance scalogram, which plots the wavelet variance as a function of wavelet width (proportional to pattern scale) by averaging over all locations, can depict the relative contribution of the decomposed spatial scales. For the simulated data presented here as an illustration, the prominent spatial scales of the pattern are 2^1, 2^4 and 2^6 (Figure 7.16), based on the positions of the peaks in the variance plot. Carl and Kühn (2010) used this multiresolution decomposition method to remove the effects of small-scale spatial autocorrelation to

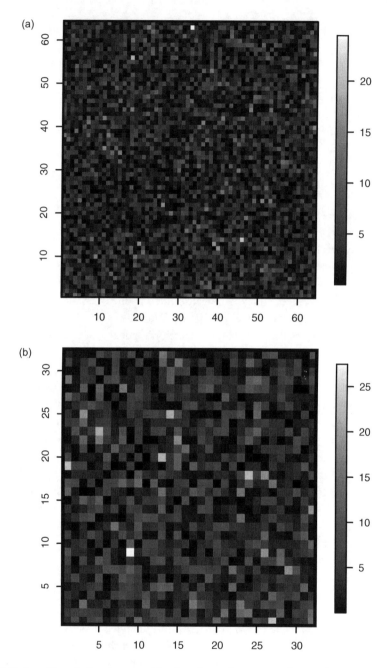

Figure 7.15 A multiresolution decomposition based on a discrete wavelet transform (DWT) of the dyadic simulated data of Figure 7.12 (128 × 128 units): (*a*) spatial scale 2^1; (*b*) spatial scale 2^2; (*c*) spatial scale 2^3; (*d*) spatial scale 2^4; (*e*) spatial scale 2^5; and (*f*) spatial scale 2^6.

7.6 Multiscale Analysis   223

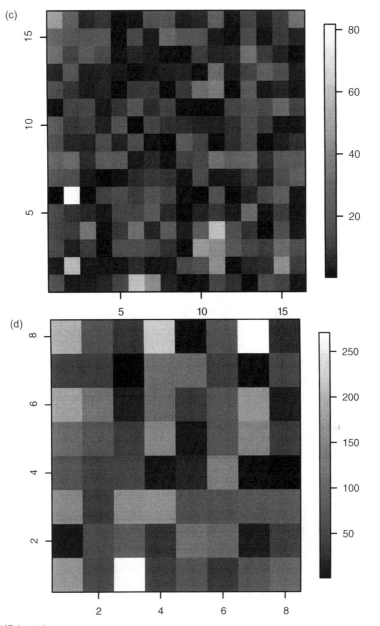

Figure 7.15 (cont.)

examine only the larger scales in a subsequent regression analysis. This decomposition procedure can be used to detrend data when the spatial patterns are not linear or not stationary.

Other wavelet methods are available, and more are being developed. Floryan and Graham (2021) described a method that integrates aspects of wavelet analysis with PCA (Principal Components Analysis), which they call Data Driven Wavelet

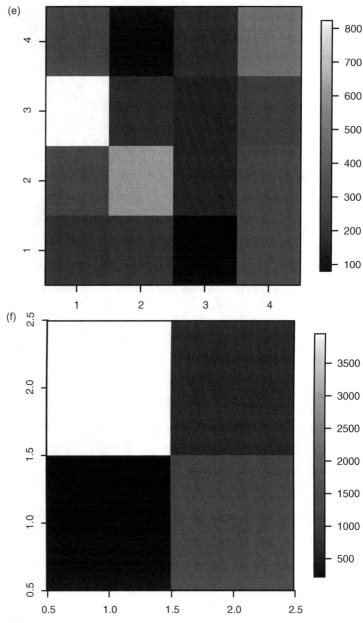

Figure 7.15 (cont.)

Decomposition (DDWD). Although their focus was originally the analysis of turbulence, the approach allows the characterization of localized hierarchical structures in multiscale systems including many natural systems such as various fluid flows as well as networks and their graphs (cf. Hammond et al. 2011). From a historical perspective, it is interesting to see the methods we used in Dale and Mah (1998) referred to as 'traditional' wavelet methods (Floryan & Graham, 2021). That is probably justified

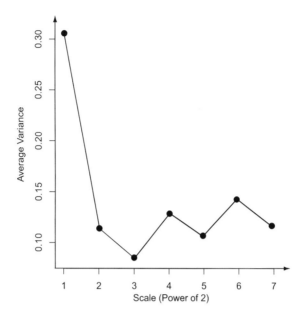

Figure 7.16 Wavelet scalogram of the simulated data illustrated in Figure 7.12. The key orthogonal spatial scales are 2^1, 2^4 and 2^6.

given the range and sophistication of the methods that have appeared since and the wavelet applications of the future.

7.7 Concluding Remarks

The areas of focus in this chapter, spatial regressions and their variants and multiscale spatial analysis, have much in common: rapid ongoing development and informative application to ecological studies, as well as the obvious overlap in the use of MEM techniques. The multiscale analysis has conceptual links to the multispecies pattern analysis described for quadrat data in Chapter 3 (Dale & Zbigniewicz 1995) and to the spatial Local Contributions to Beta Diversity methods in Chapter 9 (Legendre & De Cáceres 2013), combining ordination techniques with other analytical approaches to provide insight into spatial structure (see summary in Box 7.1).

A recent trend is to design ecological studies over larger areas to understand the processes that determine species' spatial arrangement. For new data types, novel spatial and statistical methods have been developed to evaluate and understand them. This chapter has shown how spatial analysis and multivariate analysis merge to allow multivariate multiscale spatial analysis. This is a dynamic and rapidly advancing field and more developments are expected in the years to come. We note that spatial eigenvectors, used as spatial predictors in inferential statistics, are not process-based and give little insight into underlying processes. Hence, environmental variables that

> **Box 7.1** Summary of Multiscale Methods
>
> 'Multiscale structure is all around us…' (Floryan & Graham 2021).
>
> Multiscale methods are described throughout this book, not just in this section. The multiscale feature of spatial analysis can be accomplished in several ways: changing spatial lag, changing the size of the template or search window or changing the grain of the procedure. This list gives examples:
>
> - Distances to nearest neighbours of 1-, 2-, 3-, … and k-orders give larger distances (Chapter 3).
> - Ripley's K (and variants) including multispecies: association as a function of interpoint distance (Chapter 3).
> - Circumcircle scores including multispecies: score as a function of circle area (Chapter 3).
> - Quadrat variance-based analysis, including multispecies eigenanalysis: variance changes with scale (Chapter 3).
> - Wavelet analysis based on moving template: scalogram of scores by scale, heatmap of scores by location (Chapter 3).
> - Global spatial statistics (Moran's I, Geary's c): spatial autocorrelation value estimated as a function of distance (e.g. spatial lags, distance classes) (Chapter 4).
> - Moran's eigenvector maps: scale orders the eigenvectors and the pattern of site scores varies with spatial scale (Chapter 7).
> - Spectral decomposition by wavelets: spatial patterns as a function of spatial scale and the resulting scalogram (Chapter 7).
> - Species diversity: beta diversity as a function of distance between sites (Chapter 9).

are process-based should be the priority, with spatial eigenvector predictors as secondary or as a last resource, because their explanatory power is limited to the study area extent and location. In the absence of alternatives, however, these approaches do provide ecologists with useful tools for the analysis of data in a spatial context and with a range of spatial scales. As always, the best advice is to use the range of analytical tools that are available but applied with full knowledge of the conditions of their capabilities and limitations.

8 Spatio-temporal Analysis

Introduction

For almost any ecological system, the concepts of spatial structure will include a temporal component, at least implicitly. Expanding spatial analysis to include time is not the same as adding one more spatial dimension; time is different because of its directionality. Even in cyclic community phenomena, in which building and degrading phases seem like mirror images of each other, different processes are responsible for the two phases, even if the resulting sequences appear to be merely reversed in time. Furthermore, space may not be a substitute for time as the stationarity of the processes may not hold (Damgaard 2019).

Quantifying the changes in spatial statistics calculated at sequential times is worth some discussion (Section 8.1), although it is not the recommended approach to what should be multi-dimensional analysis (Parrott *et al.* 2008). The methods in this chapter are mainly more technical and the discussion is limited to analysis not modelling because the topics like spatially explicit dynamic models are sufficiently complex to require books of their own (Otto & Day 2011; Clark 2020*b*). Our treatment of spatio-temporal analysis is broader than just the statistics for such data, but Cressie and Wikle (2011) provided excellent coverage of both statistics and models for those.

The data used for spatio-temporal analysis can be classified in several ways, but an important criterion is continuous versus discontinuous (discrete) data either in time or in space. If we are monitoring the environmental conditions in a nature reserve by an array of hygrothermographs placed throughout the reserve, the data are continuous in time but discontinuous in space. If we are studying the movements of animals using global positioning system (GPS) radio collars, which report on position once every hour, the data will be discrete in time and space. In both examples, we assume stationarity in order to interpolate between sites that have hygrothermographs, or between the animals' reported positions.

In cases such as permanent sample plots, in which tree stems are mapped and re-mapped at intervals, no interpolation may be necessary: stem No. 23 was alive in 1990, standing dead in 1998 and a downed log in 2005. The data and the analysis of spatio-temporal patterns bring us closer to observing the processes affecting the pattern because sufficient data are available to recover most of the important transitions (see Nathan *et al.* 2022).

To complete the classification of data types, it is possible to have data that are more-or-less continuous in both space and time, limited only by the resolution of the recording technology, for example, the flight path of a butterfly, although the observer might divide the movement into units for analysis. Similarly, animal tracks in the snow are also of this kind. Mark–recapture data are discrete in both space and time because they are records of animals caught, tagged for identification, released, and caught again. The space component is discontinuous by trap locations and the time by trapping session.

Spatio-temporal autocorrelation is a concept that is basic to most of the approaches presented in this chapter. It is the lack of independence between objects, events or observations due to their positions in space and in time. The simplest kind is the case of short-range positive spatio-temporal autocorrelation where samples are more similar when they are closer together in space or in time (as discussed for space alone in the previous chapters). Setzer (1985) used a Mantel test (see Chapter 7) on spatial and temporal distances between aphid galls on cottonwoods (*Populus deltoides*) and found that galls close in space were likely to suffer mortality close in time. In comparison, more complexity is likely to be found in cases involving cyclic behaviour such as diurnal migration, such as the vertical migration of zooplankton, where autocorrelation will be positive at short space and time lags, becoming negative over short space and longer time lags and then positive again over even longer time lags.

Just as there are several measures for spatial autocorrelation (e.g. Moran's *I*, Geary's *c*; see Chapter 4), there are several different indices of spatio-temporal autocorrelation. One is Griffith's space–time index (Griffith 1981; Henebry 1995) for T temporal units and n spatial units, with w_{ijt} as the weights and z as the deviation from the overall mean of the observations:

$$I_{s-t} = (nT - n) \frac{\sum_{t=2}^{T}\sum_{i=1}^{n}\sum_{j=1}^{n} w_{ijt-1} z_{it} z_{jt-1}}{\sum_{t=2}^{T}\sum_{i=1}^{n}\sum_{j=1}^{n} w_{ijt-1} \sum_{t=1}^{T}\sum_{i=1}^{n} z_{it}^2}. \tag{8.1}$$

Clearly, this measure combines evaluations of temporal autocorrelation (of z_{it} with $z_{i,t-1}$) and spatial autocorrelation at individual times (of z_{it} with z_{jt}).

If we are measuring autocorrelation for a particular separation or distance class, d, the weights might be more precisely written as $w_{ijt}(d)$. As with Moran's *I*, the expected value is a function of the number of samples:

$$E(I_{s-t}) = \frac{-(T-1)}{T(nT-1)}, \tag{8.2}$$

which is approximately $-1/nT$ for large T. For large sample sizes, the usual assumption of convergence to normality is justified (Henebry 1995).

We begin with the use of temporal changes in some of the spatial statistics already described. We then present some truly spatio-temporal methods based on join counts, cluster change detection, polygon change and movement through space. We also provide a discussion of the concept and some applications of spatio-temporal graphs in which the nodes (and possibly edges) have locations both in time and space. The final topic is the synchrony of ecological phenomena, especially the cyclic or eruptive behaviour of animal populations.

8.1 Spatial Statistics Reassessment

One of the simplest kinds of spatio-temporal analysis is to examine changes in almost any of the spatial statistics described so far, as a function of time. This general approach to spatio-temporal analysis by repeating basic spatial statistics two or more times is straightforward and can be very informative and useful. For detailed long-term studies, this approach becomes even more important, and it is not only animal movement ecology that will benefit from big data approaches (cf. Nathan *et al.* 2022). One main disadvantage is that any summary statistics and changes in them may miss some of the important details of the actual changes to individual units or particular sub-regions in the spatial structure. It is important to know the size, spacing, positions and spatial evolution of clusters, in addition to the general degree and scale of aggregation as a function of time, especially for pest management.

Averaging over space, as a summary statistic does, before examining changes in time raises concerns and Parrott *et al.* (2008) recommended that we not 'collapse' the spatial dimensions before considering the temporal aspects of the data. The reduction is analogous to following plant community succession by summarizing species composition in a single measure and following its changes through time. We can learn something important about the community dynamics, but certainly not the whole story. Chapter 9 comments in greater detail on evaluating measures of community diversity through time, expressed as temporal beta diversity.

8.2 Spatio-temporal Join Count

Methods to elucidate spatio-temporal patterns in a single factor by autocorrelation analysis consider two 'lags', intervals along the axes of space and time. One such method, Griffith's index (Equation 8.1; Griffith 1981), related to Moran's *I*, examines the values of a variable at different locations and different times and it is appropriate for continuous variables such as density. If the data are discrete count values then a Poisson space-time joint index should be used (Griffith & Paelinck 2009) and if a binary variable such as species presence–absence is considered, then join count statistics are best.

Join count statistics measure association in nominal data from a lattice or grid (Chapter 4). Consider binary data where each cell of a lattice takes one of two values, say black or white. A join is a connection at a particular lag between pairs of defined types, like black-to-black. In the spatio-temporal approach, a two-dimensional lattice represents space as one dimension and time as the other, such as a transect of n quadrats observed at m intervals. Suppose the black cells represent plant establishment, then the occurrence of plants of different ages in the same quadrat can be distinguished by placing several black cells in that column of the lattice. Join lengths are specified by a combination of the intervals: $d = 2$, $t = 3$ signifies joins of length two on the spatial axis and three in time. Spatio-temporal association can be determined by comparing the observed number in a class to the number expected from randomness (Jacquez 1996). The simple null hypothesis (H_{01}) is that the observed number can be accounted for by random occurrence. If more joins of a particular class (d, t) occur than are expected, this indicates a tendency for stems to be separated by distance d in space and t in time.

For each join class, many random 'data' lattices are generated to provide a reference distribution, possibly based on several models of randomness. The association statistic for each join class can be displayed as two-factor correlograms, using a bubble plot, with the circle size representing probability and position indicating the join class (see Figure 8.1). Little and Dale (1999) used this method to characterize the clonal expansion of *Populus balsamifera*, in an Alberta field, using three null models for comparison. This technique is a temporal adaptation of a mosaic analysis, the *random paired quadrat frequency* approach (Dale 1995; Chapter 3). As already commented, spatial analyses can provide more insight by using several null models for comparison.

8.3 Spatio-temporal Analysis of Clusters and Contagion

These methods originate with studies of disease incidence and some of that vocabulary persists, but they translate well into many ecological areas, not just disease and parasites. The simple version of the approach is *cluster detection* (not to be confused with the multivariate technique of *cluster analysis*) and the methods for detecting disease clusters, for which the 'at risk' population is unknown, are versions of familiar point pattern analysis. The epidemiological literature emphasizes approaches that include the locations of individuals at risk, as well as those affected, or at least the 'at risk' density. The question is whether the disease cases are more clustered than can be explained by local variations in the overall population (Jacquez 1996).

Wakefield et al. (2000) presented a comprehensive review of the methods for cluster detection in the general area of spatial epidemiology and Tango (2021) provided a technical treatment of this topic. The methods fall into several categories. *Traditional* methods include a simple comparison of the numbers of cases observed in different areal units (townships, counties) with the expected number (based on population and global disease rate), using a goodness-of-fit test. The *distance-adjacency* methods include Moran's I for rates in contiguous areal units and Diggle and Chetwynd's (1991) variant of Ripley's bivariate K-function analysis for point data.

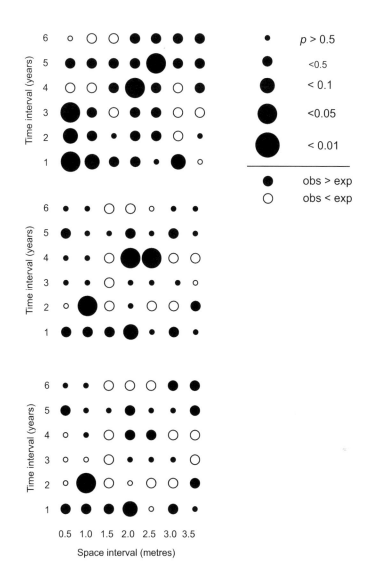

Figure 8.1 Spatio-temporal join count analysis of the *Populus balsamifera* data. At each space and time lag combination, the size of the symbol represents the significance level and white and black circles (○, ●) represent the sign of the difference between observed and expected values. Three different null hypotheses are tested for comparison (see Little & Dale 1999 for details).

Locally specific methods include the moving window approach and risk surface estimation (see Wakefield *et al.* 2000 for details). We describe two approaches for general spatio-temporal analysis.

The first set of methods is best appreciated by looking at Figure 8.2, which shows the progression of a disease (black dots) through a population (all dots) at four time periods. Clearly, the disease is spreading out from one corner of the figure. Having

Spatio-temporal Analysis

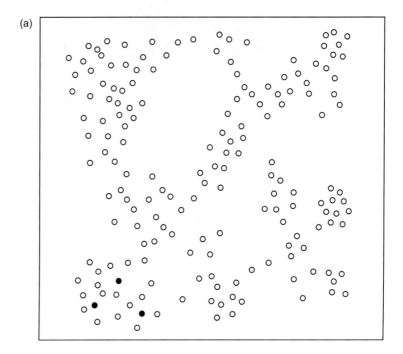

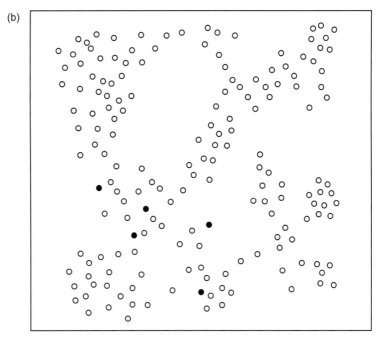

Figure 8.2 (a)–(d) Maps of the incidence of newly diseased organisms (filled circles) in a population at four different times. The spread of disease is obvious.

8.3 Spatio-temporal Clusters and Contagion

(c)

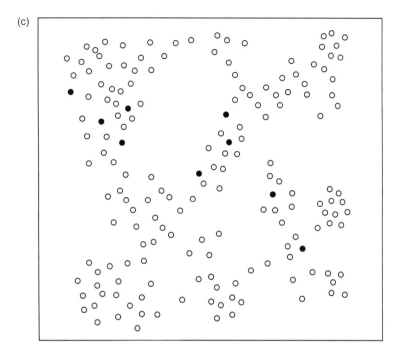

(d)

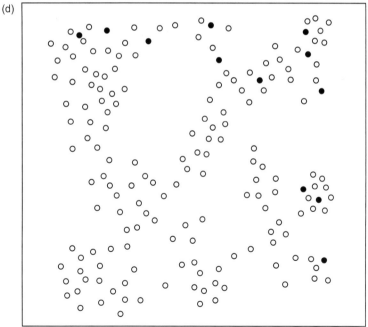

Figure 8.2 (cont.)

Table 8.1 Space–time data.

	Time	
	Near	Far
Space		
Near	114	2
Far	125	110

defined thresholds for 'near' versus 'far' in time and 'near' versus 'far' in space, pairs of disease incidences are classed as 'near' or 'far' in time and 'near' or 'far' in space to produce a 2 × 2 table. The table can then be tested with the usual statistics to determine whether incidences close in time also tend to be close in space (Knox 1964). Table 8.1 gives the near-versus-far counts from Figure 8.2, using thresholds of two-time steps and 40% of the side of the sample area.

The goodness-of-fit statistics are highly significant, indicating spatio-temporal association among the disease incidences, despite the many pairs that are near in time but separated in space (chiefly in time 4, Figure 8.2*d*).

The Mantel test was developed to determine spatio-temporal clustering by comparing two 'distance' matrices (Chapter 7), containing the spatial and temporal separations of events. In this context, the Mantel test can be seen as an extension of Knox's approach, using quantitative distances, rather than just 'near' and 'far'. Where d_{ij} is the separation of two events in space and s_{ij} is the separation in time, the basic Mantel statistic is:

$$Z_M = \sum_{i=1}^{n-1} \sum_{j=i+1}^{n} d_{ij} s_{ij}. \tag{8.3}$$

Evaluation of the test statistic is by randomization comparison. In the example in Figure 8.2, the observed statistic is $Z_M = 1{,}219.14$ which is highly significant by a randomization test and equivalent to a value very close to 1.0 in a standardized range of -1.0 to $+1.0$, indicating a strong association between temporal and spatial proximity.

In spatio-temporal networks, the dimensions of space and time are not directly commensurate. The spatial distance is the number of steps from source to sink, the sum of the edge weights, or the sum of the physical lengths of the edges. When the time dimension is divided into finite and equal intervals, the number of time steps will be determined by the duration of the intervals, called the *temporal granularity*. Temporal distance can be defined either as the number of steps or as the time elapsed from source to sink, called *latency* (Williams & Musolesi 2016). Then, the spatio-temporally shortest path is the temporally shortest path with the minimum spatial length (Williams & Musolesi 2016). This concept gives us the *spatial efficiency*, which is the average reciprocal spatial distance between pairs of nodes, and the *temporal efficiency*, which is the average reciprocal latency between pairs of nodes (Williams & Musolesi 2016).

8.3 Spatio-temporal Clusters and Contagion

Neither of these 'distance' methods use any information about the 'at risk' disease-free population, nor does the spatio-temporal version of Ripley's K-function analysis. Using only focal events, like disease incidences, Ripley's K-function counts the number of these within spatial distance s and time distance t of each event (Diggle et al. 1995):

$$\hat{K}(s,t) = \frac{|A|T}{n(n-1)} \sum_i^n \sum_{j \neq i}^n w_{ij} v_{ij} I\left(d_{ij}^S \leq s\right) I\left(d_{ij}^T \leq t\right), \qquad (8.4)$$

where w_{ij} and v_{ij} are weights for edge correction and the I's are the usual indicator functions. A and T are the area and time-span and their product is the total spatio-temporal 'volume'.

Observed and expected values are compared in the usual way and plotted as a function of distances s and t. This approach has many potential problems, the first being the incommensurability of time and space units already discussed. The second is that, unless the time series of observations is very long, temporal edge effects may be an important factor. The third concern is whether, because time is directional, a one-sided temporal search template should be used rather than a two-sided 't-bar' template.

Returning to the incidence of disease in Figure 8.2, now consider the positions of the uninfected individuals. This is accomplished using a spatio-temporal graph, in which the nodes have both spatial and temporal locations; the edges join the nodes in pairs but may not represent actual locations. The nodes are persistent and fixed, but may change categories, say from 'uninfected' to 'newly infected'. Similarly, edges can belong to categories, such as their time interval or 'potential transmission' versus 'transmission not possible'. The same set of nodes can produce at least three kinds of graphs:

(1) The first begins with simple planar subgraphs using a topological algorithm (e.g. minimum spanning tree, MST), one subgraph for each subset of nodes that are newly infected in the same time interval. The first full graph is the union of these time-interval subgraphs (Figure 8.3a); each time interval can be considered as a layer of a multilayered structure, now commonly referred to as a multilayer network (Bianconi 2018; Dale & Fortin 2021, section 4.2). In a single-layer network, the edges of each subgraph are given the labels of the appropriate time interval.
(2) A second approach is a series of graphs depicting the pattern's evolution, with the nodes labelled with their categories. Different sets of edges can be created, say Minimum Spanning Tree (MST) and Delaunay triangulation (DT), with those that are likely routes of infection changing categories as time proceeds. Using more than one set of edges for potential transmission routes allows us to compare hypotheses about the spatio-temporal characteristics of transmission (not shown).
(3) The third approach is a single graph in which each node newly infected in time t is connected to the nearest node that was newly infected in time $t-1$ (or in any previous time interval), as in Figure 8.3b. This graph is a spanning tree of the ever-infected individuals, and some of the edges have direction, from previous to

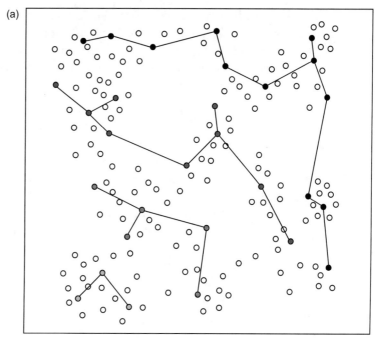

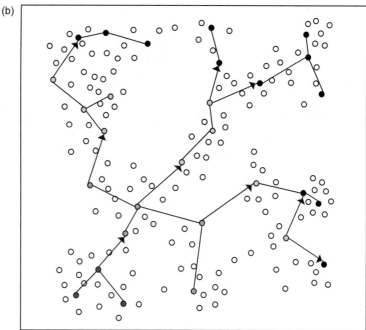

Figure 8.3 (*a*) A population with four time periods of records of new infections. Subgraphs for each time period. There are no edge crossings in this graph. (*b*). A population with four time periods of records of new infections. MST for all nodes ever infected. We can count the number of edges between nodes at different times. Some edges (those ones) have directions. There are no edges between time periods more than one unit apart.

current infection, thus producing a spatio-temporal digraph. The edges may represent likely routes of disease transmission, although the actual infection pathways may not be known. For mobile animals, one approach is to compare the network of individuals' contacts with the network of transmission, to determine their relationship. Applications of multilayer networks to answer ecological questions will be illustrated with more examples in Chapter 11.

8.4 Spatio-temporal Scan Statistics

In Section 5.6, we described spatial scan statistics used to detect and assess clusters of events, based on moving window templates. An obvious extension is to include time as well as space (Kuldorff *et al.* 2005). As with spatial scan statistics, the spatio-temporal versions involve the calculation of a probability or likelihood ratio to evaluate the result. The risk in interpreting these results associated with the potential for spatial non-stationarity is exacerbated by the possibility of temporal non-stationarity as well. On the other hand, the non-commensurability of time and space should not be an issue because a scan template need not be isodiametric. Takahashi *et al.* (2008) introduced a flexible approach to the shapes for spatio-temporal scan templates, which will be useful in ecological contexts where irregular shapes are a natural choice.

Spatio-temporal scan statistics have proven useful for ecological studies. Marj *et al.* (2006) used scan statistics to detect forest fire clusters in time and space using Florida daily fire records from the 2005 fire season. The use of these scan statistics to detect and evaluate spatio-temporal clusters of forest fires has received a general recommendation from Tuia *et al.* (2008). Bayles *et al.* (2017) provided an excellent example in their study of the invasion of the citrus psyllid *Diaphorina citri* in southern California. They combined spatio-temporal scan techniques with Moran's I and the Getis-Ord statistic (Getis & Ord 1996) to identify hot-spots of the invasion and to document its anisotropic spread. This example illustrates the value of spatial epidemiology methods for evaluating other kinds of invasive phenomena.

8.5 Polygon Change Analysis

In Chapter 3, the focus was on points or events in the plane, with little consideration of the analysis of irregular polygons. Similarly, studies of spatio-temporal analysis also emphasize point events rather than polygons, quite naturally when the subject is the dynamics of tree stems or disease foci. The analysis of polygons has greater complexity than point pattern analysis, due to the greater complexity of the data themselves.

Analysing a dynamic system of polygons is even more complex because the characteristics of the polygons can change (position, size, shape), as can their connections to neighbours, but also old polygons may disappear and new ones develop. One practical approach to analysing polygons is to calculate summary statistics for each of

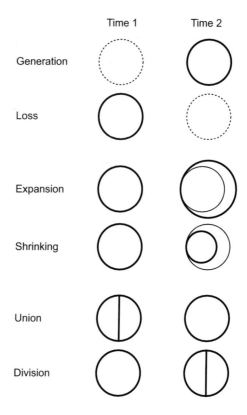

Figure 8.4 The primitive events for polygon change analysis: generation, loss, expansion, shrinking, union and division.

several observation times and then examine changes in those summaries. Using several summary statistics has advantages, but much about the characteristics of the polygons will be missed, thus causing a loss of information.

Sadahiro and Umemura (2002) developed a sophisticated approach to quantifying changing polygons on the condition that the polygons are immobile and that any change occurs in steps. They divide the stepwise behaviour of a polygon into six primitive events, illustrated in Figure 8.4:

(1) generation, the appearance of a new polygon;
(2) disappearance, the loss of a polygon;
(3) expansion, the increase in area occupied;
(4) shrinking, the loss of area;
(5) union, two polygons merging; and
(6) division, a polygon splitting into two.

The changes observed between two times can be described by combinations of these primitive events.

The two sets of polygons from the two observation times, Γ_1 and Γ_2, are overlaid to create a new set of polygons, Γ_u (Figure 8.5). These new polygons can then be

8.5 Polygon Change Analysis

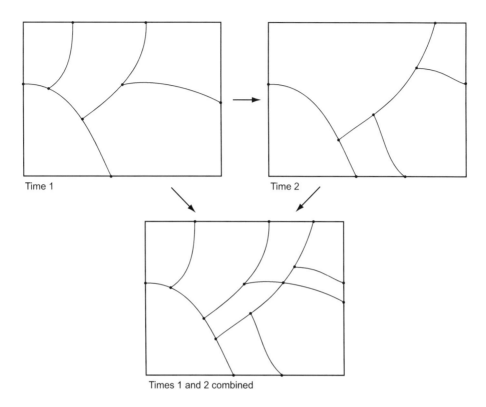

Figure 8.5 The two sets of polygons from time 1 and time 2 (Γ_1 and Γ_2, upper diagrams) are overlaid to create a combined set (Γ_u, lower diagram). There are five polygons in each of the first two sets and nine in the combined set.

classified into three groups: Ω_0, which existed at both observation times; Ω_1, which existed at time 1 but not time 2; and Ω_2, which existed at time 2 but not time 1. The arcs in the diagrams are classified into 12 groups based on the combinations of four possible states in which they existed at times 1 and 2: boundary, partition, internal to polygon and absent. Figure 8.6 illustrates these classes. The two classifications are then used to deduce possible sequences of primitive events that gave rise to the observed changes. The changes between two observation times are decomposed into the smallest number of primitive events possible, and one useful statistic is then the number of such primitive changes, M_e, which can be standardized to the total number of polygons:

$$m_e = \frac{M_e}{|\Gamma_1| + |\Gamma_2|}. \tag{8.5}$$

This approach follows that of Claramunt and Thériault (1997), who proposed a scheme of 16 primitive events, accommodating mobile polygons. Claramunt and Thériault (1997) also suggested that the events could be assigned weights appropriate to the type of event: generation and expansion events could be weighted by the area

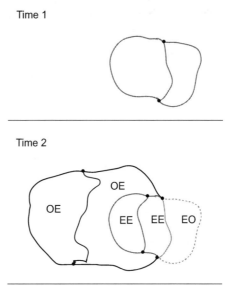

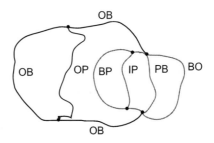

Figure 8.6 In this example, there are two polygons in the first set, four in the second, and five in the combined set. The polygons can be classified as 'EO', extant at time 1, but not at time 2; 'OE', extant at time 2, but not at time 1; and 'EE', extant at both times. The edges that define the polygons can be classified as 'BO', boundary at time 1, but extinct at time 2; 'PB', partition then boundary; 'IP', interior then partition; 'BP', boundary then partition; 'OP', partition new at time 2; and 'OB', new boundary.

gained, and shrinking and disappearance events by the area lost, with partition and union events having weight 0.

The analysis of polygon change has been extended to polygons that move, with the development of STAMP (Spatial-Temporal Analysis of Moving Polygons) methods within a GIS environment (Robertson et al. 2007), now available in R (Long et al. 2018). This extension augments the earlier basic scheme with five movement events: displacement, convergence, divergence, fragmentation and concentration. Its usefulness was demonstrated by applying it to the spread of a wildfire in northwest Montana (USA), for which the method produced a good quantitative summary of changes in size and direction of movement. Grillet et al. (2010b) investigated the spatial dynamic of malaria incidences in Venezuela. Also, Vepakomma et al. (2012) applied this polygon change analysis to quantify gap spatial dynamics in the boreal forest in Québec.

The analysis of the dynamics of delineated polygons may present a promise for future studies, but the approach based on the polygons as units may be superseded by raster methods, quantifying the relationships among the labelled cells of a fine grid of detailed data, with polygons treated as clusters of contiguous cells.

8.6 Analysis of Movement

The movement of individual organisms is of crucial interest in many ecological studies: the spread of disease vectors in epidemiology, the identification of home ranges in wildlife ecology or the invasion of a clonal plant into a new habitat. At the level of the individual, movement is also closely tied to behaviour, perception, ecological interactions and learning (Lewis et al. 2021). The quantitative analysis of movement depends on whether the movement (or its record) is continuous, like a sidewinder crossing a dune, or occurs as discrete units, like a butterfly's flight with stops at individual flowers. Even if the movement is continuous, it is often divided into units, such as the locations at 5-minute intervals. For movement analysis, we will use the steps analogy and present the material in the language of spatio-temporal graphs. The animal's observed position in space is one node of a graph, p_t, with a location and a time, and a directed edge joins p_t to p_{t+1}, although the actual route between the two locations may not be known.

With such data, movement analysis is based on the edges' lengths and the angles between them (see Figure 8.7) and it requires the correct calculation of the average of angles. For any set of angles, whether the directions of the edges, labelled α_i in Figure 8.7, or the *turning angles*, δ_i, the angles are represented by vectors of unit length and coordinates (x_i, y_i), as in Figure 8.8. The coordinates of the mean vector are:

$$\bar{x} = \frac{1}{n}\sum_{i=1}^{n} \cos(\alpha_i) \quad \text{and} \quad \bar{y} = \frac{1}{n}\sum_{i=1}^{n} \sin(\alpha_i). \tag{8.7}$$

In polar coordinates, this is

$$(\bar{x}, \bar{y}) = (r_a \cos \varphi, r_a \sin \varphi), \tag{8.8}$$

where $\varphi = \tan^{-1}(\bar{y}/\bar{x})$ and

$$r_a = \sqrt{\bar{x}^2 + \bar{y}^2} = \frac{1}{n}\sqrt{n^2\bar{x}^2 + n^2\bar{y}^2} = \frac{1}{n}\sqrt{\left(\sum_{i=1}^{n}\cos\alpha_i\right)^2 + \left(\sum_{i=1}^{n}\sin\alpha_i\right)^2}. \tag{8.9}$$

The measure r_a cannot take negative values, so it is not a correlation; it is a measure of *angular concordance* or *angular concentration*. It takes a value 1.0 when all the

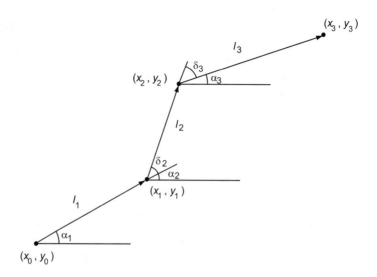

Figure 8.7 Movement as a series of straight-line units, l_1 to l_3, with known start- and endpoints. The αs are the absolute angles and the δs are the turn angles.

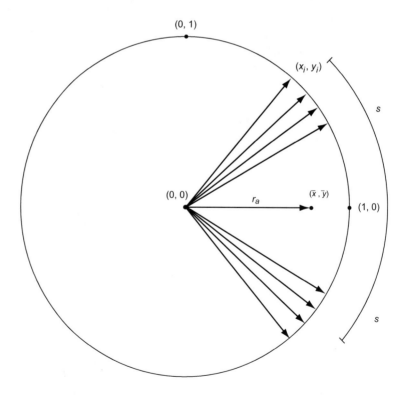

Figure 8.8 Calculation of the average of a set of eight angles depicted as eight-unit vectors (x_i, y_i). The angular concordance is r_a and the angular deviation is s.

8.6 Analysis of Movement

angles are the same and a value 0 when all the vectors cancel out. Figure 8.8 shows high concordance and r_a is 0.75. The circular equivalent of the standard deviation for linear data is s, the *angular deviation* (Batschelet 1981):

$$s = \sqrt{2(1 - r_a)}. \tag{8.10}$$

Here, $s = 0.71$, which converts to 41° (from $s \times 180°/\pi$), a relatively small variance.

One characteristic of interest in a movement path is the *directional autocorrelation* of adjacent edges, and that measure is the average of the cosines of the turning angles (Batschelet 1981), as in Figure 8.9. When the two steps are in the same direction, the correlation is 1.0; when they are at right angles, the correlation is 0; and when they are directly opposite, it is −1.0. If the turning angles are concentrated near zero, indicating a tendency to continue in the same direction, this produces positive directional autocorrelation.

For the analysis of movement data, start with calculating the edge length autocorrelation and angular autocorrelation for pairs of edges at a range of lags. Figure 8.10 shows the results of this kind of analysis for five sets of artificial data with different kinds of autocorrelation. We could also look at the autocorrelation of net displacement as a function of lag. The most difficult problem in analysing field data is usually the lack of stationarity. This potential (and probable) departure from the underlying assumptions must be accounted for both in analysis and interpretation. One way of dealing with non-stationarity is to make it explicit by allowing the data to be divided into subsets for different activities, such as searching versus feeding, with different parameters for each (the distributions of distances and angles) and probabilities of transitions between the conditions. This concept will be described in detail as the fourth alternative model in this list of possible approaches to non-stationarity.

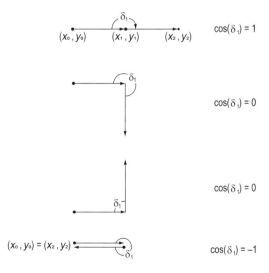

Figure 8.9 The cosine of the turning angle as a measure of angular autocorrelation.

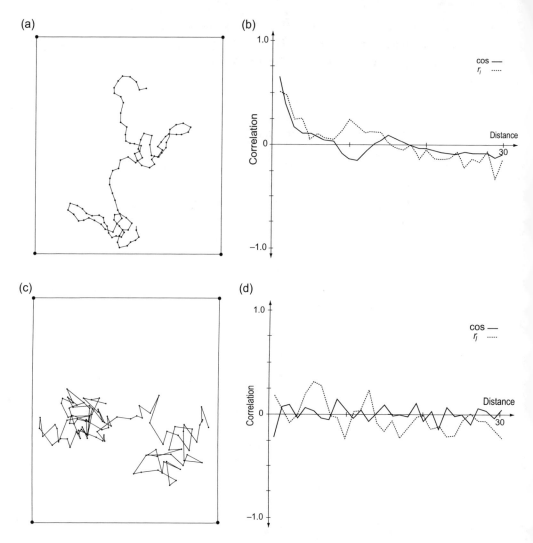

Figure 8.10 Pairs of paths and the autocorrelations of the units' lengths and angles as a function of distance or lag: (*a*, *b*) short-range positive autocorrelation in both length and angle; (*c*, *d*) angles and lengths both independent; (*e*, *f*) directional bias, but lengths independent; (*g*, *h*) directional cycles, lengths independent; and (*i*, *j*) cyclic behaviour in lengths, angles independent. The dots in the illustrations of paths are the corners of the study area.

Observed structures like these spatio-temporal graphs are usually assessed by comparison with null models and, as a null hypothesis for movement data, consider the *random walk* in which the direction of each unit of movement is randomly chosen uniformly from all possible directions with its edge length drawn from a distribution. The distribution can be determined a priori or estimated from the data. It is easy to calculate the expected displacement and other characteristics of the path, such as the distributions of angles, resulting from a random walk. However, paths of clonal plants and mobile animals generally do not match these characteristics well; angles near 180° are less common than those near 0° and net displacement tends to be greater than the random walk produces (Turchin 1998).

8.6 Analysis of Movement

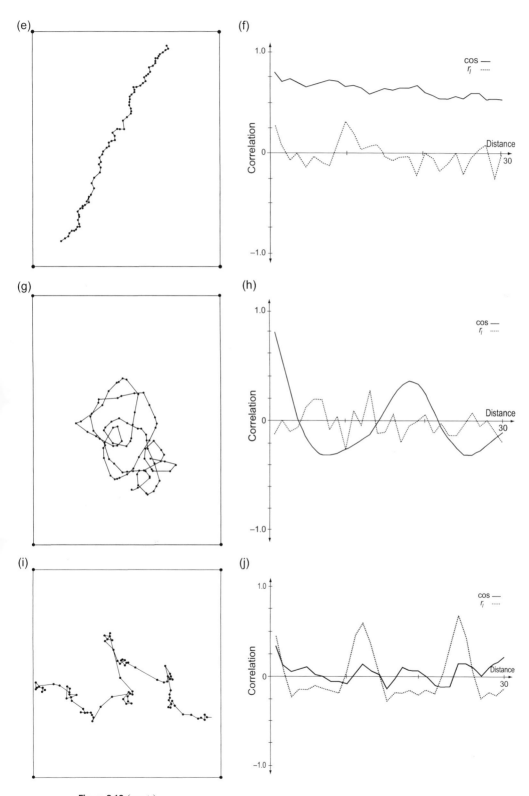

Figure 8.10 (cont.)

An alternative is the *correlated random walk* (CRW; Kareiva & Shigesada 1983) in which adjacent edges are autocorrelated for some properties but the expected net displacement is still zero. We can assess the first-order autocorrelation of length and direction of the units of movement and compare the observed and expected path directional complexity, called *tortuosity* (Wiens et al. 1993) or *sinuosity* (Sanuy & Bovet 1997). Kareiva and Shigesada (1983) have provided the derivation of the expected values for this model. The appropriateness of CRW can be evaluated by comparing the observed net displacement after n steps with the model's prediction. Many studies have used these approaches to study animal movements, including those of caribou, clover and caterpillars (Berthelot et al. 2020).

Kareiva and Shigesada (1983) have recommended that we should go beyond the simple first-order models and look at higher-order Markov models. That is, the length and direction of an edge can depend not just on the one immediately preceding it, but perhaps on the characteristics of the two, three, or more edges preceding. Figure 8.10*f* illustrates a path with regular cycles, although more realistic cyclic behaviour with a random component could be modelled by a high-order Markov model.

Another alternative is the *biased correlated random walk*, which includes a directional tendency, either absolute or relative to some habitat element (Turchin 1998; Berthelot et al. 2020), so that the expected net displacement is not zero. Rather than selecting angles uniformly from the full circle, they are drawn from a non-uniform distribution, usually unimodal. This approach is similar to adding 'drift' to the model, by including a set absolute directional component to each unit of movement (Wiens et al. 1993).

The last model we will consider is appropriate when the data seem divided into subsets related to different animal behaviours such as searching versus feeding. The idea is that the different behaviours have their own parameters for edge lengths and angles in their random walks and that transitions between the behaviours occur with fixed (but unknown) probabilities. For a two-level model of searching versus feeding, the edge length and angle parameters are λ_s and λ_f, and α_s and α_f. The probabilities of transition from searching to feeding and from feeding to searching are γ_{sf} and γ_{fs}; remaining in the current behaviour mode has probability $(1 - \gamma_{sf})$ and $(1 - \gamma_{fs})$. This is also a Markov model for which we do not know the parameters; it is a Hidden Markov Model (HMM) because we do not observe the underlying process, only its results. We must derive estimates of its parameters from the data by using maximum likelihood estimation. This approach is one of the hallmark features and a key technique of movement ecology, and it is readily available in R code (*moveHMM*; Michelot et al. 2015, 2016).

Whichever model is used, randomization techniques based on reordering the edges or Monte Carlo procedures using parameters determined from the observed characteristics of the path can be used to evaluate the results (cf. Manly 2018).

Another important theme in the analysis of animal movement data is the search to associate the locations and the characteristics of the paths with the habitat through which the animal moves and thus to evaluate differential habitat preference and use. The simplest approach is to look only at the set of locations of the animals, without

8.6 Analysis of Movement

considering movement. Indeed, one could compare the frequency of animals' presence data in particular habitat types with the availability of those habitat types in the landscape (Fall et al. 2007). This approach has some advantages, including the fact that it can be used for one or several animals. We might have some concerns about what is implicitly defined as the experimental unit and as the replicate if we include data from several animals (Hurlbert 1984). Leaving those concerns aside, this approach cannot be used as described because autocorrelation in the data has not been accounted for. Positive autocorrelation makes statistical tests too liberal, giving more apparently significant results than the data actually justify. We have discussed this at length in Chapter 9, in the context of spatial autocorrelation and statistical testing. The same considerations apply to the spatio-temporal autocorrelation inherent in animal movement data.

One assessment of preferential habitat use is based on the fact that movement in the favoured habitat is more *tortuous* with more frequent and tighter turns; yielding less net displacement and greater residence time (Turchin 1998). Any set of k edges in a movement path can be assigned an index of tortuosity or compactness. Several measures are available but we suggest one based on the 'convex hull' of that portion of the path, that being the smallest convex polygon that contains it. Where m is the diameter of the convex hull, the largest node-to-node distance in the convex hull, and L is the total path length within it, then a simple measure is L/m. Figure 8.11 shows two examples of this measure.

For each point i in the path and integer k, we can calculate the compactness for a sub-path of length k centred on i. Then, those scores can be compared for different habitat types for a range of sub-path lengths. Statistical testing can be carried out by

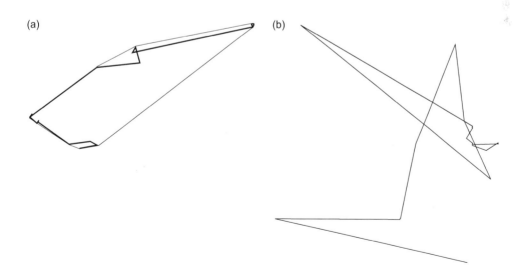

Figure 8.11 Measures of path compactness or tortuosity. (*a*) A sample of elk movement of low tortuosity: $L/m = 1.56$. The bold lines are the path and the fine lines complete the convex hull. (*b*) A sample of elk movement with higher tortuosity: $L/m = 4.12$.

superimposing the path itself, after random translation, rotation and reflection on the habitat map, say 1,000 times, and recalculating the scores. The scores from the original path can be evaluated by comparison with the values from these relocated versions.

The number of times a path crosses itself is another statistic related to path complexity. Consider that the paths in Figure 8.10; part (*e*) has no crossings; part (*i*) has many, but they always involve steps that are close to each other; part (*g*) also has many crossings but some are temporally far apart in the path. A second statistic would be a measure of whether the crossings occur between edges of the path that are close to each other in time or temporally distant; a simple measure would be the pairwise correlation of the temporal labels of edges that cross. Again, its significance can be evaluated by comparison with random relabelling of the nodes.

In complicated situations or for large data sets, this kind of analysis could be carried out for short sub-paths of the sequence, with the results recorded, mapped, or plotted in order to detect changes in time or space. Where non-zero net displacement is expected, one question of interest is how to detect important behavioural patterns from the incomplete information provided by periodic data acquisition. In Figure 8.12, we can determine whether the characteristics of the path indicate avoidance of the area marked as a shaded rectangle or unusually rapid movement across that patch.

This thinking leads to a hierarchy of hypotheses and a hierarchy of tests:

(0) *Complete randomness.* The nodes as endpoints of the edges are randomly and independently positioned in the entire area being considered.
(1) *Complete avoidance* of the gap. No nodes of the spatial graph are within a given area, here the shaded rectangle. For 28 observations and a shaded area of 14% of the total; with the usual randomness and independence assumptions, the probability that the shaded area has no nodes is $(0.86)^{28} = 0.0146$.
(2) *Few crossings* of the patch. Given the positions of the nodes, if they were randomly joined in pairs by edges of the graph, the frequency of realizations with one or fewer intersections of the patch can be shown to be significantly small. The concern might be whether completely random pairing of nodes

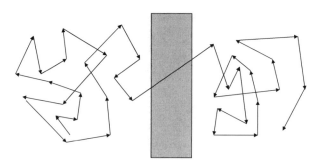

Figure 8.12 Locations of an animal tracked through a landscape at equal time intervals, with the order indicated by the edges. The rectangle represents a patch of habitat that may be avoided by the animal.

constitutes a fair test, since it will tend to produce longer edges than observed. Perhaps we should use a restricted randomization that limits edge lengths.
(3) *Rapid crossing.* Compared to other single transition path segments, it is by far the longest, indicating very rapid movement. Hence the individual is avoiding a specific landcover type. This can be approached using a randomization test, although the number of edges may sometimes be too small. Comparison with the model of a correlated random walk (CRW; see Turchin 1998) should be useful here.

Other questions that can be answered include:

(i) Is the path self-avoiding? The number of self-crossings is only five. Random pairing of nodes would produce more, but is that a fair test? Again, a correlated random walk might be a more appropriate model for comparison.
(ii) Given information on habitat classification for the entire area, not just the shaded rectangle, can we use a resource selection function or other approaches (such as a step-function, patch-focused CRW and so on) to characterize the possibly non-random use of different habitat types?

These kinds of questions can be best approached by using a spatio-temporal graph to model and analyse the essential features of the system. Other interesting examples of spatio-temporal graphs for studying animal movement to be pursued include the analysis of a group of identifiable individual animals, such as a family group or herd, to study the dynamics of their relative positions at a series of time intervals as they move through the landscape. The topic of the spatio-temporal analysis of the movement of individuals is clearly both important and worth intensive research on methods and applications and it has, in fact, given the rise of the subdiscipline of *movement ecology* (Nathan et al. 2008; Joo et al. 2022).

8.7 Spatio-temporal Networks

We have referred to spatio-temporal graphs or networks more than once in this chapter: on regarding disease spread in a stationary population and in describing animal movement. A graph is a set of nodes joined in pairs by edges. So is a network, but, in a network, the nodes usually have labels or magnitudes and the edges have categories or quantitative properties such as rates or capacities. The ecological literature now refers to networks almost exclusively, rather than graphs, and we will keep the discussion general and talk about spatio-temporal networks.

Spatio-temporal networks are distinguished by the fact that the nodes have spatial and temporal locations and the edges may or may not, and a key question is the appropriate metric of the distance between nodes. In any spatio-temporal network, the dimensions of space and time are not directly commensurate. In some applications, the spatial distance is the number of steps from source to sink, the sum of the edge weights or the sum of the physical distances of the edges. Temporal distance can be defined

Table 8.2 Characteristics of spatio-temporal networks.

Time	The original events or objects may be long-lasting or of very short duration; their records may be in continuous form or in discrete intervals (e.g. equal steps or order of occurrence; time 'slices') Continuous-time data can be converted to the discrete form by 'binning' Explicit time-course records can be converted to an integrative summary of the whole time-course Explicit time values or temporal labels (e.g. time period)
Space	The original events or objects may be point events or they may have length, area or volume; their records may be as continuous data (e.g. real number x, y coordinates) or in discrete format, such as raster, mosaic/polygon or object-referenced data Continuous data can be converted to the discrete form by 'binning' Number of spatial dimensions: 1D, 2D, 3D, as dictated by the system

Nodes	Time: instantaneous (duration close to 0), ephemeral (short duration), transient or intermittent, persistent/permanent
	    If persistent: constant or variable in characteristic of interest (e.g. size, state, quality)
	    Temporal characteristics: one-time snapshot (of contemporaneous nodes) versus integrative summary of many time periods versus sequential time course (nodes as 'ancestors' and 'descendants')
	Space: fixed or stationary versus moving
	Representing: locations, objects, events
	Properties: categorical, qualitative or quantitative labels; associated functions

Edges	Time: instantaneous (duration close to 0), ephemeral (short duration), transient or intermittent, persistent/permanent
	    If persistent: constant or variable in characteristic of interest (e.g. flow rate)
	    Temporal characteristics: a single time period versus summary graphs of many time periods indicating contemporaneity versus sequential time course (e.g. edges linking 'ancestors and descendants' or 'identity' edges linking instances of the same node)
	Space: symmetric versus asymmetric or directional
	Representing: connections, processes, relationships, objects
	    Properties: categorical, qualitative or quantitative labels (including signs); associated functions; implicit directionality from any temporal extension

either as the number of steps or as the time elapsed from source to sink, called *latency* (Williams & Musolesi 2016).

In these networks, space and time may be continuous in records or divided into discrete units. Table 8.2 provides a more complete description of the possible characteristics of these networks (see table 10.1 of Dale 2017). A particular version is one in which the data can be divided to create a spatial subgraph for each of several time periods: a multilayer network with one layer per time period. Figure 8.13 illustrates the concept, drawn with the layers aligned like the layers of a cake. Edges between nodes within layers may reference different interactions than those included as edges between layers.

Spatio-temporal networks include multilayer structures in which the layers are sites rather than times and each layer can be a time sequence or a time-only graph

8.7 Spatio-temporal Networks

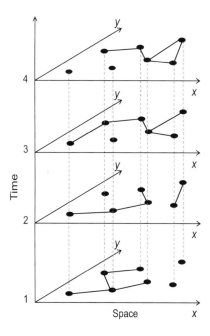

Figure 8.13 Multilayer network for spatio-temporal data. Each layer is a time slice.

(cf. Dale & Fortin 2021, figure 4.8). Timóteo *et al.* (2018) used a multilayer approach with the layers being habitats to investigate the spatial structure of seed dispersal in the Great Rift landscapes. If the layers are landscapes at different times with the intralayer edges representing the connections between the landscape patches as the nodes, the differences among the layers represent the dynamics of the landscape connectivity, an important factor determining conservation success (Martensen *et al.* 2017; Zeller *et al.* 2020). This is a promising approach and Chapter 11 examines the roles of multilayer networks in spatial and temporal analysis more thoroughly.

8.7.1 Phenology

We will complete this discussion of spatio-temporal networks with the variation of phenology in space and time. Phenology is the sequential appearance of seasonal characteristics of plants, such as leaf-out or flowering or other organisms like egg-hatching in insects. Large-scale patterns and changes in climatic conditions can be assessed from the phenology of particular species' stages in a network of sites. Phenology is highly important in the context of climate change (Wolkovich & Donahue 2021), which affects natural processes and conservation efforts and it presents some challenges to techniques of analysis.

Phenological data provide a complex network with two kinds of nodes: persistent spatial nodes, the sites and transient temporal nodes, specific seasonal events, such as the first leaf for species A, located at each site at a particular date. Spatial edges between sites can be constructed for spatial proximity, ease of dispersal, similar

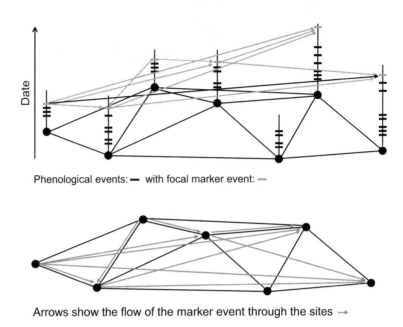

Figure 8.14 A year's phenology of seasonal events in spatial sites as spatio-temporal structure. Adjacent sites are joined in a Delaunay triangulation.

topographic or vegetation features or climatic conditions derived from continuous environmental recording. Temporal edges for events can simply tie together the events that occur in the same year at the same site. Spatial edges between events can be used to indicate instances of the same event at different sites and these may be directional to indicate temporal priority (see discussion in Dale & Fortin 2021). Phenological patterns are often discussed in terms of 'flow' and the 'same-event' edges can be compared in both space and time to evaluate variation in the flow of different events through the study region (Figure 8.14). The flow occurs in a 3-dimensional space created by mapped locations in two spatial dimensions and the temporal dimension of the seasons. The flow of phenological markers results from the interaction between the progress of environmental conditions, such as day length and temperatures, with the physiological responses of the plants, which may vary by ecotype within species (Chuine & Régnière 2017). The phenology network may be most useful when a matching network records environmental data. Such time series of environmental data can be analysed for spatio-temporal characteristics such as trends, cycles and synchrony, which can then be compared with similar patterns in the phenological data. Usually, environmental monitoring has fewer locations than the phenological network (Schröder et al. 2014) and the environmental data must be interpolated for the phenology sites, often using Kriging (Chapter 4; Gerstmann et al. 2016). Also informative in a different way, the phenological data may allow interpolation of environmental factors to sites not monitored (Ruiz et al. 2012; Gerstmann et al. 2016) and it is possible that the relationship between environment and phenology can be evaluated by co-Kriging, although the assumptions may not be fully met (Chapter 4).

The phenological network for any year (Figure 8.14) can be considered as a layer of a multilayer network, giving rise to further complexity but a wealth of information and possibilities for sophisticated spatio-temporal analysis. Multilayer networks including standard spatio-temporal ones, as in Figure 8.13, are discussed in detail in Chapter 11. The phenological networks we have described here are unusual in the fact that the passage of time in the interlayer space is continuous between the layer-to-layer steps, and in the complexity within the layers, each being 3-dimensional.

The multilayer structure makes it possible to detect and quantify the effects of environmental change (Schröder *et al.* 2014; Gerstmann *et al.* 2016) and to interpolate environmental factors to sites not monitored (Ruiz *et al.* 2012) or to place current phenological timings in the appropriate historical context (Pearse *et al.* 2017). Schröder *et al.* (2014) compared 1961–1990 with 1991–2009 for phenological data in Germany and found significant shifts in plant phenological phases of about 7 days, with an overall prolongation of vegetation growth of 3 weeks. Most changes were correlated with air temperature; for example, apple trees (*Malus domestica*) started to flower earlier with increasing mean air temperature (Schröder *et al.* 2014). Phenological networks, through spatio-temporal analysis, can also be used to study the advantages or costs of loose or close synchrony among populations and between closely related or distant species. These considerations lead naturally to a general discussion of synchrony and asynchrony of ecological phenomena.

8.8 Spatial Aspects of Synchrony

Although samples that are closer together in space or time may tend to be more similar than samples that are taken further apart, it is also possible for more widely spaced samples to be similar. In space, a simple alternation of patches of high density with gaps of low density will produce cyclic behaviour and we are familiar with many examples of temporal cycles in ecology, such as masting by trees and the famous population cycles of snowshoe hares and lynx, or small mammals such as lemmings and voles. Often these cyclic phenomena are more-or-less synchronized over large areas and a number of studies have examined the relative effects of the characteristics of population dynamics and dispersal and of external forces, such as climatic events, in causing large-scale synchrony (Guichard *et al.* 2019). Synchronous cycles are not the only form of spatio-temporal organization observed in natural systems or in the models used to investigate their properties. Expanding circles, travelling waves and spirals of high density are other possibilities, somewhat like the wave-regeneration in some forests or the development of banded vegetation in 'brousse tigrée' (tiger brush). Much of the literature on spatio-temporal organization has concentrated on the obvious cycling of certain populations because they are widespread and provide a challenge to our understanding of the factors that determine the abundances and distributions of organisms (cf. Krebs *et al.* 2001).

Given a cyclic system that is found over a wide geographic area, it is of interest to ask about the relationship between the cycles as a function of distance. Do the cycles exhibit the same periodicity or does the period change with location? If the cycle lengths are more or less the same, are the cycles in synchrony over the geographic range or does synchrony decline with distance?

For data from a set of spatial locations, several different but closely related methods can answer these kinds of questions. In general, the data for any particular location, i, will consist of a time series of population densities, $N_i = N_{i1}, \ldots, N_{it}, \ldots, N_{iT}$, often log-transformed before analysis:

$$X_t = \log(N_t + 1). \tag{8.12}$$

The method described by Hanski and Woiwod (1993) uses these log-transformed data, but removes the first-order temporal autocorrelation effects by fitting the equation:

$$X_{t+1} = a + bX_t, \tag{8.13}$$

and then using the residuals for further analysis:

$$R_t = X_t - (\hat{a} + \hat{b}X_{t-1}). \tag{8.14}$$

One approach is to calculate the cross-correlation between the two series with a time lag of zero. For two time series, x_1 and x_2, the cross-correlation is:

$$r_{12}(0) = \frac{\sum_{t=1}^{T}(x_{1t} - \bar{x}_1)(x_{2t} - \bar{x}_2)}{\sqrt{\sum_{t=1}^{T}(x_{1t} - \bar{x}_1)^2(x_{2t} - \bar{x}_2)^2}}. \tag{8.15}$$

With two time series for each of several locations, the effect of spatial distance on their cross-correlation coefficient can be assessed. Tobin and Bjørnstad (2003) presented an interesting example of this approach to study the spatio-temporal relations between a prey species, the house fly (*Musca domestica*) and a predatory beetle, *Carcinops pumilio*. Given the cross-correlation of the two time series for each of many locations in commercial hen houses, in which the fly is a serious problem, a kernel function can be used to give an estimate of the cross-correlation at any given distance. They found that, during the exponential growth phase of the fly population, the beetles were strongly negatively cross-correlated with their prey at local spatial scales.

An alternative is to compare the $n \times n$ matrix of correlation coefficients, r_{ij}, with the $n \times n$ matrix of inter-site distances, d_{ij}, using a Mantel test with a randomization procedure to test significance. Having more correlation coefficients than sites, pseudo-replication may bias the results (Koenig & Knops 1998). Mantel correlogram results may not be of ecological interest, if all it detects is a decline of autocorrelation with distance. Figure 8.15a gives an example of the correlation of tree-ring widths in *Picea glauca* as a function of geographic distance in Alberta (Peters 2003). The results for two trees per site and five sites show little evidence of a systematic decline in

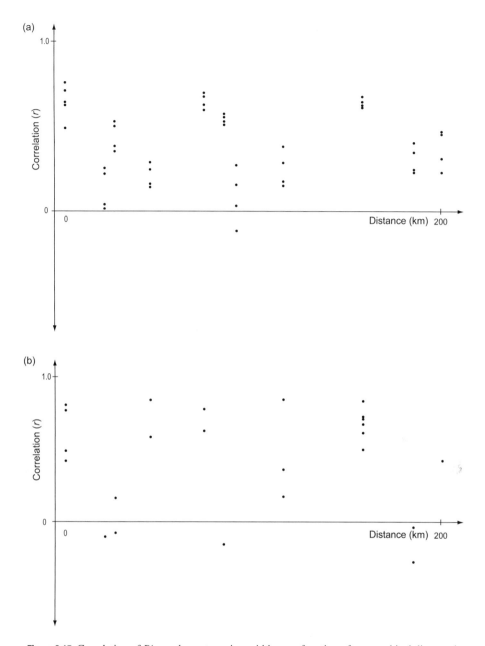

Figure 8.15 Correlation of *Picea glauca* tree-ring widths as a function of geographical distance in central Alberta. (*a*) Five sites, the same two trees per site. (*b*) The same five sites, two trees per site, but the trees were not reused (sampling without replacement).

correlation with distance, indicating that the situation is not the simple one that Koening and Knops (1998) described. In fact, the simple decline with distance may not be as common as those authors suggest, as described in Chapter 6. They recommended a 'modified correlogram' which plots the mean correlation coefficient between the time series of randomly chosen pairs of sites within specified distance classes, with each site used only once to avoid pseudo-replication. Ranta *et al.* (1997) used a randomization technique to compare the level of synchrony at any particular site with others, by choosing other sites at random, thus avoiding the same problem. Figure 8.15*b* shows a reanalysis of the same *Picea glauca* tree-ring data of Figure 8.15*a*, but using randomly chosen pairs of trees that are then not reused (sampling without replacement). The conclusions would be the same. Where each site has two time series of data, such as acorn production and annual growth in oak trees, the cross-correlation coefficient between the two series can be used in the same way (Koenig & Knops 1998). It is not clear how great a problem is the reuse of data; many exploratory analysis techniques are based on repeated use, with TTLQV being an extreme example (Chapter 3). For a wide range of methods, the trade-off is between independence and the effective use of the information available.

Bascompte and Solé (1998) described how the analysis of two data sets confirmed the predictions from spatially explicit dynamic population models of spontaneous self-organization. Several studies have demonstrated the phenomenon of travelling waves of density, for example in field voles (*Microtis agrestis*; MacKinnon *et al.* 2001) and in red grouse (*Lagopus lagopus* ssp. *scoticus*; Moss *et al.* 2000). Travelling waves are detected in the spatio-temporal data by looking for anisotropy in the spatial covariance (Lambin *et al.* 1998; Bjørnstad *et al.* 1999). A travelling wave produces cross-correlation that declines with distance perpendicular to the wavefront but not with distance parallel to the wavefront. The plot of cross-correlation as a function of distance (Equation 8.15) is divided into a few direction classes and examined for these differences. Lambin *et al.* (1998) have described a modelling method to estimate the speed and direction of a travelling wave and to determine its statistical significance.

Two related questions can be asked about (a) the consistency of spatial pattern through time and (b) the synchrony of temporal patterns in space. This section has described methods that can be used to answer these questions but answering them merely leads to more questions, now about the ecological processes that produce the dynamic patterns we detect.

8.9 Concluding Remarks

Spatio-temporal analysis, and the phenomena with which it deals, is one of the most rapidly developing areas in ecology. It is critical to a mature ecological understanding, not just for spatial patterns and temporal processes, but also for spatial dynamics and temporal patterns. Work on spatial synchrony and asynchrony is providing important insights into the basic drivers of population dynamics and community interactions. The analysis of animal movement through its habitat, and how that is related to the habitat structure, is an area of active research where we expect to see rapid developments. The analysis of polygon change is also one that deserves further work and effort. The developments in the application of spatio-temporal networks have been so rapid and important we have created a whole chapter that focuses on multilayer networks (Chapter 11).

Spatio-temporal analysis is a field that can handle and profit from rich datasets; many of the techniques described here will be most rewarding with detailed spatial information and many times of observations.

In conducting statistical tests for these studies, we must beware of spatial and temporal autocorrelation: their causes and their effects, how they are similar and how they may be different. A theme throughout this book is the lack of independence among observations. While that lack of independence causes problems for statistical tests through autocorrelation, it is also the property that makes prediction possible, and prediction (of many kinds) is crucial to the scientific value of ecology. Predictions, in general, are more powerful and potentially more useful when they are quantitative, and so large data sets of high-quality numerical data provide an important advantage. We have discussed the problems and interesting qualities of spatial and temporal autocorrelation and it is an important consideration in analysing spatio-temporal data of all kinds. The concept of time-to-independence or distance-to-independence can be mistaken and misleading. We need to learn to take advantage of that lack of independence in the data and use it for our own purposes. Therefore, it is much better to use all the information available and to evaluate the characteristics of autocorrelation in the data to be used in later analysis.

9 Spatial Diversity Analysis

Introduction

The topic of diversity in biological systems, whether called 'species diversity' or 'biodiversity', is the subject of thousands of scientific articles and is a central concept of many books (e.g. Magurran & McGill 2010; Lovejoy & Hannah 2019). Informally, diversity is related to the variety of classes or categories represented in a collection of objects (e.g. study sites); it is a key factor in the *complexity* of any system (Ladyman & Wiesner 2020) and an essential feature of what is considered ecological complexity (Riva *et al.* 2022). In ecological systems, diversity affects the coexistence of species and the structure of natural communities and it has implications for the persistence of species in disturbed systems, the conservation of organisms and the functioning of the ecosystems in which they occur. One important factor is how the diversity is organized in space: space in the sense of area and location. The spatial arrangement of diversity is a complexity that affects the dynamics of the system, contributing to its stability or its vulnerability to change.

Much discussion of diversity in the ecological literature has concerned what the concept means, or should mean, and how it is to be measured. For our purposes and the spatial analysis of a community of 'individuals' belonging to 'species', *diversity* is a measure of how difficult it is to predict a randomly encountered individual's species. (The familiar Simpson's index measuring diversity is defined as the probability that two randomly chosen individuals belong to different species.) A second version is *spatial diversity*, which measures the difficulty of predicting an individual's species, given the scale of observation and the location of the encounter. This refinement offers the choice of adapting the familiar non-spatial approaches by including the location or developing new measures that include spatial information explicitly. To begin, we describe the importance of spatial considerations for evaluating diversity in ecological systems.

9.1   Space and Diversity

Basic diversity analysis requires quantifying the density of each species in each delimited area. This treatment is initially 'aspatial', although space, as location or scale, may be included somehow, usually implicitly rather than explicitly. Much of the discussion of diversity has been clearly aspatial, in part because many organisms of interest are mobile animals, like birds, insects or fish. In addition, a spatial assessment may be limited by the data collection: traps or sampling stations may blur the spatial signal. This feature is of special concern in studies using environmental DNA (eDNA) and metabarcoding to resolve species identification. Depending on the source of the eDNA, the blurring effect may be greater because the DNA from various sources is amalgamated by sampling, requiring resolution to achieve species identification (Cantera et al. 2022). Yet, if the eDNA sampling is designed to be spatially determined, high-quality data for spatial diversity analysis can result in highly localized information (Matsuoka et al. 2019; Lamy et al. 2021; Cantera et al. 2022).

A diversity measure for highly mobile animals summarizes various spatial components: the location of the study, the study extent, and the change in the species list between locations. As the study extent increases, the number of species (richness) also increases. This is the species–area effect that is well-known and well-studied (Scheiner et al. 2000; Chase & Knight 2013; Myers et al. 2015; Keil & Chase 2019; Keil et al. 2021).

To examine spatial aspects of diversity, we can ask how standard diversity measures (like species richness or Simpson's index) change with location and how they respond to the spatial structure of the environment. We can consider how spatial locations and spatial structure are to be included in the analysis. Scale and location have already been included as aspects of spatial context, but we will add spatial heterogeneity and spatial dependence.

9.1.1 Spatial Scale

The concept of scale includes both *grain* (the size of the smallest discernible unit), and *extent* (the size of the area under consideration) (see Chapter 2). Scale may be confounded with heterogeneity and location, but it includes the total area sampling effect, shown by a 'species–area curve', plotting the number of species encountered as a function of the area searched. Most simply, as larger areas are searched, more species are found, although the variability in densities may increase.

Extent is a foundation for ecological diversity which usually recognizes three levels: α-diversity at the smallest scale, within sites, and γ-diversity for an entire region (Whittaker 1977; Figure 9.1); β-diversity is then the variation among sites within a region. This intermediate level may be of greatest interest to many ecologists, allowing tests of hypotheses about processes that affect diversity (Legendre & De Cáceres 2013).

Grain is the other aspect of scale with effects on diversity measures. Consider the organisms as random labelled point events in the plane: a larger sampling unit has an increased chance of containing more species and a greater probability of containing one event of any given label. As grain increases, given a random arrangement, the frequency distribution of the labels in a sample approaches the overall distribution. If the labels are not randomly arranged, but are localized, larger sample units have non-linear increases in richness as the units grow to include more than one

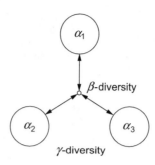

Figure 9.1 A single collection gives α-diversity and the entire area of study gives γ-diversity; the second intermediate level is β-diversity, the diversity among or between primary sites or samples.

environmental sub-area. Each sample unit is affected similarly and the aggregate effects of increasing grain follow the same pattern.

As with any kind of spatial analysis, there may be an advantage to making explicit the quantitative changes in diversity over changing extent or changing grain (see Turgeon *et al.* 2017).

9.1.2 Spatial Location and Environmental Gradients

Location has obvious effects on diversity through the environmental conditions, the species already established and species that have the potential to arrive. The environmental gradients that intersect there also affect how diversity changes through space, at local or geographic levels. The variability among sites may be directly related to site positions on spatially explicit gradients, which will be manifested in their characteristics as a function of location. Site locations can be included in a multivariate analysis of species composition by using x and y coordinates, such as north and west, as independent variables (possibly including $x^2, y^2, xy, \ldots$ as well). This approach takes account of site location and can be used to partial out the variability that is attributable to location, but it does not remove the effects of spatial dependence on statistical tests (see Section 9.1.4) and these location effects may be confounded with the effects of environmental gradients, possibly not measured or considered. Under these circumstances, the diversity among sites may be closely related to the gain or loss of species between sites (Section 9.2.2).

The location may be absolute in referring to position in a frame of reference, such as latitude and longitude; or it may be relative to other sites or objects, such as distances to other islands.

9.1.3 Spatial Heterogeneity

The effects of heterogeneity on diversity may result in a greater diversity of substrates, environmental conditions or resources that should support a greater diversity of species. Where that environmental heterogeneity has a spatial structure such as patches, it may determine the spatial structure of the diversity.

Heterogeneity among sites has several causes, including abiotic environmental factors, stochastic events like colonization and ecological similarity declining with physical distance due to spatial processes like dispersal (Soininen *et al.* 2007; Anderson *et al.* 2011; Keil *et al.* 2021). This kind of heterogeneity is documented in the large amount of literature on β-diversity (composition variability among sites) in its various versions. To be truly spatial, the positions on gradients or the relative locations and distances of the sites should be included in the analysis, in order to quantify that variability in an explicit spatial context. Otherwise, the heterogeneity occurs in a spatial context but it is not preserved in the measures of among-site variability.

9.1.4 Spatial Dependence

The implications for diversity arising from spatial dependence in the relative positions of samples and individuals may be based on the suggestion (which may not be true) that things closer together tend to be more similar than those farther apart (Tobler's Law; Tobler 1970). Spatial dependence is not restricted to simple physical proximity but may include connections between the units and the processes that make them functionally closer (dispersal routes, winds, currents and so on). This dependence can arise in the biological variable of interest, from environmental variables that affect it, or a combination of the two ('innate', 'induced' or 'double' in the language of Chapter 6). It is usually positive at short range, but it can be negative, and with patchiness, it can cycle between positive and negative with increasing distance. The often-cited decrease in similarity with increasing distance, usually as an exponential decay, is only sometimes true, because patchiness can lead to similarity that cycles with distance (see Chapter 6). As with many spatial analyses, explicit mapping of values or plotting relative values as a function of distance may be crucial to correct interpretation.

9.2 Application: Why Spatial Diversity

The basic measures of diversity are designed for a single collection, either a sample or a complete census. Those two cases are different. We must, therefore, distinguish between a *sample* of a large 'population' (in the statistical sense) and a *complete census* of all the organisms of interest in the study area. In a sample, the rarest species may be missed so that the actual species richness is greater than the species count (and unknown), whereas, in a complete census, the true number should be known. Our calculations and interpretation must take account of this difference (Scheiner *et al.* 2000; Chao *et al.* 2014). The analysis and interpretation may also differ between having discrete 'natural sampling units' such as islands, versus samples from a spatially continuous structure. The discrete sites are often portrayed as a spatial and functional network based on the distances between the sites and the movement of individuals or propagules between them (Dale & Fortin 2021). This metacommunity structure will have obvious implications for diversity at and among the sites and the way it is studied (Chapter 1).

The concept of diversity starts with a single site, but we can extend it to the higher levels of organization or greater complexity of spatial structure. To introduce the material that follows, we provide artificial data illustrating the importance of including spatial information in diversity analysis.

> **Example 9.1**
> Figure 9.2 shows three branches of a river, designated as $x = 1, 2, 3$, which are sampled with four samples each, with $y = 1$ being near the confluence and $y = 2, 3$, and 4 further upstream (Tables 9.1 and 9.2). Species A and B occur at the downstream sites while Species J occurs only at the stream heads. Species C and D, E and F, and G and H are found mainly, but not exclusively, in branches $x = 1$, 2, and 3, respectively. Without the spatial locations, the pattern of presences and absences is very difficult to interpret. The explanation appears to lie in the replacement of species along the branches; this could be tested (see Section 9.2.2.4; Matsuoka et al. 2019).

This case is not particularly sophisticated, but it mimics situations encountered in diversity analysis when the spatial structure is not at first explicitly included. We will now consider some of the technicalities of diversity measurement and how these are to be applied in spatial analysis.

9.2.1 α-Diversity

The first level of diversity for a single collection is 'α-diversity' by convention (Whittaker 1977). Site diversity can be based on a number of samples, whether

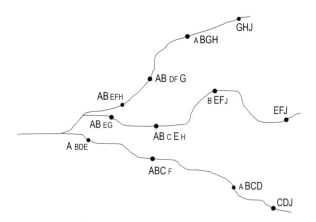

Figure 9.2 The importance of spatial structure in diversity studies: three branches of a river with community compositions for various locations. Letters for species, with the larger font representing the most abundant at each site. The positions in the drainage system help interpret the patterns of diversity.

Table 9.1 Aspatial summary

Sp.→ Site↓	B	A	E	F	C	D	G	H	J	Site Total	(x, y) Location
1	■	■	+	+			+			5	(3, 1)
2	■	■		+		+	■			5	(3, 2)
3	■	■	■		+		+			5	(2, 2)
4	+	■	+			+				4	(1, 1)
5	■	■		+	■					4	(1, 2)
6	■	+			■	■				4	(1, 3)
7	■	■	+				+			4	(2, 1)
8	■	+				■		■		4	(3, 3)
9	+		■	■					+	4	(2, 3)
10					■	■			■	3	(1, 4)
11			■	■					■	3	(2, 4)
12							■	■	■	3	(3, 4)
	9	8	6	5	4	4	4	4	4	← Species Total	

*A square symbol indicates high abundance, a plus sign indicates presence with low abundance and a blank indicates absence.

distributed through space, as quadrats for vegetation, or through time, as trapping sessions for flying insects. As described, α-diversity is affected by scale (grain of sampling and extent) as well as by the spatial intensity of sampling.

For the quantitative *measurement* of diversity, several questions arise: Is this a complete census or a sample? If a sample, what proportion of the site area is covered by samples? Are there measures of abundance or presence–absence data? How certain is the identification of the organisms and the taxonomy of their classification?

In general, assumptions for α-diversity include: (1) all identified taxa are treated as equivalent in their effect on the diversity measurement, (2) the measure of abundance is appropriate, and (3) where 'individuals' are counted, all individuals are treated as equivalent. Measures of diversity tend to combine the number of taxa, 'richness', and the equality of their representation, 'evenness', with various amounts of influence on the index value (Magurran & McGill 2010). Two such measures for abundance data are Simpson's index, D_S, based on probability, and the Shannon–Weaver index, H', based on information theory; both use the proportion of the total abundance that belongs to the ith category of the s in the classification, p_i:

$$D_S = 1 - \sum_{i=1}^{s} p_i^2, \tag{9.1}$$

and

Table 9.2 Spatial summary

x = 1

sp:	A	B	C	D	E	F	G	H	J
y = 1	■	+		+	+				
2	■	■	■			+			
3	+	■	■	■					
4			■	■					■

x = 2

sp:	A	B	C	D	E	F	G	H	J
y = 1	■	■			+		+		
2	■	■	+		■			+	
3		+			■	■			+
4					■	■			■

x = 3

sp:	A	B	C	D	E	F	G	H	J
y = 1	■	■			+	+		+	
2	■	■		+		+	■		
3	+	■					■	■	
4							■	■	■

*A square symbol indicates high abundance, a plus sign indicates presence with low abundance and a blank indicates absence.

$$H' = -\sum_{i=1}^{s} p_i \log_e p_i. \quad (9.2)$$

D_S arose as the probability that two randomly chosen individuals belong to different species. The information theoretic index has several positive features, such as the fact that it can be partitioned when applying more than one classification. Given the many possible measures of diversity, Hill (1973) proposed a scheme for some unification with a general index:

$$N_x = \left[\sum_{i=1}^{s} p_i^x\right]^{\frac{1}{1-x}}. \quad (9.3)$$

For $x = 0$, $N_0 = s$; for $x = 1$, $N_1 = e^{H'}$; and for $x = 2$, $N_2 = (1 - D_S)^{-1}$. N_1 is called the 'effective number' of species, because it is the number of species with equal frequencies that would give the same value of H'. Each of these measures has advantages and disadvantages; as described in Hill's original explication and other summaries (Jost 2007; Chao et al. 2014; Gaggiotti et al. 2018).

9.2.1.1 Spatial Structure in α-Diversity

Some studies require an analysis of diversity's spatial structure at a site, based on specific designs for sampling, such as grids of pitfall traps for beetles, or the direct mapping of organisms, such as mussels on a rocky shore. An alternative to mapping individuals is to use small sample units, either scattered or in grids, with known locations and recorded species composition. For such data, Keil et al. (2021) made a strong case for the importance of the spatially explicit measures of interspecific associations (including distance dependence) for testing biodiversity theories. For mapped labelled point patterns, Shimatani (2001) advocated for an index of diversity based on the probability that 'randomly chosen pairs' of plants belong to different species in multivariate point data, to determine how tree stem diversity changes with distance. The index described as $\alpha(r)$, diversity as a function of distance, r, resembles the transformed Ripley's K for interspecific pairs, $K_{X,\sim X}$ (Chapter 3). In our notation, it is:

$$\hat{K} = \sum_{i=1}^{N} \sum_{j \neq i}^{N} \left[w_{ij} I_t(i,j) \mid m(i) \neq m(j) \right] / N^2. \tag{9.4}$$

This is the corrected count of pairs within distance t that belong to different species. In Shimatani's version, $\alpha(r)$ is the probability that two randomly chosen tree stems, separated by a distance no more than r, belong to different species: essentially Simpson's index with a distance threshold. The two are closely related but the modified Ripley's K takes values -1 to $+1$, whereas $\alpha(r)$ takes values 0 to 1.

9.2.2 β-Diversity

The second level is β-diversity: the variation among sites within a region (Whittaker 1977). Originally, β-diversity was defined relative to γ and α, but β can be calculated independently from the variance in a site-by-species data table (Legendre et al. 2005). The discussion of β-diversity is complicated by the fact that it includes several overlapping concepts (Anderson et al. 2011; Keil et al. 2021). To begin, there is a conceptual difference between a series of sites along an explicit environmental gradient, and a cluster of sites with no gradient effective or evident.

9.2.2.1 β-Diversity without Gradient

The first version of β-diversity, M_β, comes from a partition of total diversity, M_γ:

$$M_\gamma = M_\beta + \sum_{j=1}^{m} q_j M_{\alpha j} = M_\beta + \overline{M_\alpha} \text{ so that} \tag{9.5}$$

$$M_\beta = M_\gamma - \overline{M_\alpha}. \tag{9.6}$$

Depending on the measure, say M' instead of M, the relationship may be:

$$M'_\beta = M'_\gamma \div \overline{M'_\alpha}. \tag{9.7}$$

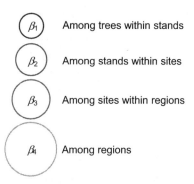

Figure 9.3 A possible hierarchy of levels of β-diversity, denoted by subscripts, based on spatial or organizational scale, such as trees within stands, stands within sites, and so on.

In Equation (9.6) all three measures have the same units, such as 'species equivalents', but in Equation (9.7), α- and γ-diversity have the same units but β-diversity has none because it is a ratio.

If the three levels of α, β and γ are insufficient, β-diversity can be subdivided. In a study of tree canopy beetles, Crist et al. (2003) used four sub-levels: β_1, among trees within stands; β_2, stands within sites; β_3, sites within regions; and β_4, among regions (Figure 9.3). Identifying sub-levels may be particularly useful for partitioning diversity related to physical or geographic structuring (Legendre et al. 2005; Legendre & De Cáceres 2013).

To include the spatial structure explicitly in the analysis, start with records for each sampled site of its location in x–y coordinates and relevant characteristics such as area, elevation and so on. One approach is then to calculate a measure of diversity for each site and compare site-pair diversity differences with their geographic distances. The weakness here is that two sites may have similar diversity values even when the species lists are very different, reminding us that α-diversity is a synthetic variable. It is better to use species composition directly, calculating a dissimilarity coefficient as a β-diversity measure of composition differences and comparing those with geographic distances by the Mantel test. The introduction of pairwise site similarities leads to the concept of species composition complementarity to be discussed in Section 9.3.

9.2.2.2 β-Diversity with Spatially Explicit Gradient

This version begins with sites arranged on an explicit environmental gradient, with some species present at the beginning of the gradient dropping out farther along (losses $= l$) and others that are absent at the beginning being added (gains $= g$). Those gained may be lost farther along, as illustrated in Figure 9.4. Gains and losses of species at sites along a gradient are referred to as 'turnover', but it is conceivable that a species with a bimodal distribution could appear and disappear more than once. The concept may be expressed as species replacement rather than turnover. The associated measures are usually the net losses and net gains from beginning to end. This turnover version of β-diversity, called β_τ, is spatially explicit. Other formulations of β_τ are possible, based on calculations comparing among-site diversity to within-site diversity. With a spatial gradient, this assessment of β-diversity is spatial analysis in a single dimension.

9.2 Application: Why Spatial Diversity

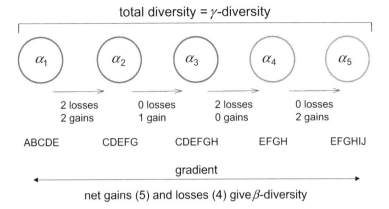

Figure 9.4 The concept of β-diversity based on species turnover along a gradient. Here are five sites with species compositions given by lists (species A to J), showing losses and gains of species between adjacent sites. Turnover β-diversity is the total net losses and net gains, each compared to the average richness (5.2) for the sites: $\beta_\tau = (4+5)/(2 \times 5.2) = 0.87$.

In comparing the no-gradient version with the explicit-gradient version, Vellend (2016) showed that if species turnover is limited to losses and gains between the first sample and the last, and if the richness in the terminal plots is close to the average, then both versions measure the same characteristic. This equivalence is dependent on the spatial arrangement of the plots which must reflect the change of species along a gradient. One consequence of the spatial structure is that the similarity between plots decreases with distance between them. This is a safe assumption for samples on an environmental gradient but less easily defended generally because the patchiness of natural systems may work against it (Nekola & White 1999). Yet, if the gradient is long enough, the ends of the gradient may appear similar because they both lack species that are common in the middle, something like the horseshoe effect in ordination (Legendre & Legendre 2012).

With a single spatially explicit gradient, it seems easy to relate changes in composition and diversity to a unidirectional change in controlling environmental factors, but natural gradients are almost always very complex, with many different physical factors changing along the gradient, some increasing and others decreasing. Likewise, the diversity of different components of the natural communities along a gradient may change in very different ways.

In spatially expressed gradients, physical closeness brings a similarity of conditions encountered without great differences in environmental conditions. In this situation, 'next to' combines spatial and environmental proximity and the connection is made easier by the gradient's single dimension of functional importance. In this way, at least, these spatially explicit gradients resemble temporal series.

9.2.2.3 β-Diversity with More Than One Gradient

The possibilities and difficulties for β-diversity analysis on a single explicit gradient are amplified when there is more than one. This is like the 'no clear gradient' version and the spatial structure of diversity can be considered as 'spatial variation in β diversity' (Koleff et al. 2003). For more than one gradient, partitioning may indicate which parts of that spatial variation are attributable to which of the gradients (Legendre et al. 2005; Legendre 2008), which will be especially useful when the contributions of individual sites to β-diversity can be mapped as a function of location and thus related to the environmental gradients (see Legendre & De Cáceres 2013). Turgeon et al. (2017) provided a good illustration of this approach in a multi-scale analysis of fish community assemblage responses to impoundments on boreal rivers. The dams caused nearshore fish community assemblages to change in composition but diversity was little affected. This approach shows much promise for future applications of spatial diversity analysis.

9.2.2.4 β-Diversity: Pairwise Relations

To compare the diversity of pairs of sites, we can use the β-diversity measures given with $n = 2$. For presence–absence data, the pair measure of compositional difference is a kind of 'turnover' (β_τ), a count of species not in common. With s_i species at one site and s_j at the other, and $s_{i\&j}$ at both, that count is:

$$\beta_\tau = s_i + s_j - 2s_{i\&j}, \tag{9.9}$$

with the proportional change being,

$$\beta_p = (s_i + s_j - 2s_{i\&j})/(s_i + s_j - s_{i\&j}). \tag{9.10}$$

The latter is equivalent to the familiar Jaccard dissimilarity measure, which is the proportion of all species found only at one site.

For abundance data, the correlation coefficient of the species abundances at one or both sites is a good candidate measure of similarity, with its complement being a measure of diversity, but there are many possibilities for measures of differences in multivariate composition (Legendre & Legendre 2012).

Baselga (2010) proposed that for any pair of sites, β-diversity, measured by the Sørensen index, β_{SOR}, can be partitioned into components attributable to turnover and nestedness. Nestedness is the tendency for species-poor sites to have only the most common species and for rare species to occur only at the richest sites. The proposed partition is:

$$\beta_{SOR} = (b+c)/(2a+b+c) = \beta_{TUR} + \beta_{NES}, \text{ with } \beta_{TUR} = \min(b,c)/[a + \min(b,c)], \tag{9.11}$$

and

$$\beta_{NES} = [|b-c| \div (2a+b+c)] \times \{a/[a + \min(b,c)]\}. \tag{9.12}$$

Consider Example 1; the intuitive assessment was that, despite some nestedness, turnover along the branches of the river was likely to dominate. Here, $\beta_{SOR} = 0.75$, which is partitioned into $\beta_{TUR} = 0.5063$ and $\beta_{NES} = 0.2437$, suggesting that turnover has twice the influence of nestedness on the observed pattern. These measures of

9.2 Application: Why Spatial Diversity

diversity are aspatial, but they can clearly have a spatial interpretation in a context like this.

In a similar application, Rouquette *et al.* (2013) compared taxonomic groups in the response of the Jaccard distance between sites to different measures of physical distance (Euclidean, connective network and flow). Based on Mantel tests, with and without removing the effects of environmental distance, they found that the taxonomic groups responded differently, with macroinvertebrates and diatoms most closely aligned with network distance. Butterflies did not respond to any of the distance measures, nor did plants once the environmental distance was removed. As a point of clarification, these authors refer to the Jaccard distance as 'species turnover', which does not match the Baselga (2010) use of the term.

In a study at a larger scale, Pinto-Ledezma *et al.* (2018) investigated the relationship between the turnover partition of β_{SOR}, β_{SIM} in their notation, and biome boundaries across North America. The spatially explicit results are geographic maps with red-to-blue coded cells indicating the relative importance of β_{SIM} and β_{NES} for three forms of diversity: taxonomic, functional or phylogenetic. The results for all three were highly congruent based on spatial regressions (Chapter 7), with β_{SIM} higher in the dry temperate biomes in the south and β_{NES} higher in the polar biomes of the north.

It seems that the most effective method to study the spatial aspects of β-diversity is to focus on pairs of sites, however designated, and then base more general interpretations on the results from these pairs. This moves beyond the derivation of β-diversity from α- and γ-diversity, but it offers many advantages for spatial analysis. One important application of *site-pair β-diversity* (SPBD) evaluation is the creation of weights for the edges of spatial graphs depicting diversity characteristics, or to look at very local patterns of diversity. Legendre and De Cáceres (2013) showed how the matrix of pairwise dissimilarity coefficients can be used as the basis for the overall evaluation of β-diversity in the entire data set. From this derives the important technique of quantifying the site-focussed *local contributions to β-diversity* (LCBD) and the equivalent organism-focussed *species contributions to β-diversity* (SCBD). The values of LCBD can then be plotted on the map of the sites (Legendre & Condit 2019) or associated with the nodes of the spatial graph. Thus, a complete picture can be achieved with LCBD scores on the nodes and the SPBD values on the edges. The same basic approach can be modified to examine and evaluate directional changes in presence–absence data (Schmera *et al.* 2022); this is an area of ongoing research and recent advances.

An alternative to assessing diversity based on pairs of sites is *zeta* (ζ) diversity partitioning, which is based on the number of species shared by several assemblages (sites); the ζ component designated ζ_i is the mean number of species shared by i sites (Hui & McGeoch 2014). The familiar pairwise β-diversity measures can be expressed by combinations of ζ_1 and ζ_2, but the higher order ζ_i components estimate the partition of species subsets shared by a particular set of sites. This approach to diversity represents species occupancy and turnover in independent samples and can be shown to scale with sample size, grain and distance (Hui & McGeoch 2014). There is a clear conceptual relationship between the number of shared species and the compositional diversity discussed in Section 9.3. Zeta diversity has been applied to study the gradients driving species turnover (Latombe et al. 2017) but more spatially explicit applications should be investigated.

9.2.3 γ-Diversity

The level of γ-diversity applies to the largest extent, the entire study region. The effect of spatial extent on γ-diversity is a frequent focus of studies, but spatial grain also affects it, especially through the ability to detect or quantify rare species. It is also affected by spatial partitioning, which at evolutionary spatial and temporal scales may play an important part in diversification and species coexistence, leading to higher overall diversity. Spatial diversity analysis can be applied to features other than taxonomic classification, such as plants' spectral response to solar radiation (Laliberté et al. 2020). Spectral diversity integrates chemical, structural and taxonomic characteristics of the plants in a regional set of communities. Data used in measures of γ-diversity can be remotely sensed and then partitioned at appropriate scales into α- and β-diversity, allowing the detection of spectrally unique areas that may be the foci for conservation or further study (Laliberté et al. 2020).

The ecological subdiscipline of macroecology has matured in recent years; it studies the large-scale relationships between organisms and environment, focusing on abundance, distribution and diversity (Brown 1995; Gaston & Blackburn 2000; Shade et al. 2018). Although there is a clear link between some aspects of macroecology and some topics in spatial analysis, such as the spatial turnover of species, there are many within macroecology that are not comparable, such as the relationships between body size and extinction. Macroecology and spatial analysis overlap most in studying the factors that determine γ-diversity and the possible roles of physical structure, such as insularity or environmental gradients, spatial extent and spatial autocorrelation. Shade et al. (2018) expressed an ambitious proposal for macroecology as a unifying science for all forms of life and possibly all combinations, which leads us to the next topic of species combinations and community composition.

9.3 Combinations and Composition: Agreement and Complementarity

The concept of spatial structure being determined by the positions of organisms of different species relates to the determination of species associations, either pairwise or multivariate. For multivariate cases with presence–absence data for s species, we examine 2^k tables to detect combinations of k of the s species that are unexpectedly rare or unexpectedly common. Although the idea of ecological diversity originated with the unpredictability of single species, it can be extended to the unpredictability of species assemblages. We can measure (a) the diversity of the *species combinations* represented by vectors of presence–absence data, such as $\mathbf{p}_j = (1, 1, 0, 0, 1, \ldots)$ or (b) the diversity of *species compositions* represented by vectors of abundances, such as $\mathbf{a}_j = (10, 12, 0, 0, 3, \ldots)$.

With *combinations* as the basis for sample diversity, we are concerned with identifiable sets of species in presence–absence data and the analysis proceeds through the mathematics of combinatorics. With *composition* as the basis, this is multivariate quantitative data and the analysis resembles multivariate techniques such as ordination. The evaluation of sample diversity makes sense both at the level of a single collection (α) and at the level of comparing several sites within a region (β).

9.3.1 Species Combinations

To introduce species combination diversity, consider the frequency of pairs or multiples of species occurring in spatial proximity. To evaluate multispecies combinations, create 2^k tables of co-occurrence counts ($2 < k \leq s$) and test for significant departure from independence with the G statistic and $2^k - k - 1$ degrees of freedom. A significant result overall can be interpreted using Freeman–Tukey standardized residuals to indicate which combinations are unexpectedly rare or unexpectedly common. For eight species, combinations such as no species or all species are likely to be rare, whereas those with four species may be unexpectedly common. The count will be sensitive to the size of the sample units because larger units tend to contain more species.

Species combinations can be presented as stacks of boxes or circles, like binary tally sticks, filled for presences and empty for absences (Figure 9.5). We can then count the number of unique combinations without concern for similarity or

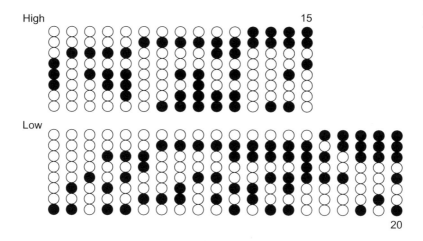

Figure 9.5 (redrawn from Dale 1999, figure 5.8c). Each column of eight circles represents the presence (solid) or absence (hollow) of each of eight species in a combination that occurs with significantly high (15 combinations) or significantly low (20 combinations) frequency based on the results of 2^k contingency table analysis. The data are from the SE Lyall moraines in the Canadian Rockies (MacIsaac 1989; Dale 1999).

complementarity among them, or we may include an evaluation of how similar or different the combinations are. Figure 9.5 (redrawn from Dale 1999, figure 5.8c) shows the 'significantly' high- and low-frequency combinations of eight species from SE Lyall moraines in the Canadian Rockies (MacIsaac 1989). The 35 apparently significant results are convincingly more than expected (see Dale et al. 1991) and should indicate real effects. The spatial relationship of complementary combinations will be of interest in the search for 'checkerboard' patterns of species occurrence, which may indicate competitive interactions as drivers of community composition. Association analysis using 2^k tables can reveal details about the spatial relationships of species not available from other methods.

For the diversity of species combinations or *combinatorial diversity*, consider these two data sets of eight samples, each having seven species (Table 9.3). In the first case,

Table 9.3 Examples of different incidence combinations

Set 1: Species × Samples $(F = -\Sigma p_i \log p_i = 1.86. \ e^F = 6.39)$

Samples: Species:	1	2	3	4	5	6	7	8	Σ
A	1	1	1	1	1	1	1	1	8
B	1	1	1	0	0	1	1	1	6
C	1	1	1	1	1	0	0	1	6
D	1	1	0	0	1	1	1	1	6
E	1	0	1	1	0	0	1	1	5
F	0	1	1	1	1	1	0	0	5
G	1	0	0	0	0	0	0	0	1
Σ	6	5	5	4	4	4	4	5	37

Set 2: Species × Samples $(F = -\Sigma p_i \log p_i = 1.91. \ e^F = 6.73)$

Samples: Species:	1	2	3	4	5	6	7	8	Σ
A	1	1	1	1	1	1	1	1	8
B	1	1	1	1	1	1	1	1	8
C	1	1	1	1	1	1	1	1	8
D	1	1	1	1	1	1	1	1	8
E	1	1	1	1	1	1	1	1	8
F	1	1	1	1	0	0	0	0	4
G	0	0	0	0	1	1	1	1	4
Σ	6	6	6	6	6	6	6	6	48

there are eight different incidence combinations (presences and absences) but only one instance of each.

The second case has only two distinct incidence vectors (the columns), but there are four instances of each. Designate the number of distinct combinations (s species in t samples) as K, with κ_m repetitions of the mth combination so that $p_m = \kappa_m/t$. Using an information-theoretic measure (like the Shannon–Weaver index, but base 2), Juhász-Nagy (1993) introduced a measure for the diversity of *combinations*:

$$F = -\sum_{k=1}^{K} p_k \log_2 p_k. \tag{9.13}$$

This considers only the number of distinct combinations, not how different they are; two very similar combinations and two very different combinations both count as distinct. It may be useful to include a measure of their similarity and one approach is to calculate the similarity between all pairs of incidence vectors, with the Jaccard coefficient, $a/(a + b + c)$, for samples or species, and then average the pairwise values for each set (Koleff *et al.* 2003; Legendre & Legendre 2012).

An extension is to evaluate a 2^s table of the frequencies of possible incidence combinations of the s species simultaneously, rather than the pairwise summaries. That table has s dimensions, one for each species. We can use the differences and similarities and the lack of independence among the observed combinations, to determine the 'effective number' of combinations; that is the number of independent combinations equivalent to the non-independent combinations observed, of which there are t. In a set that contains all 2^s possible incidence combinations, only s can be independent, so that $t' = s$, even if t is greater than s. For a set of only two combinations, $t' = t = 2$ if they are independent. The effective number of combinations should account for the similarity of combinations and their evenness: the more equal the frequencies, the greater the effective number. Based on this thinking (Hill 1973), if H is any entropy measure of species combinations' diversity, then $M_1 = e^H$ estimates the effective number.

From different starting points, Ricotta (2006) and Jost (2006) suggested a similar conclusion when their results are combined. To account for similarities among combinations, Ricotta (2006) suggested modifying the basic species combination diversity by including the dissimilarity of the mth and kth vectors as d_{mk}:

$$F = -\sum_{k=1}^{K} p_k \log_2 \left(1 - \sum_{m \neq k}^{K} d_{mk} p_m \right). \tag{9.14}$$

Jost (2006) extended Hill's (1973) concept of effective species numbers to suggest that the exponential forms of entropy-based indices provide measures of diversity as 'effective number'. Combining the two ideas, and using the Jaccard coefficient, the effective number of species combinations is M_1, calculated as:

$$F' = -\sum_{k=1}^{K} p_k \log_e \left(1 - \sum_{m \neq k}^{K} [1 - J_{mk}] p_m \right) \text{ and} \tag{9.15}$$

Table 9.4 More examples of incidence combinations

Set 3: Species × Samples $(F = -\Sigma p_i \log p_i = 1.05.\ e^F = 2.87;\ M_1 = 1.5)$

Samples: Species:	1	2	3	4	5	6	7	8	Σ
A	1	1	1	1	1	1	1	1	8
B	0	0	0	0	1	1	1	1	4
C	1	1	1	1	1	1	1	1	8
Σ	2	2	2	2	3	3	3	3	20

Set 4: Species × Samples $(F = -\Sigma p_i \log p_i = 1.59.\ e^F = 4.94;\ M_1 = 1.85)$

Samples: Species:	1	2	3	4	5	Σ
A	1	1	1	0	0	3
B	1	1	0	0	1	3
C	1	1	1	1	0	4
D	1	1	1	0	1	4
E	1	0	1	1	0	3
Σ	5	4	4	2	2	17

$$M_1 = \exp(F'). \tag{9.16}$$

This measure responds to the similarity of the combinations and to the equality of their frequencies, as illustrated by Tables 9.3 and 9.4. Set 1 gives $M_1 = 1.72$; Set 2 gives $M_1 = 1.16$, because the combinations are so similar. Set 3 gives $M_1 = 1.5$, which is between 1.0 and 2.0, as suggested. Set 4, by comparison, illustrates the property that species diversity and combinatorial diversity are quite independent; here $e^F = 4.94$ but $M_1 = 1.85$.

This 'effective number of combinations' measure has many good features: simplicity, relationships with other effective numbers and measures of species diversity and reflecting both combination similarity and evenness.

Neither the entropy measure nor the effective number transformation takes account of the locations of the combinations; they are not spatial. The locations of species combinations and the relationships between them become important in the study of the combinatorics of species co-occurrences (non-random versus random) as applied to the assembly of communities. There are at least two kinds of non-randomness to be considered: the non-randomness of the combinations observed and, given that, the non-randomness of their locations. The spatial structure of combinatorial diversity will be discussed after some further comments on compositional diversity.

9.3.2 Nested Subsets, Constrained Compositional Diversity

9.3.2.1 Compositional Agreement and Complementarity

The concept of diversity is about unpredictability, but the unpredictability of species assemblages may be constrained by factors including the phenomenon of nestedness: rare species occur only at the richest sites and the species-poor sites contain only the most common species. The species sets at depauperate sites are therefore proper subsets of those of richer sites, producing nested subsets. The usual way of depicting this relationship is by a grid of presences (1s) and absences (0s), with species ordered from most common to most rare, and sites from species-rich to species-poor. If the nestedness is perfectly ordered (Figure 9.6a), the presences will be neatly packed into the top left part of the diagram as determined by the numbers of sites (m), species (s) and total occurrences (P). The significance of departures from this order (Figure 9.6b) can be determined by a restricted randomization procedure in which the frequency of each species, p_i, is maintained and presences assigned to sites by probability weights.

The causes of this nestedness pattern relate to the total area available, the range of substrates or environmental conditions, species' responses to factors determining extinction, and the site characteristics that affect establishment or recolonization, such as island size, distance from the mainland source of colonizers and isolation from other islands. Isolation influences nestedness and is a clearly spatial factor. For

(a) Perfectly nested: nestedness discordance = 0.

Sites Species	a	b	c	d	e	f	g	h	i	j	k	l	Frequency
A	1	1	1	1	1	1	1	1	1	1	1	1	12
B	1	1	1	1	1	1	1	1	1	1			10
C	1	1	1	1	1	1	1	1					8
D	1	1	1	1	1	1			0				6
E	1	1	1	1			...						4
F	1	1					0						2
Richness	6	6	5	5	4	4	3	3	2	2	1	1	Σ = 42

(b) Less nested: nestedness discordance (1's below 0's in upper triangle) = 7.

Sites Species	a	b	c	d	e	f	g	h	i	j	k	l	Frequency
A	1	1	1	1	1	1	1	1	1	1	1	1	12
B	1	1	1		1	1		1	1	1			8
C	1	1	1	1	1		1	1					7
D	1		1	1	1	1		1					6
E	1	1	1	1			1						5
F	1	1				1				1			4
Richness	6	5	5	4	4	4	3	3	2	2	1		Σ = 42

Figure 9.6 Nested subsets. (a) The stepped diagonal indicated by a broken line separates the block of presences from the block of absences; its shape and position will change with the number of species, number of sites and their ratio. In (b), the shaded cells are those of 'unexpected' absences or presences based on the hypothesis of perfect nestedness.

example, Murakami and Hirao (2010) investigated the causes of nestedness in insects on Bahamian islands. Spatial measures related to isolation included distance to the mainland, over-water distance to mainland, distance to the nearest neighbouring island in the network and total area of neighbour islands within one of three threshold distances (250, 500 and 1,000 metres). All insect groups showed that at least one of the spatial variables had significant effects on nestedness.

The spatial aspects of nestedness will be most evident when the driving forces are spatially contagious, and the spatial pattern of nestedness will have a signature of its own. The concept of nestedness goes beyond 'nested: yes or no?' and measures of partial nestedness can be helpful. We recommend *node overlap, decrease filling* (NODF; Almeida-Neto et al. 2008). It is 1.0 when the array is perfectly nested and the value 0.0 when the array is perfectly modular with exclusive species lists for subsets of sites and exclusive site lists for subsets of species. To use this measure for insightful spatial analysis we need (1) a measure of the nestedness of any individual site compared to the whole array and (2) a measure of the relative nestedness of any pair of sites.

Box 9.1 gives the details of three ways of creating a site nestedness score, comparing the site with the whole data set, for (a) presence–absence data, (b) frequency data based on occurrence at sub-sites and (c) abundance data. The site nestedness score, thus derived, can be colour-coded for mapping and subsequent spatial analysis, with blue indicating high agreement with the overall ranking and red indicating strong disagreement.

A nestedness score for each species or each site can be calculated by the concordance measure of its abundance ranking with the overall frequency ranking. For assessing spatial relationships, a relative nestedness score can be derived for each pair of sites. One choice is an asymmetric matching coefficient, explained in Box 9.2. The overall NODF measure is essentially the average of the relative nestedness measures for all pairs of sites and all pairs of species. In a spatial graph with the sites as nodes each edge can be weighted by the relative nestedness measure of the pair of sites (Figure 9.7).

Box 9.1 Site Nestedness Scores

Presence–absence data for sites
Wilcoxon 2-sample test statistic, U, based on ranks sum R;
site score, $v_U = U/n_1n_0$, runs from 0 to 1.

Presence–absence data for sub-sites (gives site frequency data)
Count rank discordances: D; D_{max} is $s(s+1)/2$.
As a site score, $v_D = 1 - D/D_{max}$ runs from 0 to 1.

Abundance data at sites
Pearson's correlation coefficient serves as a measure of similarity and then the site score of $v_r = (r+1)/2$ runs from 0 to 1.

9.3 Combinations and Composition Comparison

> **Box 9.2** Measures of Relative Nestedness for Pairs of Sites
>
> **Unequal Site Richness**
>
> With a_{ij} as the number of species at the poorer site i also occurring at site j, $s_i < s_j$ and b_{ij} as the number at site i not at site j, a measure of relative nestedness is $m_{ij} = (a_{ij} - b_{ij})/s_{ij}$ where m_{ij} takes values between -1 and $+1$.
>
> **Equal Site Richness**
>
> For $s_i = s_j$, the measure is a symmetric matching coefficient of presences. With a_{ij} and b_{ij} as for the unequal case, because $s_i = s_j$, $a_{ij} = a_{ji}$ and $b_{ij} = b_{ji}$. A measure of relative nestedness for sites of equal richness is $m_{ij} = m_{ji}$ as defined for unequal richness where m_{ij} takes values between -1 and $+1$.

For a more complete spatial nestedness analysis, a spatial graph of any neighbour network is produced from the data (Chapter 2); each node has a nestedness score and each edge has a relative nestedness score (Figure 9.8). Spatial clustering or boundary detection methods can then delimit sets of sites by their nestedness characteristics, based on those of scores. The score map may also be used to determine 'hot' or 'cold' regions of nestedness, with anomalies indicating areas for further research or conservation efforts.

Although the nestedness concept was originally aspatial, there is obviously a close relationship between the assessment of nestedness in communities and spatial analysis. It is also obvious that this is an area of spatial theory and analysis in which there is more work to be done.

In the 'non-spatial' discussion, we described nestedness as an important characteristic of communities with overlapping species lists. Much of the interest in this topic is

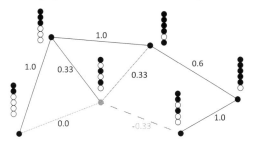

Figure 9.7 The nodes (sites) are coded by site nestedness score: black for high values (near 1.0) and grey for low (near 0.0). Weights on the edges of a spatial graph, giving the relative nestedness scores of neighbour sites. The edge depicted by a broken line is the only negative value. Some pairs are perfectly nested with a weight of 1.0.

Spatial Diversity Analysis

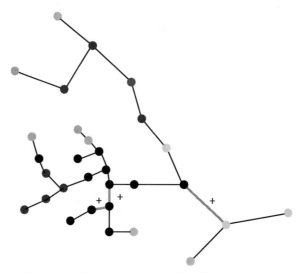

All edges negative except three as indicated by the symbol '+'
Nodes coding: black = (0.90 - 1.0);
dark = (0.80 – 0.90); light = (0.50 – 0.80)

Figure 9.8 Spatial graph of the edge of the Minimum Spanning Tree showing nestedness analysis for Tenebrionid beetles on Aegean islands (data from Fattorini 2007): nestedness scores for sites indicated by shading of nodes (black 0.90–1.0; dark grey 0.80–0.90; light grey 0.50–0.80), relative nestedness of neighbours all negative except for the three labelled with plus signs.

related to conservation for which a spatial context, such as an environmental gradient or dispersal corridors for recolonization, has important implications. In a study of orchids on Réunion Island, Jacquemyn et al. (2007) found significant nestedness but only within altitudinal ranges, indicating the importance of spatial structure for interpreting nested subsets.

One motivation for spatial analysis of nestedness is to detect the spatial 'signatures' of different processes that contribute to the phenomenon. Given an archipelago of vacant islands, sequential colonization by a species with different colonizing abilities will produce a pattern that is well-nested with high neighbour nestedness. On the other hand, given an archipelago with all islands starting with the same large set of species with different vulnerabilities to extinction, the resulting pattern will be well-nested overall if some sites retain most species, but neighbour nestedness will be low due to spatially independent extinctions. If extinctions are frequent at all sites, all species lists may be small subsets of the original and overall nestedness may be weak with low neighbour nestedness.

9.4 Spatial Diversity: Putting It All Together

One major conclusion from considering the spatial aspects of diversity analysis is that indices of diversity or measures of richness, being synthetic variables, can conceal

9.4 Spatial Diversity: Putting It All Together

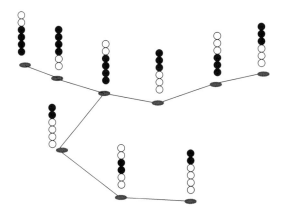

First-order neighbours are similar in richness but not in composition.
Second-order neighbours tend to have similar composition.
Average Jaccard's are 0.10 for first-order and 0.77 for second-order.

Figure 9.9 Artificial example showing the similarity of sites as spatial graph nodes. First-order neighbours are similar in richness but not in composition (average Jaccard's coefficient 0.10); second-order neighbours have similar compositions (average Jaccard's coefficient 0.77).

much important information about the spatial relationships among samples, sites or regions. It is tempting to interpret similarities in richness or diversity in a way that should be based on species composition: sites that are first-order spatial neighbours may have similar diversity or richness, although the second-order spatial neighbours have greater similarity in composition. Figure 9.9 provides an artificial illustration of this point, using richness as a simple version of diversity. Most joins are between sites with the same richness (6 of the 9), but they are of low similarity (using Jaccard's index: 0/4, 0/6 or 2/6), and all uniquely second-order joins have a higher similarity (2/4). This point about indices is not new, but the inclusion of the spatial context as an explicit part of the analysis emphasizes the conclusion. A corollary is that the spatial component of diversity analysis is often crucial to understanding the patterns in the data. The converse is that if diversity were just a number, then its spatial analysis would resemble the spatial analysis of any quantity, with trends and spatial autocorrelation. It is the complexity of the factors that contribute to diversity that makes its spatial analysis interesting. That also explains why we spent such effort on the basic analysis of species combinations and species composition diversity.

Figures 9.7 and 9.8 suggest that the best approach to spatial analysis of diversity is through spatial graphs, as previously described. Start with a set of sites each with location and a list of the species present, with or without species' abundances. These can be analysed by assessing the relationship between composition and location (latitude, longitude, altitude, etc.) or they can be treated as the nodes of a spatial graph, with edges determined in any of several ways (physical distance, topological considerations, functional connection, etc.). These edges define which pairs of nodes are neighbours and so form the basis of any analysis that involves neighbours or neighbourhoods. Consider Shimatani's approach to spatial diversity analysis (Section 9.2.1) which is closely related to Ripley's multivariate K (Chapter 4): it evaluates the probability that two randomly chosen individuals within a given distance of separation belong to different species. That is the same as taking the mapped locations of the individuals as the nodes of a spatial graph and defining its edges and the nodes' neighbours by a threshold of the distance being tested. Thus, the method uses a distance-defined spatial graph.

We can summarize the approaches described in this chapter based on several different categories as applied to the creation and analysis of a spatial graph (Box 9.3). The first category is a general description of the 'property' or ecological characteristic being investigated, including such things as neighbour diversity, diversity as a function of scale (Shimatani's K), spatial structure of richness or diversity of species combinations, nestedness and so on. The next category is the identity of the units that are nodes in the spatial graph: individual organisms, sites or samples. Having identified the nodes, the next category is the rule used to determine which pairs of nodes are joined by edges. The range is large, from the 'empty graph' with no edges to a complete graph with all pairs joined, but the most usual choices lie between the minimum spanning tree (MST) and the Delaunay triangulation (DT), with approximately two to six edges at each node (Chapter 2). Sparser graphs, such as nearest neighbours, are less useful because they are not fully connected. Another method to assign edges is with a threshold physical distance, creating edges between all pairs that are closer than that threshold. Both the nodes and the edges can have labels or weights associated with them, and the next two categories in our classification can be the node label, such as a site richness value, and the edge label, such as relative nestedness. We can include categories of overall purpose, such as cluster identification or boundary detection, and appropriate test procedures.

Comparing the results from two graphs with the same nodes but different edge rules allow us to test different hypotheses about the forces structuring diversity in the system being studied. For example, if there is a much greater similarity of species composition between pairs of first and second neighbour sites in a graph based on functional connections like river flow than in a graph based on geographic distance, this provides good evidence of the dispersal of the organisms by water rather than by air.

We have included the concept of edge diversity in the table, although we have not described it in detail. For spatial graphs, the edge labels will most often be weights, such as distance or flow capacity, or a comparative statistic like relative nestedness, but could be categorical, such as hedgerow versus stone wall for corridors between habitat patches. In either case, edge diversity may be useful in the evaluation of spatial heterogeneity or in assessing the correlation of node label and edge characteristics. When the edges are ecological interactions within communities, spatial aspects of edge diversity are covered in the assessment of the β-diversity of interaction dissimilarity (see Poisot et al. 2012). This is another topic worthy of further and ongoing investigation. It will be discussed from another angle when metawebs are related to multilayer networks in Chapter 11.

Box 9.3 Summary of Characteristics for Diversity Analysis with Spatial Graphs

Property	Node	Edge rule	Node label	Edge label	Graph character	Test procedure
Neighbour diversity	Individual	Any TR	Species	None	Species differences Location differences	Compare counts
Shimatani's K	Individual	Distance thresholds	Species	None	Diversity versus scale	Counts versus distance
Richness	Site	Any TR or complete	Site richness, s_j	β_p (site pair beta diversity for p/a)	Local variation Hot spots or regions	Index; cluster or Boundary detection; sp.ac. versus path length; LCBD evaluation
1° diversity	Site	Any TR or complete	Site diversity, M_j or LCBD	β_d (site pair beta diversity for abundances)	Among-site diversity Hot spots or regions	Index; cluster or boundary detection sp.ac. versus path length; LCBD evaluation
Composition	Site	Any TR or complete	Tally stick p-a	Difference tally	Regional differences, trends	Clusters or boundaries sp. ac. versus path length
Composition	Site	Distance threshold	Tally stick p-a	None	Distance decay	Similarity versus distance
Combinatorial diversity	Site	Complete	Tally stick p-a	None	Florula diversity	Significance of differences
Nestedness	Site	Any TR	Site score	Direction of difference? Relative nestedness	Hot spots or regions	Clusters or boundaries or trends; sp. ac. versus path length
Composition	Site	Functional connection (rivers, corridors, ...)	Tally stick p-a; Abundance vector	Direction of difference?	Neighbour similarity	Similarity versus path length; trends
Edge Diversity	Site	Any TR	Any	Category	Local heterogeneity	Category diversity index
Edge Diversity	Site	Any TR	Any	Weight	Local Heterogeneity	Weight variance

Key:
p-a = presence–absence data.
sp.ac. = spatial autocorrelation.
TR = topological rule.
Column 1: the ecological characteristic investigated.
Column 2: the units that are nodes.
Column 3: the rule for determining edges. TR = topological rule (e.g. minimum spanning tree, MST or Delaunay triangulation, DT); distance threshold rule or complete (all possible edges).
Column 4: node label, a value (e.g., species richness) or a list or vector (e.g., a species list depicted as a tally stick), LCBD = local contributions to β-diversity.
Column 5: edge label, e.g., similarity index or relative nestedness of two sites.
Column 6: the graph characteristic of focus, e.g., detecting hot spots, the scale of spatial autocorrelation or homogeneous regions.
Column 7: the test procedure to be followed.

9.5 Concluding Remarks

9.5.1 Temporal Aspects

Diversity can also be located or partitioned in time. Although temporal partitioning is uncommon in ecological studies, it is a more frequent topic in evolutionary and paleontological literature (cf. Rosenzweig 1995). Temporal partitioning is not that different from spatial partitioning, sharing the difficulties of dealing with continuous variation that can be sampled at a range of different scales, depending on the sample window that is used and whether the templates are applied with overlap, contiguously or with gaps between them. In ecological studies of plant systems, such as a study of community succession, it is much more common to track diversity as a function of time or site age, rather than partitioning total observed diversity into temporal subsets. There are many examples of this approach (Huston 1994) but here we need to acknowledge that diversity is not a real property of the community; it is a derived variable that is a short-hand and imperfect summary of more complex characteristics of the community. This is particularly true for changes in this measure through time because the components that contribute to changes in diversity may themselves change. Large changes in the species list may not produce large changes in the diversity we measure. Similarity in diversity, at the early and late stages of succession, may tell us little about the composition and functioning of the community. The derived nature of the variable does not exclude it as a characteristic of interest or importance, particularly for changes in diversity measures through time and the factors that affect them. Other related characteristics, such as nestedness, are not static variables either and their dynamics should be included. For example, in a study of the nestedness of archipelago floras in the Bahamas, Morrison (2013) found that nestedness changed little over time, despite changes in the species composition itself.

9.5.2 Complexity

As already stated, diversity is a key factor in the complexity of the systems we study. Ladyman and Wiesner (2020) have suggested some general features that can determine the complexity of a system. The list includes (a) structural features, such as many units; unit type diversity or disorderliness; and nestedness and modularity in the system's network; (b) short-term dynamics including feedback loops (positive or negative), non-linear responses and non-equilibrium or conditional stability; and (c) long-term dynamics such as spontaneous order or self-organization, adaptive behaviour and system memory. All these features are part of, or contribute to, biological diversity of different kinds in ecological systems (see also Landi *et al.* 2018). Clearly, the spatial analysis of diversity is itself complex but a key part of the understanding of ecological systems. We have shown that including the spatial component of diversity will provide a clearer understanding of this characteristic and its measurement. Of course, that may not be easy.

9.5.3 Space and Time

The final comment here follows a trend throughout this book: the recommendation to move from space-only analysis to analyses that consider both space and time. Obvious cases of this trend in the context of β-diversity are Turgeon *et al.* (2017), Brice *et al.* (2019), Legendre (2019) and Legendre and Condit (2019). We expect to see more such studies based on tracking changes in spatial measures through time, the dynamics of spatially explicit localized measures, the spatial structuring of temporal statistics and measures and statistics that are truly spatio-temporal (Chapter 8). The last may involve the handling and analysis of very large (and, yes, complex!) data sets, but there are now resources and techniques available for such challenges, some of which we will discuss in the context of multilayer networks (Chapter 11).

10 Points and Lines, Graphs and Networks

Introduction

In ecological studies, we often analyse the spatial characteristics of structural units by reducing their complexity, representing them with simplified forms such as dimensionless points, straight or curving lines or simple polygons. Obvious examples are tree stems in a forest represented by points in a plane, or riverine systems represented by branching linear structures. Points and lines can be studied in separate analyses or they may be considered together, as in a study of clonal plants, with points representing erect stems and lines representing rhizomes or stolons. In some cases, points still represent the structural units, but the lines between them now indicate relationships, not physical elements of the system, usually functional or relational interactions between the units. To describe and analyse that kind of system, we use graph theory.

Graph theory is a field of mathematics in which structural units are represented by points and relationships between units are represented by lines joining the points in pairs (Harary 1969; Dale 2017). For clarity, we refer to points in a graph as *nodes*, and lines between them as *edges* (see Figure 10.1a). The nodes may have additional qualitative and quantitative characteristics, like species labels or stem diameter. The edges may have directions (Figure 10.1b), signs (plus or minus, Figure 10.1c) or other properties such as weights or capacity (Figure 10.1d). Graph theory has been applied to a range of ecological phenomena and it provides an open and flexible conceptual model that can contribute to understanding the relationship between structure and process, including configuration effects and compositional differences.

In ecological studies, individual organisms, populations, and habitat patches are commonly objects of interest, while behavioural, physical, functional and dispersal activities are processes that link these objects. For decades, *graphs* and *networks* have been used to model and analyse organisms and their interactions (Proulx et al. 2005; Mason & Verwoerd 2007; Dale & Fortin 2010; Dale 2017). In this context, these were *aspatial graphs*, where 'aspatial' means that the placement of the nodes and edges is determined by convenience; there is no spatial information from the original system however the graph is derived. In its purest form, a graph is a combinatorial entity, depicting only structure. Networks are graphs that include quantitative information, usually representing a real system (Figure 10.1d), but in many biological applications

Introduction 285

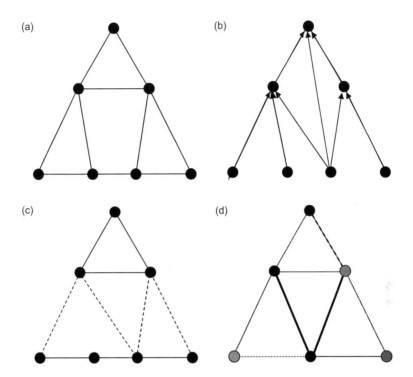

Figure 10.1 How points and lines form graphs. (*a*) A graph consisting of nodes, representing objects, joined by edges, representing relationships between them. (*b*) A digraph with edges that have directions. (*c*) A graph in which the edges have signs. A solid line for positive and a dashed line for negative relations. (*d*) A network with labelled nodes and weighted edges.

'network' and 'graph' are now treated almost as synonyms (Dale 2017; Dale & Fortin 2021), although sometimes it makes sense to talk about the graph (the abstracted structure) of a network (the system from which it is derived).

Graph theory is not new nor recently discovered. In the eighteenth century, Euler laid the foundation for graph theory in resolving the 'Seven Bridges of Königsberg' problem (see Biggs *et al.* 1976). A century later, Cauchy and L'Huillier formalized the basic concepts of topology, as an approach to 'the mathematics of position' (Biggs *et al.* 1976). In the 1950s and 1960s, Erdös and Rényi introduced probabilistic graph theory and the formal study of random graphs (Erdös & Rényi 1960). Probabilistic graph theory has been extended to models with various forms of dependence in the generation of random graphs and networks, such as 'small world' and 'scale-free' networks (Barabási 2009). Of special interest is the ongoing research on spatial graphs and networks (Barthelemy 2011, 2018), including random spatial networks constrained by the positions of the nodes in dimensional space (Barthelemy 2018).

Graph theory has been applied in food web studies (Pascual & Dunne 2006; Lurgi *et al.* 2014), conservation ecology (Keitt *et al.* 1997; Fletcher & Fortin 2018) and epidemiology (Godfrey 2013; Kinsey *et al.* 2020). Graphs are commonly used to depict functional connections among organisms (predation, pollination, competition, Bascompte 2009; Kéfi et al. 2015, 2018), within spatially structured groups of local populations (metapopulations; Grant et al. 2007). Although many of the phenomena that inspired the development of graph theory were spatial (paths, maps, electric circuits), graphs originally were not, simplifying problems by removing the spatial detail.

The focus of this chapter is spatial graphs and these can be introduced by starting with spatial sets of point events and spatial sets of lines and combining them or by beginning with graphs as abstract structural depictions with nodes and edges, and then embedding them in a spatial context. Either can work, but we will begin with the spatial sets of points and lines. We have already described the analysis of sets of point events in Chapter 3, and we will look next at the parallel analysis of sets of lines (Section 10.1). The treatment of points and lines together will follow (Section 10.2) and then various aspects of spatial analysis with graphs and networks (Sections 10.3–10.5). The further extension of the same concepts to spatio-temporal analysis by stacking up spatial graphs from different dates to create a multilayer network will be detailed in Chapter 11.

10.1 Lines Alone: Fibre Pattern Analysis

Just as we may reduce spatial objects to dimensionless points representing locations, spatial objects can also be reduced to lines or polygons, depending on their characteristics. Many linear objects are of interest in biological studies, including physical objects like roots and derived objects like dispersal trajectories. Unlike the edges of a graph which require nodes to join together in order to exist, these linear structures are the primary entities of interest. Such linear events can be studied with fibre pattern analysis (Stoyan et al. 1995; Benes & Rataj 2004), as a set of one-dimensional events produced by a 'fibre process'.

Many ecological phenomena can be treated as fibre processes: rhizomes or stolons of clonal plants, elongated burrow systems, animal migration routes and so on. In the simplest case, the fibres are straight lines, for which we can determine scales of clustering and repulsion, as in point pattern analysis. These fibres differ in their angles of orientation and those angles can also be the subject of spatial analysis. For example, Buettel et al. (2018) investigated fallen trees in an Australian eucalyptus stand to determine whether the trees tend to fall downhill and whether slope steepness enhances the effect. The answer was yes to both questions. Given these angles, we can also answer questions about relative angles as a function of distance between fibres or about angles as a function of location, as in the Eucalyptus example. In the more general case (Figure 10.2) the fibres are not identical and have quantitative characteristics, and we can test hypotheses about length, thickness, number of crossings and so on, using restricted randomization for each characteristic and then for combinations of them. Similarly, if the fibres belong to several classes, such as species, we may be interested in the equivalent of join count statistics and proceed by evaluating the number of crossings of fibres that belong to the same or different categories.

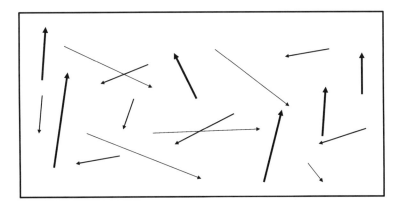

Figure 10.2 Fibres with length, direction and width. The three characteristics do not seem to be independent.

10.1.1 Aggregation and Overdispersion of Fibres

The analysis of fibre patterns can follow the methods originally designed for point patterns (see Chapter 3). Some of the most common methods for point pattern analysis are based on Ripley's K function (Ripley 1988; Chapter 3). This starts with the concept that, for λ events per unit area, the expected number of events in a circle radius t centred on a randomly chosen event is $\lambda K(t)$, where $K(t)$ is a function that depends on the pattern of events: if the events are over-dispersed, $K(t)$ will be close to 0 for short distances and increasing for larger distances (Chapter 3).

Modifying this approach for fibres requires several adjustments. First, any statistic should be based on the length of lines in a particular area, rather than on counts. Point pattern analysis uses circles as templates; linear features use a template of all points of the plane at distance t or less from the focal line. For straight-line segments, this produces an oval: a rectangle with semi-circular ends. The area of the template for line i and distance t is $a_i(t) = 2tl_i + \pi t^2$ and the length of line j that lies within it is $l_t(i,j)$. Where L is the total length of lines, the expected length within an envelope of radius t centred on a randomly chosen line is $\lambda_f K_f(t)$, where $K_f(t)$ is a function that depends on the pattern of the fibres and λ_f is the length of line per unit area. Then $K_f(t)$ is estimated by:

$$\hat{K}_f(t) = A \sum_i^n \sum_{\neq j}^n h_i(t) l_t(i,j) / n^2 L. \tag{10.1}$$

The weight $h_i(t)$ corrects for edge effects: the reciprocal of the proportion of the template within the study area. The expected value of $K_f(t)$ under complete spatial randomness is the average of the h-weighted areas of the templates for radius t: $a_\bullet(t)$. For easy comparison, we suggest plotting the statistic:

$$\hat{L}_f(t) = \sqrt{a_\bullet(t)} - \sqrt{\hat{K}_f(t)}. \tag{10.2}$$

As for point pattern analysis, negative values indicate clumping at scale t and positive values indicate overdispersion. Examples of this analysis are given in Figure 10.3, showing its appropriate response to the patterns presented.

The point-event K-function analysis can produce a map of localized scores, to evaluate place-to-place variation in deviations from the expected (Getis & Franklin 1987). However, because a fibre event does not have a single location as a point does, local scores for individual fibres would be spread along their lengths, making interpretation difficult. This is a topic for further investigation that may be worth pursuing.

10.1.2 Fibres with Properties

A fibre process may produce qualitative (such as living or dead) or quantitative (such as thickness) properties for the fibres. These characteristics can be included in the analysis by modifying the method, similar to the modifications of the univariate K-function for bivariate or multivariate data, or for including quantitative values for mark correlation (Chapter 3).

For categorical variables, we can explain the bivariate case in detail, with the multivariate case created by extension. For two kinds of fibres, to find the length of Type 2 fibres in Type 1 t-neighbourhoods and the length of Type 1 fibres in Type 2 t-neighbourhoods, calculate:

$$\hat{K}_f(t)_{12} = A \sum_{i}^{n_1} \sum_{j}^{n_2} h_i(t) l_t(i,j) / n_1 n_2 L_2 \quad \text{and} \tag{10.3}$$

$$\hat{K}_f(t)_{21} = A \sum_{i}^{n_1} \sum_{j}^{n_2} h_i(t) l_t(j,i) / n_1 n_2 L_1. \tag{10.4}$$

Then,

$$\hat{L}_f(t)_{12} = \sqrt{a_\bullet(t)} - \sqrt{[n_2 \hat{K}_f(t)_{12} + n_1 \hat{K}_f(t)_{21}]/(n_1 + n_2)}. \tag{10.5}$$

Negative values indicate aggregation and positive values indicate segregation. For multivariate data, we can consider a variety of 'interspecific' statistics of the form $K_f(t)_{I,\sim I}$ which can be evaluated on their own or compared with 'intraspecific' measures of the form $K_f(t)_{I,I}$, as described in Chapter 3 for point pattern analysis.

For C categories, we can evaluate the number of crossings of fibres that belong to each of the $X = C \times [C-1]$ categories created by the vertical order of the crossings; considering a log of Species A crossing on top of a log of Species B as different from B crossing on top of A. Where the order cannot be recovered, the number of categories for the fibre crossings is halved and, in some cases, it may be desirable to use summary classes of like-with-like versus unlike pairs of fibre crossings. We suggest

10.1 Lines Alone: Fibre Pattern Analysis

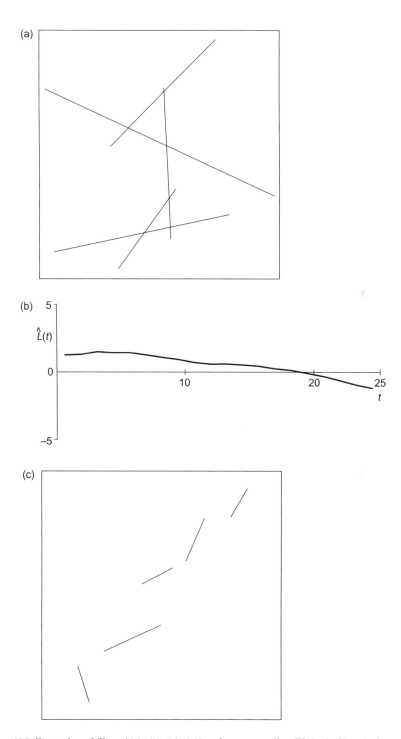

Figure 10.3 Examples of fibres [(a), (c), (e), (g)] and corresponding Ripley's K analysis [(b) (d), (f), (h)].

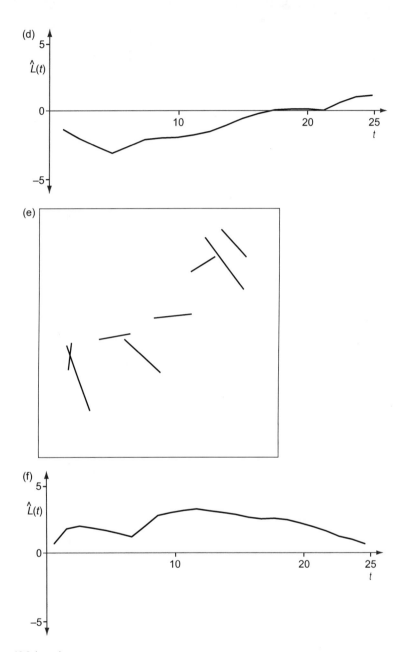

Figure 10.3 (cont.)

that restricted randomization techniques are the best approach to evaluating these count statistics (Chapter 2).

When the fibres have quantitative attributes, we calculate:

$$\hat{K}_f(t)_m = A \sum_{i}^{n} \sum_{\neq j}^{n} h_i(t) l_t(i,j) m_i m_j / n^2 L. \qquad (10.6)$$

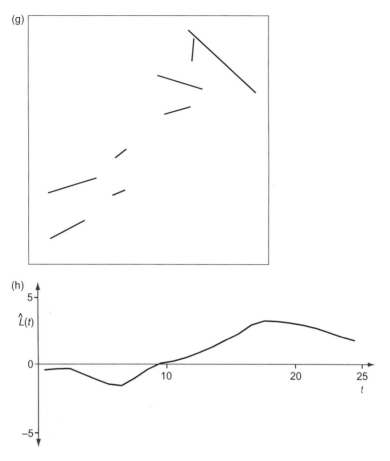

Figure 10.3 (cont.)

Then, where μ is the overall mean of the quantitative attribute, plot:

$$\hat{L}_f(t)_m = \sqrt{a_\bullet(t)} - \sqrt{\hat{K}_f(t)_m/\mu^2}. \tag{10.7}$$

Positive values indicate overdispersion of the marks, and negative values indicate their aggregation (positive correlation). This approach is often referred to as *marked fibre process analysis* (Stoyan *et al.* 1995).

10.1.3 Curving Fibres

So far, we have been concentrating on fibres that are finite straight-line segments, but some applications will require curving fibres. Curving finite-length fibres can be produced in several ways, but one practical approach is to use Bézier curves. This technique cannot produce all possible curves but it is very flexible and very complex curves can be built up from Bézier subunits by a spline technique. A curve can be

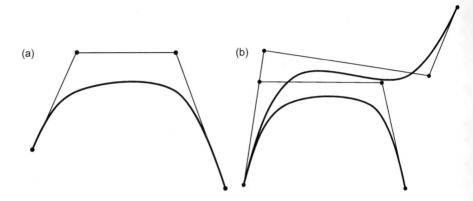

Figure 10.4 Bézier curve example, (a) simple and (b) single branching.

created of any order but cubic is most common, based on four points at positions q_1 to q_4. The parametric equation of the curve is then:

$$B(z) = (1-z)^3 q_1 + 3z(1-z)^2 q_2 + 3z^2(1-z) q_3 + z^3 q_4. \qquad (10.8)$$

An example is shown in Figure 10.4a.

This approach has many advantages, one being that transformations such as change of scale, position or angle can be generated easily for the curve by applying the transformation to the four 'control' points (Bézier 1977).

Although Bézier curves have been used in applications ranging from automotive design to the creation of type fonts, we have found no discussion of their use to create branching curves. This is easily done for cubic curves of this family by adding control points (Figure 10.4b). Combining simple branching curves allows the creation of multibranched structures resembling trees, or anastomosing systems of curved fibres forming networks. Again, this area shows promise for the spatial analysis of ecologically important objects yet to be pursued.

10.2 Points and Lines Together

Point events and lines as objects can be combined in several ways:

(1) They can be treated as two separate processes (like logs versus tree stumps on the forest floor) with the simplest null hypothesis being complete independence within and between processes. For a non-ecological example of examining the relationship between a set of points and a set of lines, Berman (1986) looked at the point locations of copper deposits (the events) in relation to linear geological features, called lineaments (the lines) in Queensland, Australia. In this example, the orientation of the lineaments was also considered (see also Foxall & Baddeley 2002). Berman's approach was a simple test statistic using the nearest

point-to-line-neighbour distances: D_i is the distance from the ith event to the nearest part of any of the lines and the statistic is their sum:

$$S_n = \sum_i^n D_i. \qquad (10.9)$$

Smaller statistics indicate more closely associated processes. Significance can be evaluated by a numerical procedure, Monte Carlo or randomization, using the lengths (and possibly the angles) of the observed fibres and the number of point events.

Foxall and Baddeley (2002) recommended the use of a variant of the empty space function (van Lieshout & Baddeley 1999), described for point events alone (Chapter 3). It compares the function, F_Y, based on the average shortest distance from a random point to an element of the set of fibres, Y, with the 'nearest neighbour function', G_{XY}, based on the average shortest distance from a random event in set X to an element of the set Y. Then, for a radius limit, r, the H-function that compares them is:

$$H_{XY}(r) = \frac{1 - G_{xy}(r)}{1 - F_Y(r)}. \qquad (10.10)$$

Again, the statistic is evaluated by a Monte Carlo or randomization procedure.

(2) In many ecological examples of points and lines, the point events are constrained to be located on the fibres, such as stems arising from rhizomes, foraging episodes on an animal's path of movement or waterfalls on a river system. This constraint gives rise to a hierarchy of hypotheses, such as distinguishing between 'Are the events over- or under-dispersed compared to CSR?' and 'Given the lengths and positions of the fibres, are the events over- or under-dispersed?' This situation is illustrated in Figure 10.5, with the events in the upper part of the figure being clumped on some of the fibre branches, whereas they are over-dispersed in the lower fibre system, although appearing to form an open clump in ordinary 2-dimensional space. These objects are constrained asymmetrically, with the point events occurring only on lines, such as waterfalls on rivers, which are 'the given' or put in place first; the points on linear structures can be analysed using a modified version of the K function (Baddeley et al. 2021). The procedure is to count the number of events within distance t of any event, with distance measured along connected fibres. The average observed count $K(t)$ is compared with the expected, again by Monte Carlo or randomization techniques; for details see Okabe and Yamada (2001). Spooner et al. (2004) provided an example of an ecological application of this technique, in both univariate and bivariate versions, studying *Acacia* tree populations on a road network in Australia. The analysis showed significant clustering over a range of scales, with the suggestion that road maintenance activities may be more important than environmental factors in determining population dynamics.

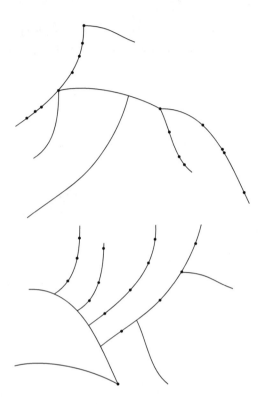

Figure 10.5 Examples of the overdispersion and clumping of point events on fibres.

One application where the 'point events on a line' analysis seems not to have been fully utilized is in the evaluation of data from line transect sampling, where individual animals are recorded on a sampling line as a way of determining population numbers and spatial dispersion (see Hedley & Buckland 2004). More can be learned, it seems, from these data sets by further analysis using the 'point events on lines' approach.

(3) In some other ecological examples, the points are 'the given' and then the lines are inserted so that they occur only at points, as with roots growing out from plant stems. This situation may be less frequently encountered, but one example of obvious ecological importance would be the growth of rhizomorphs as agents of fungal infection, such as *Armillaria*, from one tree stem (or stump) to another.

(4) A last possibility is the mutual constraint of point events and lines, with the constraints being possibly sequential: first a point event, then some lines constrained by that point, then more points constrained by the lines, and so on, as with plant shoots and their rhizomes (figures 10.13 and 10.14 of Dale 2017). Ecology has many examples of this kind of spatial structure and the events do not need to be physical objects; consider the linear sections of animal movement, with feeding episodes being the point events. This creates a mutual constraint between the elements of a point-and-line depiction of this spatial phenomenon (see comments on animal movement analysis in Chapter 8).

10.3 Points and Lines: Spatial Graphs and Spatial Networks

10.3.1 Spatial Nodes

In Chapter 2, we described several algorithms for determining which pairs of points in the plane are neighbours. Each algorithm results in a mathematical object, a *graph*, consisting of a set of points, the *nodes*, joined in pairs by lines, the *edges*; this is a *spatial graph* because the nodes have spatial locations. The edges then have physical lengths determined by the distance between the nodes, whether or not the location of the edge is explicit. For instance, we can draw an edge between the position of a tree and that of one of its known seedlings, indicating the relationship but without knowing the real route the seed travelled. On the other hand, the edges may have physical positions, such as a well-worn foraging trail between feeding locations. The nodes may have characteristics, such as the size and shape of habitat patches; so may the edges, such as the characteristics of dispersal corridors that enhance or impede dispersal (see Fall *et al.* 2007; Urban *et al.* 2009).

We will begin with an example that applies the concepts of graph theory to a spatially explicit network. Dupont *et al.* (2014) used the approach to study the spatial structure of a plant–pollinator network of marsh thistle flowers (*Cirsium palustre*, a clonal species) and bumble bees (eight species of *Bombus*: four parasitic and four nest-building). They mapped the flowering plants and determined several characteristics related to the position, such as the number of neighbours, and they marked and tracked the bees, including flower visitation. They determined the modularity of the spatial structure, which is created and enhanced by groups of nodes with more edges within the groups than between. This structure has weakly linked but strongly internally connected subsets, and this is a common feature of many ecological networks, spatial or otherwise. Here, the modules of plants and pollinator movement were connected by the tall plants in the centre of the plant patches and by the behaviour of the individual bees. The authors concluded that linking spatial ecology to network theory, especially in individual-based and animal-centred networks help answer a range of questions across taxa and ecosystems (Dupont *et al.* 2014).

Clarification: the spatial network literature uses the term 'modularity', as we have done here, but the 'modules' thus identified are often referred to as 'communities' (Barthelemy 2011), potentially confusing in an ecological context!

For spatial graphs in ecology, *modularity* is one of the three most important concepts; the other two are *connectivity*, which describes whether it is possible or how easy it is to get from one node to another, and *centrality*, which measures the importance of a node in paths through the graph's structure. These will be described and discussed in the sections that follow. The characteristics of connectivity, modularity and centrality are also key concepts for the analysis and understanding of ecological interaction networks (Dale & Fortin 2021), just as they are for the spatial networks we focus on here. Yet, some of the characteristics of interaction networks of any kind, such as the distribution of node degrees (numbers of edges per node) and their assortativity (like-to-like versus like-to-unlike) or clustering coefficient (proportion of triads with two edges that have three) are not of interest in spatial graphs. Barthelemy (2018) referred to them as 'irrelevant measures' for spatial networks; degree distributions tend to be narrow, because it is physically difficult for a node to have many edges and clustering coefficients tend to be large merely because of the constraints imposed by the spatial context. The typical examples of spatial networks are roads, railways and subways, which tend to be approximately planar with few edge crossings (Barthelemy 2018). Airline networks are atypical and tend to look more like complex networks with hubs and spokes and the potential for some nodes of high degree (Barthelemy 2011). A last variation on spatial networks to be mentioned in this context and a possibility worth exploring are multilayer spatial networks, in which different layers represent different organisms, different factors or different kinds of data (Chapter 11). The same landscape of patches could have network layers associated with genetic data, observed individual movement, satellite radio locations and structural considerations (Costa et al. 2019; Foltête et al. 2020). Of greatest interest, of course, is the phenomenon of the coupling of the layers, even if there are only two! (Barthelemy 2018, section 8.5; Bianconi 2018). An obvious application for multilayer spatial graphs is when the layers represent the same spatial system studied at different times (Legendre 2019; Legendre & Condit 2019), which may provide important insights into the forces that determine the evolution of the spatial structure (Barthelemy 2018). We will explore this approach at greater length in Chapter 11.

10.3.2 Neighbour Networks

In Chapter 3, we considered spatial sets of dimensionless point events as the objects in 'point pattern analysis' (Dale 1999; Illian et al. 2008; Diggle 2013). Their locations arise from a 'point process' such as complete spatial randomness (CSR). Other processes result in the clumping of events or in overdispersion. Testing for CSR is usually too simplistic and analysis should be designed to determine the scales of clustering or repulsion (Chapter 3). Examining a range of scales is like hierarchical hypothesis testing and the results are not all independent, so determining the statistical significance may be best evaluated using randomization procedures (Chapter 2).

One approach to evaluating a spatial point pattern is to identify the pairs of events that can be considered spatial neighbours. Joining neighbour pairs with lines creates a neighbour network which is also a spatial graph, with the events as nodes and edges being those joins. Neighbours can be defined in several different ways, each producing its own network for the same point pattern. In Chapter 2, we described a hierarchy of neighbour networks from mutually nearest neighbours (MNN) to the Delaunay triangulation, summarized in Table 10.1.

This hierarchy of neighbour networks can be used as the basis for multivariate point pattern analysis and for the analysis of marked point patterns (Chapter 3), counting the number of like-like joins (multivariate classification) or calculating correlation coefficients (quantitative marks) in the hierarchy of networks. In most applications, a 're-labelling' randomization will be an obvious and effective technique to evaluate the significance of any result (Chapter 2).

10.3 Points and Lines: Spatial Graphs and Networks

Table 10.1 Measures related to ecological connectivity for spatial graphs (revised from table 9.3 of Dale 2017).

Measure	Details	Comments
Connectedness: connected versus not connected	All node pairs are joined by a path versus some pairs have none. Digraphs can be 'strongly' (either way by directed edges), 'unilaterally' (only one way by directed edges) or 'weakly' (some path edges reversed) connected.	Connectedness allows inter-patch dispersal and recolonization after extinction; disconnected does not. Digraphs are useful for flowing systems.
Connectivity	# independent paths between A & B, κ: κ_n = minimum # nodes or κ_e = minimum # edges removed to disconnect.	Shows robustness of network to loss of patch or inter-patch link; may enhance dispersal.
Cut-nodes and cut-edges	# cut-nodes # cut-edges	Larger numbers show greater vulnerability to disconnection and potential for reduced dispersal.
Connectance	Proportion of edge positions with edges: $c(G) = 2m/n(n-1)$	Density of edges related to ease and probability of movement between patches.
Clustering coefficient (Triad closure) Node degree	Frequency of ($e_{ij} = 1$ if $e_{kj} = 1$ & $e_{ik} = 1$) d_i = # edges attached to node i (local measure). The number of neighbour nodes at $\delta_{ij} = 1$	Measures local edge redundancy Higher degree gives more links for possible dispersal; random walks suggest more organisms arriving.
Average node degree	$d.(G) = m/2$ (global measure)	Higher value suggests more dispersal possible (depends on degree distribution).
Variance of node degrees	$s^2(d_i)$: uneven to uniform degree distribution (global, but could be mapped)	High variance indicates 'hub and spoke' configuration & rapid dispersal; low value indicates homogeneity & slower dispersal.
Binned neighbour counts	The proportion of nodes as neighbours at $\delta_{ij} = 1, 2, 3, \ldots$ (local or global). Both means and variances can be used (Labonne et al. 2008).	Higher means at shorter distances with more source nodes close; variance reflects spatial heterogeneity
Components	K components: $X_1, \ldots, X_K$, $K = 1$ means graph connected; $K > 1$ means disconnected. Related: *algebraic connectivity*; K is the number of 0 eigenvalues of the Laplacian matrix = degree matrix − adjacency matrix; $L = D - A$.	More components mean more and smaller isolated subgroups of patches.
'Size'	# nodes ('order'): X_k has n_k nodes # edges: X_k has m_k edges Total of patch area: $A_k = \sum_{i \in X_k} a_i$	Related to pool of organisms for inter-patch dispersal and to ease of dispersal. Related to number of organisms available for dispersal.

Table 10.1 (*cont.*)

Measure	Details	Comments
Path length Graph diameter	δ_{ij} is geodesic path length = # edges $d_p(v_i, v_j)$ is the sum of physical edge lengths Max(δ_{ij}): geodesic path length Max$(d_p(v_i, v_j))$; sum of physical edge lengths	Affects dispersal between patches i and j. Smaller diameter may allow faster dispersal throughout landscape.
Component diameter, X_K	Max$(\delta_{ij}|$ both nodes in X_K): geodesic path length or Max$(d_p(v_i, v_j) \mid v_i \& v_j$ in X_K): sum of physical edge lengths.	Smaller diameter may allow faster dispersal within component. Shorter steps may facilitate movement.
Algebraic connectivity	This is the second smallest eigenvalue of the Laplacian matrix (the number of 0 eigenvalues is the number of components). (L = D − A).	Larger values indicate greater robustness and fault tolerance; also, the ease of synchronization in a dynamic network (Chung 1997).
Spectral gap	The difference between the first & second eigenvalues of the adjacency matrix indicates when nodes have robust connections to other nodes.	Larger values suggest robust structure even for sparse graphs; in random walks, faster convergence to limiting distribution.
Characteristic path length	$C_g = \frac{1}{n_k(n_k-1)} \sum_{i \in X_k}^{n_k} \sum_{j \in X_k}^{n_k} \delta_{ij}$ $C_p = \frac{1}{n_k(n_k-1)} \sum_{i \in X_k}^{n_k} \sum_{j \in X_k}^{n_k} d_p(v_i, v_j)$ Geodesic and physical distance versions. If $K = 1$, then X_k is G, the whole graph.	Shorter average path lengths suggest better dispersal between any pairs of habitat patches.
Path directness	$D = \frac{1}{n_k(n_k-1)} \sum_{i \in X_k}^{n_k} \sum_{j \in X_k}^{n_k} d_{ij} \div d_p(v_i, v_j)$ Average of straight-line distance over path physical length.	Values near 1 show that the dispersal routes are spatially efficient; low values indicate indirect routes with detours, which increase risk and reduce dispersal.
Betweenness centrality (node)	Local measure: proportion of shortest paths that go through v_i.	High centrality nodes are more important for dispersal routes and for conservation design.
Variance of node betweenness centrality (global)	Global measure of differentiation among nodes for this property.	Low variance: all nodes of equal importance. High variance: some nodes are in key locations and highly important.
Closeness centrality (node)	Local measure: mean geodesic distance of v_i from other nodes. Its variance is another global measure, as above.	High closeness centrality nodes are more important because they are closer to others; high variance indicates greater differentiation for this property.
Eccentricity (node)	Local measure: maximum distance of v_i from any other.	Indicates the node's isolation for potential recolonization.

Table 10.1 (cont.)

Measure	Details	Comments
Modularity	$M = \sum_{S=1}^{N} \left[\frac{m_s}{m} - \left(\frac{K}{2m} \right)^2 \right]$ N = # modules; m_s = # edges in module s K_s = # all edges of nodes in module s	Highest when most edges are within modules and few between modules.

The neighbour networks can also be used to transfer analysis methods from lattice systems to irregular arrangements of observations. For example, the multivariate analysis of clustering creates groups of observations with greater similarity within groups than among groups. In 'spatially constrained clustering' the grouping of similar objects is conditional on being adjacent in space. That spatial adjacency is usually defined on a lattice or grid but can be defined by one of the neighbour networks for irregular patterns of events. Spatially constrained clustering has many applications in ecological research and is obviously closely related to techniques for determining spatial boundaries among observations, such as delimitating landscape habitat patches (Chapter 7).

One suggestion is to apply the same analysis using the six basic neighbour networks (Chapter 2) because the differences in moving up or down the hierarchy can provide valuable insights. Applying only some of these networks may be sufficient; the MNN network may be too sparse for some analyses and the DT too dense with strong dependence among comparisons (all those triangles!). Dale (2017) has provided an interesting example with MST and the GG assessing the relationships of diversity, species composition and nestedness for Tenebrionid beetles on islands in the Aegean. The focus was on the autocorrelation of community characteristics between neighbouring islands (details in Dale 2017, see figures 9.8 and 9.9).

In addition to the network hierarchy described here, many other neighbour networks are available for application, including the Ulam tree and networks defined by the 'β-skeleton' rule (see Dale & Fortin 2014). We have also described some based on physical distance thresholds in Section 2.8.2, and those based on distance and direction in Section 2.8.3. Many other rules for neighbour networks remain to be explored in detail, but the general principles for application will probably resemble those discussed for this 'topological' hierarchy.

10.3.3 Signed and Directed Graphs and Networks

The edges of a graph can have weights of different kinds including signs for positive or negative (+ or −, a binary weight) indicating the nature of the relationship; these are *signed* graphs (signs indicated by edge colour or by solid versus dashed, recall Figure 10.1c). Edges can have numerical weights such as measures of similarity or capacity.

For asymmetric relationships, we use edges that have direction (creating a digraph). This suggests one form of hierarchical analysis, comparing the results with and without edge directions included (see Fortuna *et al.* 2006). Asymmetric relationships are sometimes signed positive or negative, such as the nurturing effects of nurse plants or the 'interference competition' of chemical allelopathy, producing edges with signs and directions. An obvious application of digraphs is in analysing movements through time, where the nodes are locations of an individual object, and the edges recording the displacement between subsequent observations. The set of edges for one individual in such a *displacement digraph*, when it is also a spatial graph, is both the graph-theoretical path and a diagram of the actual path of movement.

A *bipartite graph* consists of two subsets of nodes with edges only between subsets, not within; consider the relationship between pollinators (subset 1) and plants (subset 2) with edges joining each plant to its pollinators. Bipartite networks also apply to plants and their herbivores, or in other parts of food webs. Plants, herbivores and carnivores may form a tripartite graph of the three separate levels, although omnivores may lead to a blending of the levels (see Dale & Fortin 2021, figure 4.1). Because of the way the node subsets are defined, these network graphs are not often spatial, but may provide non-spatial summaries of spatial processes as in pollination studies (such as Dupont *et al.* 2014; but see a spatial version in Pasquaretta *et al.* 2019).

A *graph of graphs* consists of graphs at two levels or orders, such as a spatial graph of sites in which each site has its own graph of its trophic network (Figure 10.6).

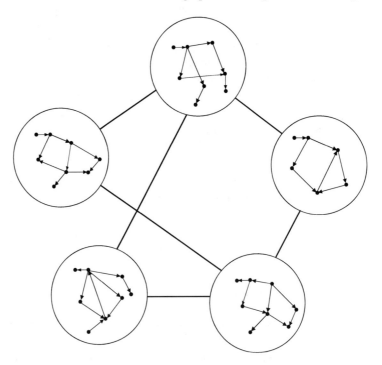

Figure 10.6 A graph of graphs (meta-network): each node of the main graph has a second-order graph associated with it. Usually, the main graph will be spatial, showing locations and neighbours, with the inserted node graphs depicting some structure or set of functional relationships for each location.

Fortuna and Bascompte (2008) showed a graph of a metacommunity consisting of nodes, each of which contains a graph of multispecies interactions. Melián *et al.* (2005) compared two complete graphs of five nodes representing habitats, with edge thickness depicting the number of shared trophic modules; each habitat node has an associated tri-trophic digraph showing a simple food chain or a food web with omnivory. These structures can also be considered as multilayer networks (Chapter 11).

10.3.4 Creating Subgraphs

Subgraphs were introduced informally in the discussion of the hierarchy of neighbour networks (Chapter 2). A subgraph of a graph has a subset of the nodes or a subset of the edges. In ecological studies, we are often interested in finding those subgraphs in which the ratio of edges to nodes is much higher than in the rest of the graph; these are *modules* and the property is *modularity*, described in Section 10.3.1. The existence of modules may reflect important substructures, as in the thistle-bee pollination example (see Pasquaretta *et al.* 2019).

There are several algorithms for dividing a univariate graph into highly connected subgraphs (Newman *et al.* 2006). This division of the graph into modules is much like spatial partitioning which splits the nodes of networks or similarity graphs into subsets, often based on the boundary detection algorithm called triangulation-wombling (Chapter 5; Fortin 1994; Jacquez *et al.* 2000). The algorithm identifies edges between pairs of nodes that are most different, indicating the steepest local changes; these edges are then candidates for a boundary between regions of high internal similarity. The technique is related to the converse of spatially constrained clustering in which the most similar nodes are joined into clusters only if they are adjacent in space (Chapter 5).

10.4 Network Analysis of Areal Units

One obvious extension to point pattern analysis is to consider events with non-zero area, such as habitat patches, as the objects for spatial analysis, prompted by the many graph theoretic studies of landscape connectivity (Urban & Keitt 2001; Fall *et al.* 2007; Foltête *et al.* 2020; Godet & Clauzel 2021). That series of papers brought renewed attention to graph theory for evaluating ecological structure, and graph theory is obviously a favourite topic of the authors of this book (Dale & Fortin 2010; 2021; Dale 2017)!

A basic assumption in landscape ecology is that a landscape can be represented by a set of identifiable patches. Studies of fragmentation consider patches of habitat suitable for the focus organisms situated within a matrix of unsuitable landscape elements. In the graph-theoretic approach, the habitat patches are the nodes of a graph and the edges are connections between the patches as determined by ecological processes. The landscape is therefore represented as a functional network, with colonization or dispersal being one primary process in conservation-oriented studies, the other being local extinction.

Several graph theoretic properties should be noted here. A graph is *connected* if there is at least one *path* (a sequence of nodes joined by edges) between any two nodes in the graph. The *degree* of a node is the number of edges that are attached to it. In a connected graph, a *cut-point* is any node, the removal of which causes the graph to be no longer connected. Similarly, any edge, the removal of which disconnects the graph, is called a *cut-edge* or a *bridge*. Obviously, a complete graph has no cut-points and no cut-edges. A graph that is not itself connected will be made up of several connected subgraphs, called *components*; a connected subgraph is a component if it is maximal: that is, if it is contained in no larger subgraph that is also connected. The *order* of a component is just the number of connected nodes it contains. For a connected graph, two measures of how strongly it is connected should be considered: (a) *node connectivity* is the minimum number of nodes that must be removed to disconnect the graph, and (b) *edge connectivity* is the minimum number of edges that must be removed to disconnect the graph. Figure 10.7 illustrates these various terms. There is an obvious analogy between the connectivity of a spatial landscape graph and the number of dispersal routes, made up of corridors and patches, available in the landscape it represents.

Graph theory is about the structure of connections and, generally, the edges of a graph do not have properties but, in some applications, the nodes and edges have properties of their own. Each edge in the graph may have a length associated with it, $d(e_k)$, which could be the physical distance between nodes that have locations, as drawn on a map, or some other property, such as cost of transport or resistance to movement. The graph theoretical *distance* between two nodes in the graph, δ_{ij}, is the minimum path length (smallest sum of edge lengths) of any path between the two nodes. For many applications in landscape ecology, the edges have the length of the (Euclidean) distances between the nodes they join, thus forming a spatial graph. Whatever the measure of length, the *eccentricity* of any node, $\varepsilon(v_j)$, is the maximum graph theoretical distance to any other node in the graph. Last, the *diameter* of a graph is the maximum eccentricity of any node in the graph.

Urban and Keitt (2001) examined the properties of landscape graphs by considering the effects of removing edges from the graph and of removing nodes, analogous to the loss of dispersal corridors in the first case and of habitat patches in the second. The characteristics they suggested for evaluating edge removal are purely graph theoretic: the number of components that result, the diameter of the largest component, and the order of the largest component. The procedure they suggested is to start with a graph with all pairs of nodes joined by edges (a *complete* graph) and then to use a series of threshold distances to remove longer edges, leaving only those shorter than that threshold, creating a series of threshold distance networks. The response of the graph properties of these networks to threshold distance provides an evaluation of the patch structure.

As an example, Figure 10.8a shows the approximate sizes and locations of 21 lakes in an extensive peatland near the Alberta–Saskatchewan border (55°45′N, 110°45′W)

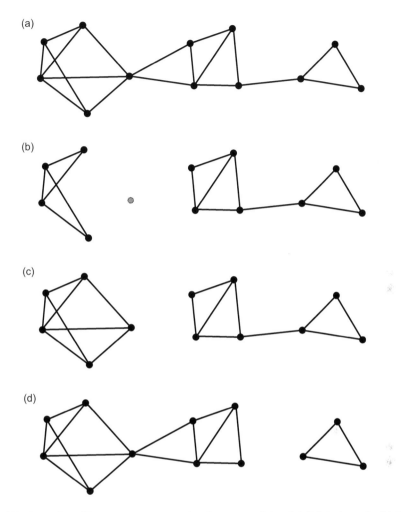

Figure 10.7 A graph to illustrate some terms related to connectivity. (*a*) Original graph, (*b*) the point now isolated is a cut-point because its removal disconnects the graph, (*c*) two lines removed disconnect the graph but (*d*) the line removed is a cut-edge because its removal disconnects the graph. The edge connectivity is 1 and the node connectivity is also 1.

and their distance-based Minimum Spanning Tree. No direct permanent surface-water features connect these lakes and a reasonable working hypothesis is that dispersal between them is inhibited by distance. Figure 10.8*b* shows their Delaunay triangulation. Following Urban and Keitt (2001), Figures 10.8*c* and *d* show threshold distance networks for 7 km and 4 km. At 7 km, the graph is a single component with no cut-points or bridges. Reducing the threshold slightly to 6 km (not shown), the graph is no longer connected, although the largest component includes 20 of the 21 lakes. The change from 7 km to 4 km for the threshold is dramatic, giving nine components, of which the largest has only five lakes. The abrupt transition from a few components, some of which are large, to many small ones is consistent with Urban and Keitt's (2001) observations on a hypothetical landscape.

304 Points and Lines, Graphs and Networks

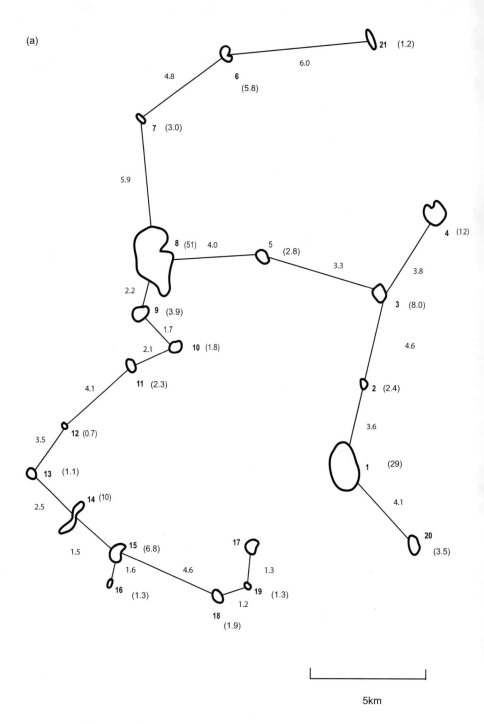

Figure 10.8 Illustration of connectivity of lakes from a peatland near the Alberta–Saskatchewan border. (*a*) The Minimum Spanning Tree for 21 lakes in a peatland, with approximate distances between them (in kilometres) and their relative sizes (arbitrary scale). (*b*) Delaunay triangulation graph of the same set of lakes. (*c*, *d*) Network of lake neighbours using thresholds of 7 km and 4 km, respectively. Different distance thresholds result in different graphs, but a graph with a smaller threshold is a subgraph of any with a greater threshold.

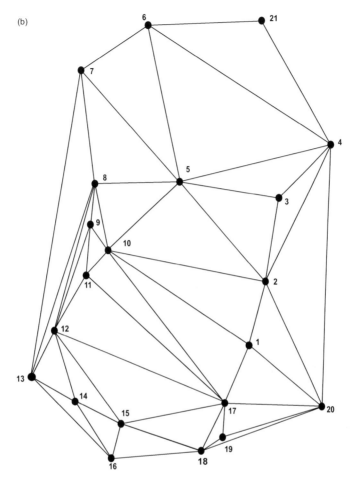

Figure 10.8 (cont.)

How can we measure the importance of an edge? We suggest that the importance of e_{ij} is the minimum 'cost' of its removal, either in the length of edges required to reconnect v_i and v_j, or as a proportion of its length:

$$c_d(e_{ij}) = l(e_{ik}) + l(e_{kj}) - l(e_{ij}) \tag{10.11}$$

or

$$c_r(e_{ij}) = \frac{l(e_{ik}) + l(e_{kj}) - l(e_{ij})}{l(e_{ij})}. \tag{10.12}$$

For example, in Figure 10.8b, the DT, the shortest replacement path for e_{35} is through v_4, giving

$$c_d(e_{35}) = l(e_{34}) + l(e_{45}) - l(e_{35}) = 3.8 + 6.1 - 3.3 = 6.6;$$

$$c_r(e_{35}) = 6.6/3.3 = 2.0.$$

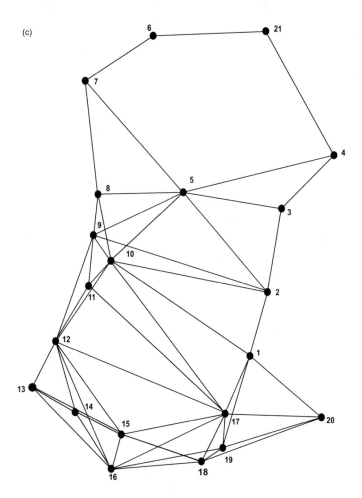

Figure 10.8 (cont.)

For $e_{9,11}$, the cost is much less:

$$c_d(e_{9,11}) = l(e_{9,10}) + l(e_{10,11}) - l(e_{9,11}) = 1.7 + 2.1 - 2.2 = 1.6;$$

$$c_r(e_{9,11}) = 1.6/2.2 = 0.73.$$

The criteria that Urban and Keitt (2001) suggested for node removal are more explicitly ecological:

(1) a recruitment index, R (a sum of patch areas weighted by patch quality);
(2) an index of dispersal flux, F (a sum of patch areas weighted by the probability of dispersal from the focal patch); and
(3) an index of traversability, T (the diameter of the largest component after node removal).

They examined the iterative node removal from the complete graph by three different procedures: (1) random choice for removal; (2) removal of the node with the smallest

10.4 Network Analysis of Areal Units 307

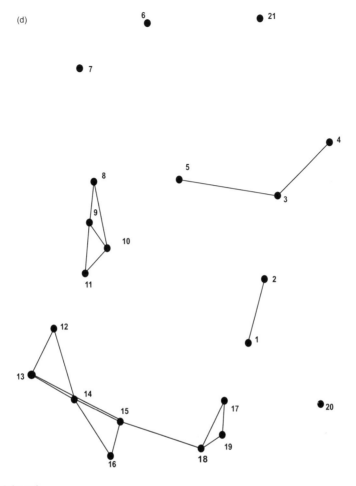

Figure 10.8 (cont.)

patch area; and (3) small-leaf removal where the smallest area node attached only to one other node in the current MST is removed (e.g. Lake 4 in Figure 10.8a). Not surprisingly, they found that small-leaf removal degraded the characteristics of the network less rapidly than did random removal. The same three measures, R, F and T, were then used to evaluate the importance of individual nodes according to each criterion. This was done by examining the difference in the measure before and after the node's removal. Because this evaluation was based on the complete graph, the evaluation of importance may be very different from one based on the DT or on the MST itself. For a complete graph of the landscape of lakes, removing Lake 8 has little effect on graph diameter, but removing Lake 21 makes a big difference. In the MST of the same landscape, the removal of Lake 21 reduces the graph's diameter somewhat, but the removal of Lake 8 has a profound effect on the graph's characteristics.

If we use a complete graph of the landscape and treat it as a purely combinatorial structure, no 'topological' distinction can be made among the points or among the edges. Once the graph is embedded in the plane, as a map, the points are differentiated, with some more central in location and some more peripheral. We can define the *perimeter* nodes as those that are in the convex hull (Lakes 21, 4, 20, 18, 16, 13, 7 and 6), with others being *interior* nodes. The importance of a node is related to its position in the graph (perimeter or interior), its degree and the distances to the neighbours to which it is joined. The removal of a perimeter node reduces the network extent ('footprint'). The removal of an interior node reduces the number of alternate paths between other pairs of nodes, a feature related to the concept of node centrality, which we now describe in detail.

The past two decades have seen many studies applying graph theoretical properties to evaluate landscape connectivity, often inspired by conservation goals (Urban & Keitt 2001; Godet & Clauzel 2021). The role of a single node in connectivity is its *centrality* which measures the importance of an individual node to the shortest paths between other nodes in the network. Centrality has many measures (Dale & Fortin 2021, table 2.1), but they can be highly correlated (Table 10.1 gives only two). Some authors have advocated combining different centrality measures to increase sensitivity for a specific purpose (Gouveia et al. 2021), although care must be taken to ensure that the 'shortest' paths are calculated appropriately and correctly (Costa et al. 2019). Centrality can be related to the potential for dispersal; a study of the landscape genetics of the marten, *Martes pennanti*, in Ontario, showed that higher proportions of immigrants were correlated with patch centrality, suggesting high patch quality (Garroway et al. 2008). Pereira et al. (2017) described the importance of assessing multi-node centrality properties in landscape graphs for the conservation of bird species in NE Spain.

Because connectivity is such a key property in these applications, and because many measures relate to it, we provided a summary with some comments in Table 10.1; it does not include all the measures from the publications cited (include Pascual-Hortal & Saura 2006; Labonne et al. 2008; Baranyi et al. 2011; Rayfield et al. 2011; Yu et al. 2013; Saura et al. 2014; Zeller et al. 2018; Foltête et al. 2020; cf. Dale 2017, table 9.3). The information in Table 10.1 should help inform the choice, but we recommend running random walks of the system's graph to understand the properties (see Dale 2017, section 9.3.3, 'A long detour: Random walks on graphs' to learn more). For a different but related application of random walks in the spatial context, Hirt et al. (2018) showed a possible relationship between individual movements and community diversity when the random walks are allometric, based on body mass and mode of locomotion.

This is an interesting area of research and more work is needed, especially on evaluating the features of habitat networks that are the most critical for the dispersal of at-risk organisms.

10.5 Spatial Analysis of Flow

A more detailed treatment of the analysis of spatio-temporal data, including animal movement, has been provided in Chapter 8, but it seems appropriate to complete the introduction to spatial graphs and networks to include movement as flow.

Graphs with directed edges (digraphs) require some modification of the concept of connectivity. A digraph is *strongly connected* if there is a path between any two nodes that follows the direction of all the edges in it. A digraph is *unilaterally connected* if there is a path between any two nodes, but not always in both directions. To be *weakly connected* is to have some pairs of nodes joined only by paths in which the edge directions must be ignored. The terms 'strongly' and 'weakly' as descriptors may be confusing and 'bilaterally' might be a better choice for the first, and 'non-directionally' for the second.

Structural characteristics affect the movement of organisms between habitat patches in a landscape network and connectivity can be related to 'least-cost' links. The costs of movement are determined by weights on the network edges, but several edges or several paths may link any pair of patches (which have shape as well as location). The assumption is that organisms move preferentially among those nodes using the paths that are least costly and there should be a relationship between the least-cost links and the paths followed between patches. Applications of such spatial graphs include the design and effectiveness of natural reserves as described by Fall *et al.* (2007) among many others. The choice of resistance values in least-cost path analysis can be subjective, however. Rayfield *et al.* (2011) showed how the selection of resistance values influences the position and length of the least-cost paths in fragmented landscapes.

Another approach that evaluates all possible paths between any pair of nodes is the method 'circuitscape' which assesses connectivity using an algorithm in the spirit of electrical circuit flows (McRae *et al.* 2008). Thus, more than just the single 'easiest' path is considered, and several parallel paths between two patches may provide a corridor that allows greater flow than a single path (Figure 10.9). This leads to considering the sum of all possible paths between patches. Several ecological studies have used this approach to understand species' movement in fragmented landscapes or to identify areas to be protected as corridors (Koen *et al.* 2010; Pelletier *et al.* 2014). Rogers *et al.* (2019) applied similar thinking to the dispersal of individual propagules, producing the concept of the 'total dispersal kernel' that encompasses all possible modes of transport. Zeller *et al.* (2018) compared connectivity models using both cost distances and circuit theory and more research is needed to evaluate the graph-based methods available. Connectivity based on the circuitscape theory can be estimated using the Julia package Omniscape.jl (Ladau & Eloe-Fadrosh 2019).

The movement of animals, such as migration between landscape patches, may not be a simple 1-dimensional path, but rather a corridor which is an elongated 2- or 3-dimensional spatial structure. This is most obvious for some linear structures such as hedgerows or ditches in an agricultural landscape, but it is equally true in less obvious cases where the path follows routes of less exposure or greater safety. A corridor could

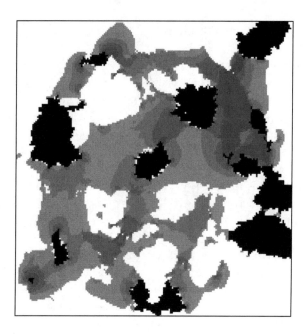

Figure 10.9 Example of connectivity analysis using circuitscape algorithm. Map of the potential probability of movement between the nodes (black patches). The probability of movement between patches is in grey (the highest probability of movement is in dark grey and the lowest in light grey).

consist of the path itself together with a buffer zone around it or a set of comparable least-cost paths (Pinto & Keitt 2009) like the anastomosing game trails across the tundra. The effects of corridors and their characteristics on the dynamics of metapopulations and meta-communities are well documented (Leibold & Chase 2018), but corridors also enhance genetic resilience for many species of different population sizes and dispersal abilities (Christie & Knowles 2015).

In spatial graphs, the positions of the nodes represent actual locations in space and many kinds of rules can determine which pairs of nodes have edges (Chapter 2). Consider data consisting of the physical locations of an animal at intervals in time. An obvious rule is to create edges between nodes joining successive locations, thus tracing the path of the animal's movement approximately and creating a connected digraph (Chapter 8).

In some cases, the nodes are not the GPS locations of animals at fixed time steps but rather fixed observation sites, traps or cameras where animals move in their vicinities. Hence, animal movements are 'captured' at more than one site and the resulting graph has edges joining nodes whenever both sites captured the same individual. With some assumptions about the animals' spatial behaviour, such as a Gaussian activity profile around a central location, the sites' locations can be included to improve estimates of density and other population characteristics (Strampelli *et al.* 2018; Jiménez-Hernández *et al.* 2020). Euclidean distance is the basic measure for comparing locations and movement in such applications, but network distance, travel 'cost' and other measures can be used (Royle *et al.* 2017).

A common approach assumes that the animals' activity centres follow a uniform distribution in the plane. Then, for individual i, it is assumed that its K observations at trap j follow a binomial distribution, $B(K, p_{ij})$, with p_{ij} determined by the Euclidean distance, d_{ij}, between the individual's activity centre and the trap:

$$p_{ij} = p_0 \exp(-\alpha_1 d_{ij}^2), \qquad (10.13)$$

where parameter α_1 controls the shape of the activity distribution around its centre, such as $\alpha_1 = 1/2\sigma^2$ for a Gaussian response. Where $\alpha_0 = \text{logit}(p_0)$, the values of α_1 and α_0 are to be estimated from the data by maximum likelihood (Strampelli *et al.* 2018) or Bayesian methods (Jiménez-Hernández *et al.* 2020). Several helpful software packages are available for this analysis, including the package *secr* in R. The literature also contains refinements of the basic method, such as allowing for missing observations (Jiménez *et al.* 2020) or using least cost path analysis instead of Euclidean distance.

A final comment on animal movement concerns transportation networks created or maintained by the animals themselves. Many studies have investigated animal transportation networks with semi-permanent structures like trails, galleries or tunnels (Perna & Latty 2014). In such cases, the investigation of spatial aspects may be easier, but costs are difficult to calculate because transportation itself is not the only cost to be considered. Costs to the animal are also associated with creating and maintaining the network. Assessing these systems should include understanding the substrate for the network and how it interacts with network development.

10.6 Testing Hypotheses with Spatial Graphs

Using graphs to test ecological hypotheses often compares the ability of different graphs to explain the observed structure in the data. With a single variable, we can determine which graph, of several choices, produces the highest autocorrelation of the node or edge values, indicating the best explanatory structure for the pattern of observations. If the similarity between sites decays with distance, the first-order neighbour autocorrelation would be significantly lower in a complete graph (all nodes are first neighbours) than with short-distance edges (only nearby sites are neighbours).

With predictor (independent) and response (dependent) variables, we can evaluate which graphs best explain the dependent variables of nodes (e.g. genetic diversity) or of edges (e.g. site similarity), as a function of independent variables such as patch size (nodes) or inter-patch distances (edges). The graph with the best-supported hypothesis is the one that produces the strongest regression or the largest correlation coefficient for the relationship between the explanatory and dependent variables at adjacent nodes or edges.

We have already mentioned, as an example, the relationships among diversity, species composition and nestedness for Tenebrionid beetles on islands in the Aegean, focusing on the autocorrelation of community characteristics between neighbouring islands (details in Dale 2017, see figures 9.8 and 9.9). In that example, two sparse topological network rules were investigated, MST and the Gabriel graph, and even the single comparison provided insights that would not otherwise be achieved.

In general, we can use a range of appropriate pairs of null and alternate hypotheses, each based on characteristics of nodes or edges, or both, to evaluate explanations of observed structures in ecological systems. In this way, graphs, with the great variety of rules available to create them, provide ecological researchers with a rich and sophisticated approach to hypothesis testing.

10.7 Concluding Remarks

We provided two conceptual developments that lead to spatial graphs and networks. One begins with patterns of point events put together with linear objects by the rule that the lines can only be placed where they join two points. The second begins with graphs as mathematical objects depicting structure in the abstract and then embeds them in space, usually two dimensions, thus forcing their spatial nature. The two developments have different intuitive appeals and the spatial graphs thus produced have a broad range of applications in spatial analysis and, undoubtedly, many more will be developed.

Ecological connectivity and its relationship to the wide range of graph theory measures (Table 10.1) will continue to be both a subject of interest and a topic worth further investigation. That relationship has profound implications for the functioning of natural systems and conservation and re-introduction programmes.

11 Spatial and Temporal Analysis with Multilayer Networks

Introduction

Modifications of network structure to accommodate ecological complexities such as disease spread, metawebs or multipartite structures are best represented by *multilayer* networks like those in Figures 11.1 and 11.2 (Boccaletti *et al.* 2014; Kivelä *et al.* 2014; Pilosof *et al.* 2017; Bianconi 2018; Frydman *et al.* 2023). As the name suggests, a multilayer network consists of several layers, each with its own set of nodes and a set of intralayer edges and possibly with interlayer edges to join nodes in different layers. Figure 11.1 illustrates a spatio-temporal network that maps the onset of phloem necrosis in a plantation of *Hevea brasiliensis* over 4 years, with one layer per year. The nodes are individual stems and edges join neighbours in the same onset year using a Minimum Spanning Tree algorithm (Chapter 2). Figure 11.2 is also a multilayer spatio-temporal network: the arrangement of pools in a wetland (nodes) and the surface water that flows between them (edges). The layers represent the initial structure and its seasonal changes, with layers for drought, flood and post-flood rearrangement. Interlayer edges joining instances of the same node in different layers (*replica* nodes) are *identity edges* (Bianconi 2018).

In this chapter, we will focus mainly on cases where the layers are spatial networks representing different times. Using spatio-temporal multilayer networks allows ecologists to address a series of questions related to the dynamics of an ecological system. Here are some questions that could be analysed with this framework:

- How do we define epidemiological clusters of nodes and change through time?
- What is the best way to control for and estimate the statistical relationship between the spatial and social networks to understand the ecology of a particular disease?
- How do we measure the spatial diversity of metawebs?
- What determines metawebs' variability across space?
- How do the constraints of metawebs spatial context affect their dynamics?
- How do the spatial and temporal contexts affect the relationship between spatio-temporal and social disease networks?
- Does multiple species' connectivity vary through time?
- What is the effect of multispecies interaction on connectivity and dynamics?

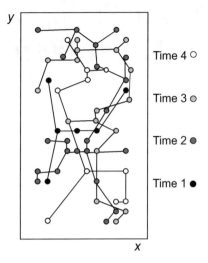

Figure 11.1 *Hevea brasiliensis* necrosis spatial distribution at four time periods depicted as spatio-temporal multilayer network, with one layer per period. The nodes are individual tree stems as mapped in two-dimensional space, and they appear only in the layer corresponding to the year in which the disease appeared. Within a layer, neighbouring stems are joined by edges according to the Minimum Spanning Tree algorithm.

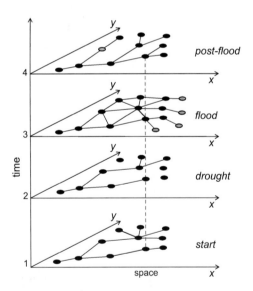

Figure 11.2 Spatio-temporal multilayer network of pools in a wetland; grey indicates pools that were temporary during the flood. Identity edges between layers are indicated by dashed lines.

Before addressing such specific questions, we describe some of the technical details of multilayer networks, to help with access to the broad literature available, which derives from many disciplines.

11.1 Multilayer and Multiplex Networks

In general, a multilayer network is a family of M network graphs, $M = (G, C)$, where G is a layer graph for $u = 1$ to M (Boccaletti *et al.* 2014; Bianconi 2018) and C is the set of interlayer edges.

$$G = \{G_u | G_u = (V_u, E_u)\}, \tag{11.1}$$

V_α and E_α are the nodes and edges in layer α, and

$$C = \{E_{uv} \in (V_u \times V_v) | u, v = 1, \ldots M; u \neq v\}. \tag{11.2}$$

The term *multiplex* applies to multilayer networks in which nodes are the same for all layers ($V_1 = V_2 = \cdots = V_M$), so that only the edges vary among layers and interlayer edges may join only instances of the same node, *replica* nodes (Boccaletti *et al.* 2014; Figure 11.2). Where the interlayer edges may join different individual species or nodes, it may be called an *interconnected network* (De Domenico *et al.* 2015; Kinsey *et al.* 2020). For some multilayer networks, differences among the layer node sets may be the important feature but, in multiplex networks, the relationships among the edges are paramount.

We can aggregate the layers to create the *projection* of a multilayer network, $P(M)$ (Figure 11.3), as a single layer with the union of the layer node sets and the union of all edges between different nodes. Consider a summary of an airline's successful flights in a full week as a projection of the spatial networks of successful flights for all 7 days. The projection aggregates all layers following one of several possible rules: (a)

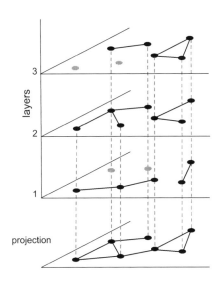

Figure 11.3 Projection of a multilayer network. Nodes in the projection are those that occur in any of the numbered layers above. Edges in the projection occur wherever there is an edge present in any of the numbered layers. Nodes coded in grey are absent from the designated layer. The dashed lines indicate identity edges between layers.

an edge occurs at any position that has an edge in any layer; (b) weighted by a count of edges in that position; or (c) weighted by the sum or average of corresponding edge weights.

The choice of the layer aggregation rule, like the choices for other forms of aggregation in ecological data (e.g. taxonomic aggregation), can have profound effects on outcomes and interpretation, especially for measures of robustness and resilience (Pilosof *et al.* 2017).

11.2 Multilayer Metrics for Emergent Properties

Standard network metrics may need modification for the characteristics of multilayer networks, but many of the challenges are already resolved, based on accommodations for temporal graphs, spatially referenced networks or spatio-temporal networks (Dale 2017; Martensen *et al.* 2017). One approach is to apply local network measures within the layers (like time slices) and complete network measures for the combined network of all layers. This is not a universal solution and some familiar metrics still require adjusting (Kivelä *et al.* 2014; Bianconi 2018), but it is a good start.

11.2.1 Node Degree and Related Measures

A node's *intralayer* degree is the number of within-layer edges attached to it: $k_u(i)$. Any replica nodes can have different intralayer degrees in other layers, $k_v(i)$ (examine the fourth node in Figure 11.2, indicated by the identity edges). The *interlayer* edges of node i in layer u to nodes in layer v are counted by $k_{u,v}(i)$. The total degree of node i in layer u is the sum intra- and inter-layer degrees:

$$K_\mathrm{u}(i) = k_u(i) + \sum_v k_{u,v}(i). \tag{11.3}$$

Some caution is needed here in how the concepts of node and degree are used. In a multilayer network of mapped locations at different times, each site is its own node in each time slice and Equation (11.3) applies; consider the stems in Figure 11.1 with interlayer edges between nearest neighbours afflicted by the disease in adjacent time periods. In multiplex networks, the sites can be considered nodes, each with several instances, one per time period, so that a node is a multilayer entity. Then, a node's multiplex degree is a vector:

$$\mathbf{k}(i) = (k_1(i), k_2(i), \ldots k_M(i)). \tag{11.4}$$

That difference needs to be considered in the interpretation of total node degree, especially as the terms are not consistently applied.

With that understood, a *clustering coefficient* or *triad completion* can be defined as the proportion of node triplets with at least two edges that have all three to complete the cycle and it is easy to distinguish between the intralayer measure of triad completion and its all-layers measure (cf. Boccaletti *et al.* 2014). Figure 11.4 illustrates the

11.2 Multilayer Metrics for Emergent Properties

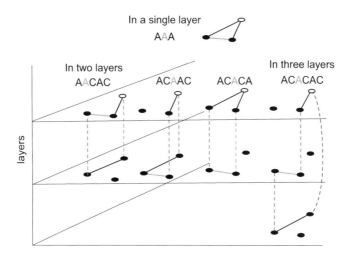

Figure 11.4 Triad completion in a multilayer network, showing the many ways that this can occur using both inter- and intralayer edges. A indicates an edge of the triangle, and all are within a layer; the grey coded A completes the triangle. Interlayer edges of identity are dashed and coded as C in the node sequences.

multilayer possibilities in a multiplex network, based on the interpretation that the replica nodes on different layers are instances of the same node (see Cozzo et al. 2015).

For a multiplex network with time slice layers, consider the correlation of the degrees of replica nodes in different layers, $k_u(i)$ and $k_v(i)$, using any one of several measures of correlation (Pearson, Spearman or Kendall). Focusing on edges instead of nodes, we can look at the *edge matching coefficient*; this is the number or proportion of edges in common between two layers. For layers u and v this coefficient at node i is $o_{uv}(i) = \sum_j a_u(i,j) a_v(i,j)$, where a is 0 or 1 to designate adjacency in that layer. The coefficient including all nodes is $O_{uv} = \sum_i o_{uv}(i)$ (Boccaletti et al. 2014). Edge matching contributes to measuring the similarity between layers in a multiplex network and, under the name 'link overlap', either global (over all locations) or local (at a focal node and its neighbours) layer by layer (Bianconi 2018). For a spatial network, the local form provides direct spatial analysis because the matching scores can be mapped onto the network's spatial context.

11.2.2 Walks and Paths

For multilayer networks, walks and paths can be based on the set of all edges. A *walk* is an alternating sequence of nodes and edges, with each node attached to the next in the sequence by an edge. A *path* is a walk that uses no element twice. A *cycle* is a path that begins and ends at the same node. If there is a path between any two nodes, the network is *connected*. The *length* of a walk or a path is the number of edges, or, if the edges are weighted, the sum of the weights. Interlayer edges can be included either as equivalent to intralayer edges and counted similarly, or in a separate category to give a different kind of distance (makes sense for layers as time slices).

In any spatio-temporal network, the dimensions of space and time are not directly commensurate. In some applications, the spatial distance is the number of steps from source to sink, the sum of the edge weights or the sum of the physical distances of the edges. When the time dimension is divided into finite and equal intervals, the number of time steps will be affected by the *temporal granularity*, which is the duration of the intervals. Temporal distance can be defined as the number of steps or as the time elapsed from source to sink, called *latency* (Williams & Musolesi 2016). Then, the spatio-temporally shortest path is the temporally shortest path with the minimum spatial length (Williams & Musolesi 2016). This definition can be generalized for multilayer networks: the shortest path between two nodes has both the shortest interlayer length and the shortest intralayer distance. For the spatio-temporal version, this concept is known as the *spatial efficiency*, which is the average reciprocal distance between pairs of nodes, and the *temporal efficiency* is the average reciprocal latency between all pairs of nodes (Williams & Musolesi 2016).

Simple edge counts of any category (Euclidean, geodesic, least-cost distance) or sums of inverse weights are the most straightforward assessment of interlayer paths (Boccaletti *et al*. 2014; Kivelä *et al*. 2014). Another emergent measure for multilayer networks is *interdependence*, the proportion of shortest paths that include at least one interlayer edge (Cozzo *et al*. 2015).

Much information about network structure can be based on the characteristics of random walks, their lengths and the number of intra- and interlayer edges (see Figure 11.5*a*). Random walks indicate the flow of information or influence through the multilayer structure within and between layers; in temporal networks, random walks are constrained to respect the direction of time (see Bianconi 2018). Figure 11.5*b* illustrates a walk (from node 1 to node 7) which takes place step by step over several intervals, as in the multi-generational dispersal of a species among sites (see Boulanger *et al*. 2020).

11.2.3 Centrality and Node Ranking

Another key property of networks is *centrality*; it measures the importance of a node, related to how the node's removal will affect network structure and function. Layers can be compared using within-layer measures from monolayer networks, including node degree, but many measures have been proposed for this property (De Domenico *et al*. 2013). A multilayer network centrality measure can be created from a weighted sum or an average of the metric's values for centrality in the individual layers (see table 3.1 and figure 3.8 of Dale (2017) for a summary of these metrics). This aggregated measure may be easy to calculate but it may not match the desired concept. De Domenico *et al*. (2015) introduced an alternative multilayer metric, similar to centrality, called *versatility*. This measure summarizes the centrality values of each node *calculated over the whole structure*, not layer by layer. They explain the logic of this approach with an authorship example (laid out in their figure 2) and provide two versions: *eigenvector versatility* and

11.2 Multilayer Metrics for Emergent Properties

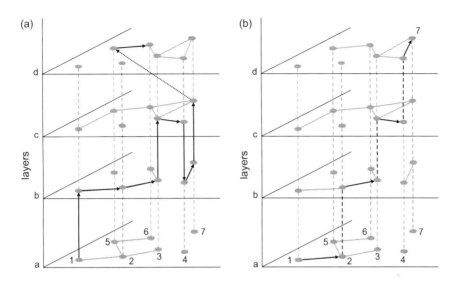

Figure 11.5 (*a*) Random walk on a multilayer network, with possible 'teleport' jump (dotted edge). (*b*) Random walk as multigenerational dispersal.

PageRank versatility. Lv et al. (2021) similarly provided two approaches, eigenvector- and PageRank-based, for multilayer temporal networks.

The six layers of Figure 11.6 represent stages of community development and the edges are positive correlations of spatial proximity between pairs of the five species (A, B, C, D and E) present at all stages. We rank the species by centrality in each layer, highest as rank 1 to lowest, with nodes of equal centrality receiving tied ranks. When the layer ranks are summed, the consensus is that node C has the lowest centrality. When the layers are aggregated, the aggregated network is a complete graph in which all nodes have equal ranks. Using the entire six-layer network and weighting the layers by the number of edges, node C has the highest multilayer centrality (versatility), as detailed in Figure 11.6 (see also Dale & Fortin 2021).

Timóteo et al. (2018) measured versatility in a multilayer network to investigate the spatial structure of seed dispersal interactions in the Great Rift, with the layers being identifiable habitats. They found that the importance of dispersers to this system was captured by multilayer versatility (a version of the De Domenico algorithm) but not by the more standard metrics from networks.

11.2.4 Clustering

Determining modularity and identifying clusters are important aspects of network analysis in any ecological application, whether the nodes are sites, species or individuals. (In the network literature, clusters of nodes are called 'communities' but, to avoid confusion with the ecological term, we will use 'cluster'.)

Figure 11.7 illustrates one cluster detection method for multiplex networks of time-slice layers with nodes arranged in space. In each layer, the maximally connected

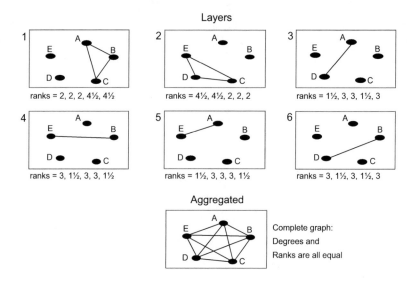

Figure 11.6 Versatility versus centrality in a multilayer network. The five labelled nodes have edges as indicated in the six layers of the multilayer structure. In each layer, the nodes are ranked according to their centrality from most central to least. Nodes that are tied for centrality within a layer receive equal ranks, with the rank total always 15. Based on individual layers, node C has the lowest centrality, but in the multilayer equivalent, *versatility*, in which the layers are weighted by the number of edges, C has the highest (see text).

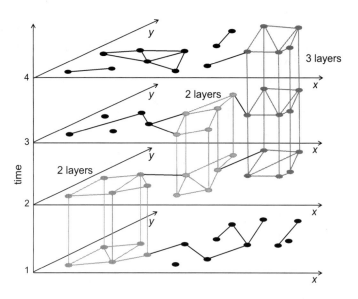

Figure 11.7 Changes in clusters according to the number of layers analysed in a multilayer network. According to how many and which layers are included in determining multilayer clusters, they can be found in any number of layers (2-, 2- and 3-layer clusters shown).

subgraphs are identified in an iterative process and compared with clusters identified in the adjacent layers. The three multilayer clusters indicated in Figure 11.7 occupy two, two and three layers, respectively.

Kim and Lee (2015) provided a useful guide to multilayer cluster analysis based on a survey of 18 publicly available data sets and a range of methods. They designated seven desirable properties for methods and we have selected three as the most important for ecological applications:

- Application to more than two layers: Most ecological networks that are multilayer have more than two layers.
- Algorithm insensitivity: The ability to apply different algorithms to improve cluster detection is highly desirable.
- Independence from the order of layers: For time-slice networks, sensitivity to the order is appropriate but, in general applications, independence is preferable.

Based on this list and Kim and Lee's (2015) table 3, the most highly recommended methods for clustering are from spectral analysis of the adjacency matrix or of the Laplacian (Dong *et al.* 2012). DeFord and Pauls (2019) provided a commentary on spectral clustering designed specifically for multiplex networks, which is an area of ongoing research.

11.2.5 Spectral Properties

The spectral properties of multilayer networks contain much information about their structure and are usually based on matrices specific to multilayers such as the *supra-adjacency matrix*. This is a 2-dimensional matrix of matrices, giving the intralayer adjacencies in sub-matrices along the main diagonal and the interlayer adjacencies in the blocks off the main diagonal (Boccaletti *et al.* 2014). If the supra-adjacency matrix is $\mathbf{A_M}$, and the diagonal supra-matrix of the nodes' intralayer degrees is $\mathbf{D_M}$, a *supra-Laplacian* can be derived as $\mathbf{L_M} = \mathbf{D_M} - \mathbf{A_M}$. The dynamics of layered networks are closely related to the spectral properties of the supra-Laplacian (Boccaletti *et al.* 2014; Kivelä *et al.* 2014). For example, the second smallest eigenvalue, the *algebraic connectivity*, and its eigenvector give important information about the network topology and structure (Kivelä *et al.* 2014). Other spectral properties provide information on connectedness: the spectral radius of a strongly connected network decreases when nodes or edges are removed; this has implications for the spread of an epidemic through the multilayer system (Kivelä *et al.* 2014).

Given the complexity inherent in the layered structure of these networks, the pursuit of spectral properties is not easy, and we will refer the reader to Boccaletti *et al.* (2014), Kivelä *et al.* (2014), Bianconi (2018) and DeFord and Pauls (2019) as starting points for further investigation. We suggest that the properties and dynamics of multilayer networks, *even with only two layers*, will be quite different from the dynamics of otherwise similar monolayer networks. Zeller *et al.* (2020) demonstrated the insights available from two layers in investigating the dynamics associated with landscape connectivity, using layers for structural and functional connectivity.

11.2.6 Resilience, Robustness, and Fragility

For properties related to resilience, robustness and fragility, going beyond monolayer can change things radically even with only two layers. ('Getting out of Flatland ... can radically change the game', Dale & Fortin 2014). Think of the change of dynamics and tactics for tic-tac-toe in three dimensions rather than just two. The effects of topology on network resilience can be unexpected (Boccaletti *et al.* 2014); for example, a failure in one layer can produce a cascade of failures in other layers. Work by Kéfi *et al.* (2018) suggests that specific combinations of interactions in ecological multilayer networks can improve species persistence, productivity and community resilience. The dynamics of these networks can be related to percolation or the spread of disease (Kivelä *et al.* 2014; Hutchinson *et al.* 2019). Increasing the layers in a multiplex percolation network can increase fragility but, depending on the model, the addition of layers sometimes increases robustness by creating redundant interdependencies among layers (Radicchi & Bianconi 2017). Huaylla *et al.* (2021) found that increasing layers from two to three (pollinator–plant–herbivore) had profound effects on how modularity and centrality identify keystone species, and we would expect the same for the keystone modules of multi-resource omnivory (McLeod & Leroux 2021). The overall effect of multilayer structures is that the layering has important consequences which are not currently well understood (Kivelä *et al.* 2014; Hutchinson *et al.* 2019). This topic is the subject of active research from which important and new insights will result.

11.3 Null Randomization Procedures

In Section 11.2, we described several metrics related to the structure of multilayer networks, especially those suitable for spatio-temporal applications. It is usually best to evaluate these measures by randomizing the networks themselves, generating null models with which the observed values can be compared. The multilayer structure makes possible a range of methods for randomization including some with restrictions. Bianconi (2018) provided a useful summary of three of those available and suitable for ecological data analysis.

11.3.1 Replica Node Randomization

The same network is maintained in each layer, but the one-to-one mapping of node-to-node between layers is randomized, as if by a random re-labelling of the nodes within each layer. The effect is to remove the interlayer degree correlations.

11.3.2 Independent Layer Randomization

The method is to randomize the networks within the layers independently, using a simple swap algorithm. Randomly choose two edges; say they are A–B and C–D. Swap edges to create A–D and C–B, provided neither already exists in the layer. Iterate this process until the characteristics no longer change. This algorithm removes the *link overlap* or *edge matching* between the layers while maintaining the within-layer degree structures.

11.3.3 Randomization Preserving Multidegree Sequences

The third method is more complicated, relying on nodes as multilayer entities; that is, each node appears in every layer. For a multiplex network, the intralayer edges can be summarized in the form of *multilinks*. The multilink for nodes i and j is a vector of M elements m_{iju} which take value 1 if i and j are joined by an edge in layer u and 0 otherwise (Bianconi 2018). Thus, any pair of nodes, i and j, is connected by a single multilink which is a yes/no vector coding the presence of an edge between i and node j through the layers from 1 to M. For M layers, 2^M multilink patterns are possible. For multilink randomization: randomly choose *two* non-zero multilinks, joining say A–B and C–D. Swap these multilinks to be A–D and C–B, except when either or both is already non-trivial. Iterate the swap process until the network measures stabilize. This randomization preserves the interlayer degree sequences and the edge matching (link overlap) of the original.

Further choices are available for null models of multilayer networks and these depend on the kind of data and the questions of interest. This is especially true when the edges have labels, weights or directions. Instead of randomizing the existing network derived from data, a different approach is to randomize the data before a network is created; see Farine (2017) for a useful discussion of the process.

11.4 Getting the Most from Multilayer Networks

As the examples that follow will show, multilayer networks can be applied to spatial and spatio-temporal analysis of ecological structures and properties and they also belong in complex system modelling to study and predict ecological dynamics. Some of the best examples are related to epidemiology, metapopulation dynamics and metacommunity structure. We have selected three thematic areas of disease ecology, metawebs and multispecies connectivity. The purpose is to provide guidance on what has already been achieved and to suggest further ways in which this approach can be used.

11.4.1 Behavioural and Disease Ecology

Spatial, social and spatio-temporal aspects of behavioural and disease ecology are often analysed using multilayer networks, with the layers depending on the situation, but frequently integrating physiological, social behavioural and ecological processes (Finn et al. 2017; Silk et al. 2017).

Kinsey et al. (2020) advocated the use of multilayer networks for studies in veterinary epidemiology, examining direct and indirect contact patterns among swine farms to answer questions such as how to define epidemiological clusters of nodes. In this example, the farms are nodes in two network layers based on (a) spatial distance and (b) animal movement. They examined degree- and eigenvector centrality, as well as node clusters (using Infomap; Farage et al. 2021). They found that eigenvector centrality and spatial proximity were not the same and were able to identify epidemiologically linked farms as clusters for multimodal transmission. This study has obvious parallels in wildlife ecology and demonstrates the importance of linking social and spatial network analysis for epidemiological insights.

Albery et al. (2021) also considered the relationship between spatial and social disease networks in animal populations, emphasizing the importance of collecting spatial and social network information simultaneously to unify the two approaches to disease ecology. The intent was to answer the question of how best to control for and estimate the statistical relationship between the spatial and social networks to understand the ecology of a particular disease. They discussed the various methods of analysing the structure of those social networks at node and edge (dyadic) levels and the use of exponential random graph models (ERGMs; Evans et al. 2020) and latent space models, emphasizing the importance of understanding the modes of disease transmission. For our discussion, their commentary on spatial confounding for social network analysis by permutations echoes our emphasis on the importance of restricted or constrained randomizations for spatial analysis (Chapter 2).

Robitaille et al. (2021) studied the effects of scale on multilayer networks, based on the social networks of caribou on Fogo Island (Newfoundland, Canada). They included social scale, spatial scale and time scale in the forms of social distance, landcover resolution and time windows or total observation numbers. They also used three habitat layers: foraging habitat, forest and open. Network measures examined included unweighted and weighted degree centrality within layers and node multi-degree and edge matching (*overlap*) across layers. The results indicate an optimal social scale of 20–100 metres, matching the 50-metre buffer often used. Increasing the spatial resolution decreased the foraging habitat area and the edge matching coefficient for that habitat but had little effect on edge-matching in the forest or open habitat. The temporal windows showed variability for several characteristics, with edge matching being lower in the June to October windows and higher in the November to May windows. Including more observations increased the connections among individuals and decreased the variance in the network metrics; both edge matching and multi-degree measures increased with increased numbers of observations. The authors concluded that multilayer networks provide a powerful approach to answering questions in behavioural ecology and provide a list of such questions that relate social networks to spatially or temporally specific associations.

11.4 Getting the Most from Multilayer Networks

11.4.2 Metawebs

Metawebs are multilayer by their very nature and the layers may be defined in several ways: by spatial location, by time or by interaction type (Pilosof et al. 2017). Consider Figure 11.2 with pools in a wetland at four times of observation. The spatial structure affects the system dynamics, including local extinction and dispersal, and the existence or loss of species interactions. Considering the food web at each pool gives a metaweb for the whole collection of pools as species enter and leave individual sites (Figure 11.8; cf. Poisot et al. 2012, figure 1). Combining the multi-slice spatio-temporal network of sites with site interaction networks creates a network-of-networks or metanetwork (Figure 11.8). Note that only the first-order network (the pools) is multiplex or multi-slice, with the same set of nodes in each layer; the nodes in the second-order networks (of the species within a pool) may not be the same among pools in space or within a pool at different times. In the 'diversity' chapter (Chapter 9), we speculated about possible measures of diversity at the edges of networks. In the metaweb context, Poisot et al. (2012) provided exactly that, in the form of β-diversity for interactions between and among sites as measures of network dissimilarity. As we described for species diversity in a spatial network, in our network of interaction networks, we suggest a map of the sites with edges between sites labelled or colour-coded for the interaction similarities between them in the same spirit as coding the relative nestedness of sites on spatial neighbour network edges (Chapter 9, Figure 9.8). Because we are dealing with a multilayer network, with the potential for interlayer edges of identity for the sites at different times, these can be labelled or

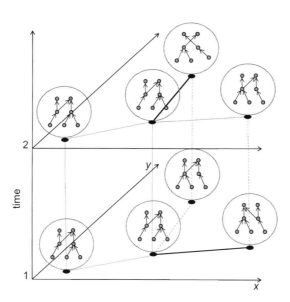

Figure 11.8 Metaweb of pools with inter- and intralayer similarities among pools coded by colour: from grey for most similar, to dark green for most different.

colour-coded for the interaction β-diversity or similarity of individual sites between adjacent time layers (Figure 11.8).

Following further the theme of spatial networks of interaction networks, Lurgi et al. (2020) used a multilayer approach to study the geographic variation of marine intertidal communities. Each site produced a multiplex network of interactions with the same set of species in each layer; the layers were (1) trophic interactions (the metaweb), (2) positive non-trophic interactions and (3) negative non-trophic interactions. They used nine network metrics as appropriate to the kinds of interactions and found that the network structure was affected by local species richness but modified by environmental factors such as sea surface temperature and upwelling. An interesting observation was that non-trophic negative interactions are more sensitive to spatial structure variation than trophic or non-trophic positive interactions. Their analysis was detailed and extensive, including structural equation models (SEMs; Grace et al. 2012) to investigate the pathways by which geographic environmental variability affects the structures of these networks, such as modularity and the number of links per species (see their figure 2). More could be done with the spatial aspects of their results, perhaps by spatial maps of neighbouring site comparisons, as we have suggested for other aspects of spatial analysis.

The spatial resolution and location were found to affect the group structures of marine food webs by Ohlsson and Eklöf (2020). Here the layers of the network are the 25 subregions of the Barents Sea recognized in their study. The group structure is essentially a form of intralayer modularity but produces clusters with ecological interpretations (Allesina & Pascual 2009; Baskerville et al. 2011). For multilayer network analysis, they compared the group structure for the entire data set with those of the subregions. Their spatial analysis found that the subregional webs have similar group structures for nearby regions, even if the species lists are not fully consistent (their figure 2).

For the question of how spatial constraints may affect metaweb dynamics, we refer to the work by Barter and Gross (2017), modelling with random geometric networks and studies of networks-of-networks (Gao et al. 2014; Gravel et al. 2016). The idea is to create a network-of-networks by combining a multi-slice spatio-temporal network of sites with the sites' interaction networks (Figure 11.8). Traditionally, the nodes would be species with edges of identity joining replica nodes between times or between sites. The alternative is to have the ecological interaction edges become second-level nodes with identity edges joining instances of those between times or sites (see Dale & Fortin 2021, figure 4.13). Figure 11.8 illustrates the concept focusing on four pools and two time periods from Figure 11.2, as shown, and a simple food web of seven species. Even if the species are consistent among sites, the first level edges of the interaction subnetworks, the species' interactions, may not be. These become second-level 'nodes' which can be joined by edges of identity. Diversity is a synthetic variable and, for spatial and temporal analysis of the multilayer network of the metaweb, the identities of the interactions that are gained or lost between locations or between times may be the most important details. The identity edges for the

11.4 Getting the Most from Multilayer Networks

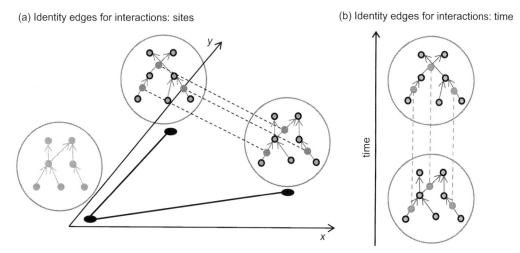

Figure 11.9 Metaweb with species constant (a multiplex network) but interactions varying, showing edges of identity for *interactions* (a) between sites and (b) between times.

interactions, as second-level nodes, provides that detail. Their consistency across time can serve as an indicator of the resilience of site network dynamics.

This conceptual approach may be pursued both by elaboration with models and by exploring further the data sets already available as metawebs or more diverse interaction networks. Labelled edges of identity for the interactions in the subnetworks should provide a range of interesting insights (Figure 11.9). Some of the spatial constraints on the interaction dynamics will act through the connectivity among the sites in space and time and, therefore, understanding that connectivity should also benefit from a multispecies network approach.

11.4.3 Multispecies Connectivity

The conceptual richness of the multilayer approach may be useful even if it is not fully exploited. The concept of a metaweb is based on the movement from one area of habitat to another, so that each species lives as a metapopulation and each set of species as a metacommunity. This view of the natural world then requires connectivity between patches to allow the movement of individuals or propagules. In assessing any area or landscape for connectivity, the multilayer approach is an obvious choice, but the layers are themselves objects of choice, with many possibilities. The layers could be times, as in Figure 11.2, where the connectivity is assessed at different seasons or under different conditions. The layers could be species; connectivity for waterfowl could be quite different from connectivity for fish. The layers could be sub-habitats, such as lentic versus lotic in the system of ponds.

To investigate connectivity in pond networks, Godet and Clauzel (2021) used a three-layer network, where the layers were three methods for defining landscape graphs: a land cover map with nodes and edges defined by expert opinion, a habitat suitability model and a hybrid method with a habitat suitability model defining the nodes and land cover maps defining the edges. One result is a set of species × method maps, colour-coded for the potential for local connectivity between habitat patches (nodes) from null to maximal. The three methods gave different maps, although all have similar nodes, and comparing these three layers and their properties provides the potential for more multilayer spatial analysis. With the methods treated as the layers, measures of edge-matching redundancy and variability can be assessed by looking at the factors affecting the discrepancies. What can be done in addition, however, is to consider different species as the layers, looking at six amphibians and one reptile, giving one 7-layer network per method. Then, in addition to comparing the results for the seven organisms, a true multilayer analysis can combine the results for metacommunity outcomes and the development of priorities to provide an examination and measurement of the multilayer centrality (versatility) that would be particularly useful. Assessing the locations and participants in edge matching (overlap) would also provide helpful insights. We will discuss other examples of similar studies, with similar comments and suggestions.

Mimet et al. (2013) studied functional connectivity for three groups of bird species, based on their habitats (farmland, forest or generalists). They examined the composition and configuration of landscapes and compared 1982 with 2003, using all three groups of species. As they suggest, since some landscape types were favourable for several species groups, those could serve as junction landscapes for networks accommodating a variety of ecological requirements. This provides a background for further studies of corridors and stepping-stones using graphs (Saura et al. 2014), as they suggest, but ideally takes advantage of three groups of species and two time periods to explore the multilayer network approach to the spatial and temporal analysis of landscape connectivity.

The multilayer approach might have been helpful in further analysis of the spatial structural and functional connectivity, especially where interactions between the species may affect the connectivity properties. We can make the same suggestion for other studies of connectivity which are essentially multilayer, but not explicitly analysed using the full set of multilayer network techniques available (Jennings et al. 2020; Zeller et al. 2020). We note that these analyses are available in R (*multinet*), Python (*Pymnet*) and Julia (*MultilayerGraphs.jl*) and that Frydman et al. (2023) have provided guidelines for an ecological multilayer network package in R (EMLN).

Multispecies connectivity studies all contribute to our understanding of the dynamic relationship between spatial structural connectivity and functional connectivity networks at the metacommunity level. Multispecies connectivity has obvious analogies with multispecies occupancy models (see Rota et al. 2016) and the spatio-temporal patterns of species assemblages based on assembling the layers of predictions for individual species with constraints from macroecological models (Guisan & Rahbek 2011). These also have conceptual ties with the spatial structure effects on metaweb dynamics discussed in Section 11.4.2 and with the spatial guilds in food webs such as those studied by Baskerville et al. (2011) in the Serengeti. For each of these approaches, multilayer networks have insights to offer.

11.5 Multilayer Networks and Spatio-temporal Analysis

As seen from the examples given, many characteristics of multilayer networks can be approached using graph theoretical properties for monolayer networks, but it is necessary to define the emergent properties of multilayer structures and to develop measures for them (Kivelä et al. 2014). Consider the spread of disease through a sessile population with the individuals as the nodes and time-labelled edges for the date of the nodes' contemporaneous disease appearance; the orderliness of disease spread can be indicated by initially short edges and few cross-temporal edge crossings (Figure 11.10; Dale 2017). For any network in which the layers represent time slices, differences between layers can be interpreted in the context of change or ecological memory within the system, with emergent properties to match. Given the complexities and sophistication of layered networks, more emergent properties will surely be discovered and described.

While the development of truly multilayer (multiplex or multi-slice) networks for ecological applications may sometimes seem daunting just from the data collection, data management and computational complexity aspects, ecologists need to be thoughtfully aware of the possibilities afforded by only two or three layers. Bianconi (2018) provided suggestions similar to those related to the example of Fraser Fir disease just described. Consider the multiplex (duplex) disease spread among humans with one network layer of physical contact and transmission and one layer of information-sharing about the disease; the interplay between the two affects both in obvious and subtle ways. Consider also the duplex spread of two viral diseases in the same population and the implications for population dynamics (Sahneh & Scoglio 2014; Bianconi 2018). The two diseases have their own layers but they are not independent and the network is spatially embedded by the locations of individuals.

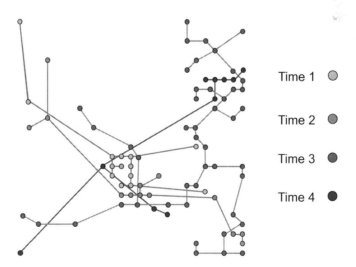

Figure 11.10 Fraser Fir mortality as time-labelled map. Multilayer spatio-temporal network of tree nodes mapped in two-dimensional space. Nodes are tree stems coded for time of disease as indicated; edges join nodes of the same time period using the Minimum Spanning Tree algorithm.

An ultimate question for ecologists may be predicting the climate's future behaviour at a range of spatial and temporal scales based on the analysis and interpretation of spatio-temporal data from climate networks. A climate network consists of a set of observation sites as its nodes, with edges determined by the similarity between sites. The similarity can be an assessment of any of several variables over a range of spatial and temporal scales (Donges *et al.* 2009). Near-surface, high-resolution monthly mean air temperature data are commonly chosen (Fan *et al.* 2018) producing a mono-layer network, but multilayer versions may include other data such as sea level pressure, sea level temperature and horizontal wind speed (Fan *et al.* 2021). A multilayer climate network represents the opportunity for great insight as well as computational challenges (Steinhaeuser *et al.* 2010).

11.6 Concluding Remarks

Clearly, multilayer networks have many potential applications in ecology because the approach can help clarify the complexities of network data when appropriate layers can be assigned. Multilayer network analysis allows us to formulate and evaluate complex systems while being explicit about processes both within and between layers (Pilosof *et al.* 2017). We have only begun to explore the complexities and sophistication available by adding directions, signs or weights to the edges, in both intra- and interlayer. In the literature, implications of social structure and network layers on the spread of disease are closest to strictly ecological research, but many others will prove to be closely analogous to studies we might perform, such as economic or transportation networks. Spatial and temporal analysis using multilayer networks is one of the most promising areas for sophisticated ecological analysis and for our understanding of population dynamics and community interactions.

12 Closing Comments and Future Directions

12.1 Reminders and Challenges

In the second edition, this section was called 'Myths, Misunderstandings and Challenges'. The myths and misunderstandings have been addressed, here and elsewhere, to the extent that what we provide now is best described as 'reminders'; the challenges remain.

Ecological studies commonly recognise the importance of the spatial aspects of ecological systems and include them somehow in their design and analysis. Spatial effects are various and may act simultaneously or interdependently, and include spatial scale, autocorrelation, locational effects, functional connections and the interaction between spatial pattern and temporal process. This complexity has given rise to some confusion and we will give reminders of solutions with clarifications acknowledging the challenges of including the spatial context in ecological studies.

We begin with the following considerations:

- Without spatial pattern there is no spatial scale. A homogeneous or totally random system provides no landmarks that create spatial scale or from which scale can be detected.
- Any observed pattern may result from several different processes (Figure 1.2). A central question concerns the relationship between observed pattern (spatial structure) and the processes that generate it or arise from it. It is generally agreed that we cannot safely deduce process from pattern, in part because different processes can produce similar spatial signatures (Figures 2.1 and 1.2). Issues related to spatial causal inference will be discussed in Section 12.6.3.
- The processes causing pattern(s) act together, with interactions reinforcing or interfering (possibly with spatial or temporal lags) or they may be fully confounded and impossible to disentangle.
- Ecological questions lacking an explicit spatial component should still account for spatial context if only to rule out possible spatial effects. For example, in studying predator–prey dynamics of two territorial populations, any spatial effects should be removed before non-spatial questions are addressed.

It may be difficult to distinguish among the various forms of spatial dependence (see Chapter 6), so understanding the ecological properties of the system

(Chapter 1) may help determine the cause(s) of any dependence detected using the range of methods available to detect and evaluate it (Chapters 6 and 7). Spatial dependence allows us to generate hypotheses about processes and pattern–process interactions from spatial patterns, but it can affect statistical tests through biased parameter estimates and reduced effective sample size. The outcome is the interesting tension, expressed as the question: 'trouble or new paradigm?' (Legendre 1993).

12.1.1 Reminders

Given these considerations, we should clarify what is important in spatial analysis in the set of related concepts; these are not independent but we have picked a few to address explicitly.

12.1.1.1 Distance and Similarity

Tobler's Law says that near things are more closely related than far things and nearby locations are more similar and interact more than distant ones (Chapter 1).

Reminder: Sometimes this is true, but ecological data are often patchy, producing cycles in similarity as a function of distance. Secondly, physical connections can outweigh distance; rivers, currents, winds and the movement of animals and human transport can enhance the similarity of distant sites. If Tobler's Law were universally true, the asymptotic variogram would be a good general assumption but, for patchy ecological data, the autocorrelation may cycle between positive and negative with distance, producing a cyclic variogram (Figure 12.1a and b).

12.1.1.2 Distance to Independence (three potential versions)

Version A: If your samples are far enough apart, they could be as good as independent.
Version B: If autocorrelation at a given distance is non-significant, it can be ignored.
Version C: If the data are sparsely subsampled to remove nearby locations, spatial autocorrelation could disappear.

Reminder: Cyclic behaviour of variables can make more distant sites more similar than close ones and functional connectivity plays a part. In some cases, it may be tempting to discard data to increase the distance between samples but that is very wasteful of information and not recommended (Legendre & Legendre 2012).

Watch for the asymmetry of significance. When the observations provide sufficient evidence to reject the null hypothesis, we cannot conclude that the process is truly random. In most circumstances, the total autocorrelation over all distances is the best quantification of its overall strength (Chapter 6). Also, the detection of 'significance' depends on the number of observations at any lag distance; larger distances tend to have fewer observations and are less likely to show significance (Chapter 2). Significant autocorrelation is more difficult to detect at larger scales, but it cannot be ignored.

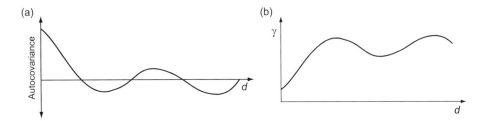

Figure 12.1 (*a*) Autocorrelation as a function of distance showing cycles in positive and negative autocorrelation resulting from patchiness (the alternation of patches of high density and gaps of low density) that is typical of many kinds of ecological data. (*b*) A variogram showing the same phenomenon, as evaluated by a different statistic.

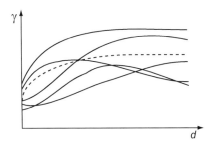

Figure 12.2 The observed variogram (dashed line) is an average over locations at which the spatial structure may be different because of inherent variability or non-stationarity.

12.1.1.3 Ecological Influence and the 'Range' of Spatial Statistics

One might hope that the zone of ecological influence in a biological system can be determined from spatial statistics' *range*, the distance at which spatial dependence approaches zero.

Reminder: Spatial characteristics such as range are *averages* over locations and directions and their values do NOT match the characteristics at actual locations. Many spatial techniques average an index over all spatial locations (whereas non-stationarity or variability may create differences; Figure 12.2) and over all directions (whereas anisotropy or variability may cause directional differences Figure 12.3). Simple inherent variability is always a factor.

12.1.1.4 Diversity Is a Variable Like Any Other

One misunderstanding about diversity is to consider it as a real characteristic of a site, like temperature, that can be measured; its statistics may be complicated, but they capture a genuine characteristic of the assemblage of organisms at a location.

Reminder: Diversity is not a variable like temperature or elevation; it is a synthetic or 'constructed' variable and is highly scale-dependent and non-stationary both spatially and temporally. So, spatio-temporal approaches are necessary for any real

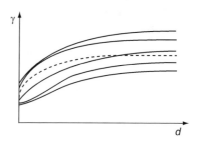

Figure 12.3 The observed variogram (dashed line) is an average over directions for which the spatial structure may be different because of inherent variability or anisotropy.

insight even into the richness component which should be simpler to predict (White et al. 2010). The misunderstanding is one of reification: if we have measures of diversity, there must be a real thing, diversity, for them to measure! There is not really, is there?

12.1.1.5 Spatial Statistics Avoid Difficult Assumptions

One optimistic view is that spatial statistics do not have assumptions like inferential statistics which require independence and asymptotic normality, and so there are none to be violated.

Reminder: Spatial statistics do have assumptions; they may not be as obvious or as restrictive but they exist implicitly or explicitly. The most important one is that the process that generated the pattern is stationary. Then, the presence of significant spatial autocorrelation affects the *effective sample size* (n') which can be smaller than n (Chapter 6).

12.1.1.6 The 'Shotgun' Approach

The suggestion is that the more methods you use in analysis, the more you learn about the data.

Reminder: There are close relationships, both conceptual and mathematical, among many spatial analysis methods we have described and discussed. Many sets of methods are only somewhat complementary and so, while using only one method will not give all the answers that the data can provide, some judgement will help choose the set of methods that provides the best range of insights.

12.1.2 Challenges

The following list provides some general categories of the kinds of difficulties encountered both frequently and in a broad range of study systems.

12.1.2.1 Forms and Sources of Dependence

We have explained spatial autocorrelation in much detail, but autocorrelation is not the only form of lack of independence. Dependence occurs when spatial coefficients are computed at several lags because the same data are used repeatedly in calculations.

Local statistics at location (x, y) will not be independent of those at location $(x + d, y + d)$ for lag r, if $d < r$. Dependence in ecological data arises as spatial dependence; spatial autocorrelation; temporal autocorrelation; phylogeny and familial relatedness (Roberts *et al.* 2017); lack of independence among descriptors (think of forest canopy height and total biomass); and lack of linear independence among variables, producing non-zero covariance. Physical and other functional connectedness can also introduce a lack of independence in a variety of ways and manifestations.

12.1.2.2 Distinguishing the Key Functional Connections

The previous entry leads to the important challenge of discerning the most critical forms of functional connectivity in the systems being studied. Think of the many choices for the links to include in a spatial network (Chapter 10). The correct choice may not be obvious but will affect the outcome of spatial analysis and interpretation or subsequent spatial modelling and parameter fitting.

12.1.2.3 Distinguishing the Key Spatial Scales

Ecologists now deal with a larger range of scales than in the past and we need to determine the key scales at which processes occur (Fortin *et al.* 2012b; Estes *et al.* 2018). Different processes, and different kinds of processes (biotic interactions vs. abiotic conditions), can act at different spatial and temporal scales. It is important to distinguish these, but not easy. Confounding and technical trade-offs present challenges for disentangling scales and determining their relative importance and their contributing factors. Two families of methods have been proposed to determine the important scales: one for sampled locations, based on eigenvalues (such as MEM, Dray 2011; see Chapter 7), and one for fully censused grid data, based on wavelets (Keitt & Urban 2005; James & Fortin 2012; Chapter 7). Further work in this area is required to clarify the best practices for ecological studies.

12.1.2.4 Local and Global Evaluations

Most spatial statistics were originally in a global form, to summarize over the entire area, but almost all can be re-formulated in 'local' versions, to assess subareas. This provides flexible and sophisticated spatial analysis enabling the evaluation of non-stationarity: compare local statistics with the global and compare local statistics at different locations. This approach is related to scan statistics (Chapter 4) and will have relevance for matters of diversity and conservation (Chapter 9). Local statistics can be used as the basis for global evaluation or for the comparison of mapped data. For grid data, Jones *et al.* (2016) described the use of neighbourhoods (like kernel functions) to calculate local measures of similarity between maps based on local comparisons of means, variances and covariances. They provide an example of this approach to compare spatial structures of distribution data of group and singleton sperm whales in the Mediterranean (Jones *et al.* 2016).

Local spatial statistics have advantages for ecological data, even when the questions are global in scope. Some of the difficulties of spatial autocorrelation can be addressed by its localized evaluation. The trade-off is that the more localized the analysis becomes, the less precise any evaluation can be, because of the reduced sample size. This question of precision becomes acute when trying to distinguish local variability from true non-stationarity. In addition, because multiple tests require Bonferroni-type procedures to adjust significance levels, which can be challenging for large numbers of local statistics, they should be used only for exploratory analysis. This commentary has focused on spatial variability, but it applies to temporal variability and spatio-temporal analysis (Chapter 8).

12.2 Back to Basics

We favour a visual approach to problems, as is evident from the number of figures. We advocate visual evaluation for every step of the analysis process. Draw a map of the locations and values; draw a spatial graph or spatial network of the data; draw the time series when you can. Plot the data, plot the results of analysis and, when fitting a model, plot the residuals.

The first step in any analysis is to plot the data. Many problems can be avoided by this simple step. A common mistake is that the x- and y-coordinate axes are used in the wrong orientation or order. Plotting the data can provide guidance in the choice of spatial statistics and exploratory methods. This early evaluation is useful for interpreting the results and understanding the patterns and the processes that generated them.

Having analysed the data, plot the results. A global analysis may detect little pattern in non-stationary data, but the non-stationarity could be revealed by explicit local analysis. For point patterns, for example, Getis' method of plotting the scores as maps may reveal spatial trends (see Chapter 3).

When fitting models, plot the residuals or map them where appropriate. The residuals are often assumed to have a normal distribution and to be independent, but that should be checked; Figure 12.4 shows why: the residuals are clearly not independent; autocorrelation is evident.

The selection of the appropriate spatial extent and grain can be difficult without prior knowledge, but a pilot study can provide information about the spatial and temporal domains of the process, as well as spatial and temporal response scales.

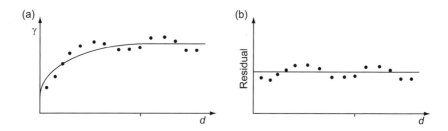

Figure 12.4 The importance of plotting residuals. (*a*) The asymptotic variogram (the line) fitted to the data (the dots) explains much of the observed variability, but it is a poor description in some ways, because it does not detect the cyclic behaviour in the data. (*b*) That behaviour becomes obvious when the residuals are plotted.

Other sources may be useful, such as aerial photographs, remote sensing images, maps and charts, as well as knowledge found in reports and colleagues' expertise.

Larger sample sizes provide more options and more power, especially when the data do not allow parametric analyses and randomization methods are the best choice. In some circumstances, a sample size of 100 may be too small (Chapters 2 and 6). Modelling the spatial structure may require bigger sample sizes and using many more smaller sample units will help.

Studies in landscape ecology use aerial photographs or remote sensing images, which give very large data sets (tens of thousands of pixels). Because the data usually cover a large area, non-stationarity should be assumed (see Chapter 2) and we recommend local statistics, at least as a first step. Global analyses will apply only if the data can be partitioned into spatially homogeneous subareas (Chapter 5). The practical issue of very large data sets is not unique to ecology but is now common in many sciences and the challenges of 'so much data!' at a more general level can be addressed elsewhere.

12.3 Numerical Solutions: Software Programs and Programming

For students in almost any branch of ecology, we recommend acquiring at least some programming skills (in languages like C#, Python, R and Julia). This may sound old-fashioned, given the wealth of software available, but that wealth is part of the reason for the advice. You need to know and understand the details of the analysis being applied.

The second reason for the advice 'Learn programming!' is that the skill allows a researcher to explore methods or modifications without relying on others, providing greater flexibility. In addition, there may be a delay between the development of a method and its general availability. The ability to write or modify programs will allow the researcher to implement the most up-to-date methods.

Detailed knowledge of data storage and the algorithm applied is critical. In many software packages, the details of analyses are hidden from view. Using programs in which the calculation details are not made explicit or cannot be understood can be misleading. Understanding programming will make you more alert to potential mis-coding, especially when using shared code. For this book, we used both software packages, like PASSaGE or BoundarySeer, and programs created in programming languages like Basic, R or Python. The software R now offers both readily available and easy-to-use libraries, the flexibility of user modification and many helpful sources (Bivand *et al.* 2008; Fletcher & Fortin 2018; Plant 2018; among others). It would be a good place to start.

Another theme for this book is the usefulness of numerical (versus analytical) solutions to methodological problems. Consider the edge correction for Ripley's K-function analysis when the study region is not a simple regular shape; a numerical approach like that described in Chapter 3 is an obvious solution. A similar numerical approach is the 'model and Monte Carlo' method for dealing with spatial autocorrelation (Chapter 6). Undoubtedly, reliance on numerical methods to solve complex problems in spatial analysis will continue to grow.

12.4 Statistical and Ecological Tests

We do not have to be expert statisticians to be good ecologists, but we do require a solid understanding of the relationship between the results of statistical analyses and their ecological interpretation. The relationship between the statistics and ecological meaning is not always direct and it may be difficult to determine the most appropriate statistic for a particular ecological question. Statistical significance does not always result from a significant ecological process and so cannot always be interpreted as ecologically important. What is important to know is the strength of the effect detected or the *effect size* (Wasserman 2004). On the other hand, a non-significant statistical result can have important ecological implications, and it is the role of meta-analysis to evaluate sets of outcomes even when non-significant (Gurevitch et al. 2018). The probability values derived from statistical tests do not tell us everything, but they are still important when used correctly as part of making decisions (Mayo & Hand 2022); they are most useful when combined with other measures such as effect size (Wasserstein et al. 2019).

We should add a reminder that spatial analysis is more than just spatial statistics and statistical tests. Exploratory analysis, model development and parameter estimation are all important components. In fact, we have colleagues who argue that statistical tests are only a half-step to knowledge, which really requires modelling and then parameter estimation; they may be right.

12.5 Complementarity of Methods

The methods for spatial analysis described in the preceding chapters are related either conceptually or mathematically (Dale et al. 2002; see Boxes 3.1 and 12.1). Understanding these relationships allows us to choose methods that are complementary in the characteristics they detect or in the range of treatment, such as global versus local. Methods may also complement each other by the inter-conversion of data type (e.g. point events versus sampling units) or by combining single-time procedures with an assessment of temporal change (e.g. patch detection and polygon change). We cannot give an exhaustive list of all useful methodological combinations, but only some examples to illustrate the concept and to provide some guidance.

Consider data consisting of a series of events in a single spatial dimension, like waterfalls along a river. To analyse such data, use the statistic W_m to detect non-randomness and then h_m to detect clumping of events (see Chapter 3) or the one-dimensional Ripley's K-function analysis to detect scales of over- or under-dispersion. If the events are clumped, plot the Ripley's scores to find the locations of greatest

> **Box 12.1** Relationships among the Spatial Analysis Methods Modified from Dale *et al.* (2002)
>
> There are close relationships among methods within each subsection of the table, with the closest relationships by similarity or by derivation indicated by the lines between the names of methods. Those relationships are described in detail in the appropriate sections of the text.
>
> (*a*) Relationship among variance:mean methods.
> (*b*) Relationship among population data methods derived from Ripley's *K* function to estimate spatial association from point data.
> (*c*) Relationship among population data methods derived from block quadrat variance method to estimate spatial structure from continuous quadrat data.
> (*d*) Relationship among population data methods based on continuous quadrats to estimate the spectral structure of the stationary (Fourier-based methods) or non-stationary (wavelet-based methods) data.
> (*e*) Network connectivity is used both in join count and spatial autocorrelation methods.
> (*f*) Relationship among sample data methods using network or Euclidean distances to compute spatial autocorrelation or spatial variance.
> (*g*) Relationship between Mantel tests and other methods that estimate spatial autocorrelation based on sample data methods.
> (*h*) Relationship between the circumcircle count method and other population data methods.

clumping. Whichever approach is chosen, the analysis uses complementary methods that detect different characteristics.

As another case, consider a mapped forest plot with the positions, species and diameters of all the tree stems recorded. The modified Ripley's *K*-function analysis will determine the scales at which the stems (of any species and any size) are aggregated or over-dispersed. Condit's Ω analysis, based on rings rather than circles, will determine whether there are any distance classes of particular interest. If the overall pattern is patchy, Getis' score mapping or circumcircle score mapping can examine the data for non-stationarity and identify the positions of patches and gaps. Mark correlation analysis can determine the aggregation or segregation of tree sizes as a function of distance (see Figure 12.5).

Given that several species are recorded, any of the multivariate approaches described in Chapter 3 can assess interspecific associations. Having already used distance-based analysis, a complementary approach would be to use the neighbour networks to look at the join counts for species pairs, comparing observed and expected counts. Lastly, the point data could be converted to raster format of quantitative data, with square units of a size determined by the previous analyses and subjected to multiscale ordination (MSO) to investigate the existence of multispecies pattern (Dray *et al.* 2012). This analysis scheme is illustrated in Figure 12.5.

Closing Comments and Future Directions

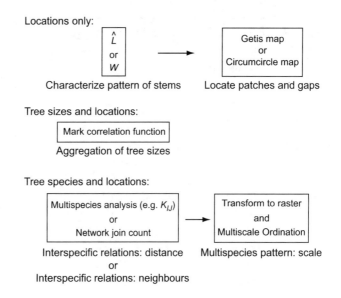

Figure 12.5 Complementary approaches to the analysis of marked event data (stems) in two dimensions: Ripley's K-function or Condit's Ω to characterize the non-randomness of the events' positions, with the Getis' method or circumcircle scores to plot the locations of the centres of patches or gaps for a given scale.

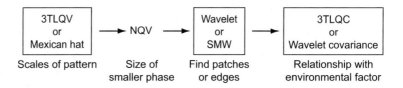

Figure 12.6 Complementary methods for analysing density data in a string of contiguous quadrats: 3TLQV or the Mexican hat wavelet analysis to detect the scales of pattern in the data, with Galiano's NQV to detect the size of the smaller phase and local wavelet analysis or a split-moving window to detect patches or edges. If an environmental factor is also recorded, 3TLQC or wavelet covariance analysis can be used to detect the scales of positive or negative association of density with that factor.

For single-species abundance data from a transect of contiguous quadrats, 3TLQV or the Mexican hat wavelet could be used to determine the scales of pattern in the data, followed by NQV to determine the sizes of the smaller phase. If the abundances are patchy, wavelets or a moving-split window (MSW) can find the edges between the regions of high and low density. If an environmental factor like altitude was recorded, any of the covariance methods described in Chapter 3 (3TLQC or wavelet covariance) could determine the scales at which abundance covaries with the environmental variable (see Figure 12.6).

For the quantitative single-variable data from spaced locations, several methods can evaluate different spatial characteristics (Chapter 4). An omnidirectional correlogram will reveal the isotropic autocorrelation structure as a function of distance. LISA's will

12.5 Complementarity of Methods

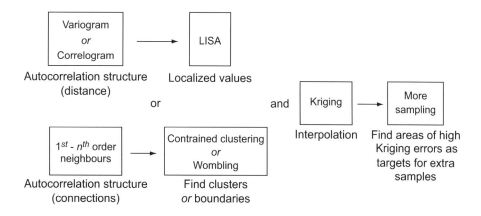

Figure 12.7 Complementary analysis of irregularly spaced records of a quantitative variable: variogram or correlogram analysis to characterize the overall autocorrelation structure as a function of distance, with LISA methods to detect local characteristics. Neighbour networks could be used to characterize the autocorrelation structure, based on connections rather than distance, with constrained clustering to find local clusters of similar values or wombling to detect boundaries between regions of different values. Kriging can be used to locate areas with high estimated variances, indicating a need for greater sampling intensity.

plot localized areas of high and low spatial association. A complementary approach is to assess the correlation of first-order neighbours in a spatial neighbour network, then second-order neighbours and so on. Spatially constrained clustering can identify aggregations of similar values or triangulation-wombling can detect boundaries (Chapter 5). Depending on the purpose, interpolation by Kriging can estimate the variable where it is not measured. Lastly, if that analysis indicates areas of high variance, subsidiary sampling may be indicated to improve the quality of the interpolation. This sequence of analyses is shown in Figure 12.7.

As a last case, consider the hourly position records for a radio-collared animal and a habitat map for those locations. A first analysis would be to quantify the radial (distance) and angular autocorrelation as a function of lag (Chapter 8). Local autocorrelation scores or local tortuosity measures would detect non-stationarity. If stationarity is a reasonable assumption (or at least piecewise stationarity), we can model the data as a basis for Monte Carlo generation of artificial 'data' for comparison (cf. Chapter 6). We could then compare the actual habitat use with either these Monte Carlo 'data' or with randomized positions of the original path of movement on the habitat map to evaluate habitat use. This analysis scheme is illustrated in Figure 12.8. If several members of the animal's herd are radio-collared, a multilayer network analysis is appropriate with one layer for each animal (see Chapter 11).

We cannot describe all combinations, but these examples illustrate the concept and procedures of complementarity. Figure 12.9, based on the 'relationship' diagrams in Dale *et al.* (2002), shows the groupings of methods we have described. It makes it obvious that many other methodological combinations could be pursued. Often the choice of subsequent analysis will depend on the results of the preceding step.

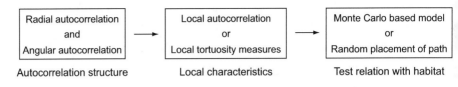

Figure 12.8 Complementary methods to analyse radio-collar position data referenced to a map of habitat types: radial and angular autocorrelation analysis will characterize the autocorrelation structure, which might then be modelled. Local scores of measures such as tortuosity can then detect areas of behavioural intensity. The relationship between the path characteristics and the habitat can be tested by comparison with paths generated by a Monte Carlo method based on the model generated in the earlier steps or by randomization of the path's position relative to the habitat map.

12.6 Looking Ahead

12.6.1 Ongoing Development

We have already pointed out many areas of methodological research that deserve further effort. For ecologists, the 'bottom line' is to clarify the relationship between process and pattern. This may be in the form of the challenge to find and make explicit the right ecological hypotheses to test and to find the right spatial measure to test them. In many instances, our concerns about methods and their weaknesses could be solved by knowing more about the biology of the system. In particular, very detailed knowledge about spatial processes would enable us to refine the methods we use. This parallels the suggestion that pilot studies and other prior information can help us make better decisions in designing surveys or experiments.

Another area to be resolved is the testing of spatial indices and measures for statistical significance. The search for significance tests is complicated by the various forms of lack of independence, both in the data and in the calculations of the statistics. Restricted randomizations seem to be part of the solution, going beyond the simple null hypothesis of 'there is no pattern'. Similarly, we want to compare the spatial pattern at two different areas or a single area at different times; in the latter case, it is more likely that the same underlying process is at play. Moreover, the pattern may be 'significantly' different at two locations, despite being the result of the same process. Comparing spatial patterns can only be partially addressed by means of stochastic spatial modelling (Remmel & Fortin 2013). Statistical tests may be viewed as the first step in a process that leads to a greater understanding through subsequent modelling and parameter estimation.

In Chapter 6, we described several models of spatial autocorrelation, in part as a learning tool. While patchiness, which is common in ecological data, can be modelled in the AR–MA structure, it is not clear how biologically realistic those models are. The question is what ecological processes would produce spatial autocorrelation resembling the models. In most statistics textbooks, the analysis of variance is presented with a model such as:

$$X_{ij} = B_i + T_j + \varepsilon_{ij}. \tag{12.1}$$

The dependent variable (perhaps crop yield) is interpreted as the sum of a block effect, B_i, a treatment effect, T_j, and an error term, ε_{ij}. When the error term is attributed

12.6 Looking Ahead

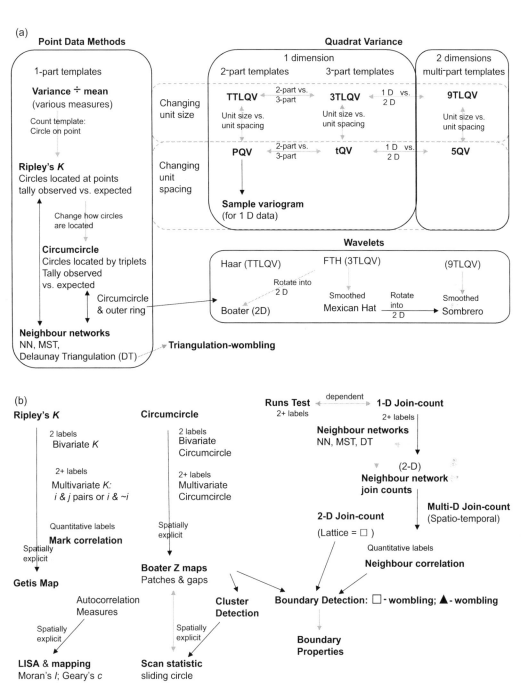

Figure 12.9 Relationships among the spatial analysis methods combining subsets of previous figures. Relative positions reflect the degree of similarities among the methods.

to variation in soil nutrients, soil moisture and light availability, the autocorrelation in yield is very similar to the induced structure described in Chapter 6. In a well-known plant competition experiment, Franco and Harper (1988) found that the sizes of first neighbours were negatively correlated while those of second neighbours were positively correlated. Large plants had smaller neighbours and small plants had larger neighbours, interpreted as the result of competition. In this case, a first-order autoregressive model (Chapter 6) with $\rho < 0$, would be a biologically realistic model of the spatial structure. In many ecological examples, the variable of interest may have both inherent and induced autocorrelation, but the biological and physical processes producing it may be unclear, and so realistic models remain a challenge.

A last comment is a reminder that inherent versus induced autocorrelation may not be distinguishable in the data for the dependent variable (cf. Chapter 6); finding a study design that provides this distinction is worth the effort. Figure 12.10 illustrates the interactions of three forms of autocorrelation in the predictor (x) and response (y) variables (none, global, localized) and the effects of their combinations on autocorrelation. A sampling design that allows separate evaluations of the spatial structure of predictor and response variables will provide a better understanding of the system. In essence, this is approaching spatial causal inference, which we will discuss in Section 12.6.3, but that should be preceded by a discussion of a Bayesian approach to spatial analysis.

12.6.2 The Bayesian Approach

Bayesian methods are increasingly popular in spatial analysis and we will provide some details of their usefulness and potential. We begin with a short general commentary on these methods and then provide a discussion of hierarchical models and their application for spatial analysis.

12.6.2.1 Bayesian Analysis

The Bayesian approach has some notable differences from the classical frequentist statistics usually taught to undergraduates. The concept of 'probability' is more closely associated with degrees of belief or credibility than with long-term averages from repeated trials or the physical properties of simple systems like tossed coins. Instead of calculating the probability of events from hypotheses or models, such as the probability of a fair die showing a 1, the emphasis is on how the observed events affect the probabilities of the hypotheses or model parameters. Consider an apparently standard die that has produced six 1's in a row: how to determine the probability that the die is indeed fair? The Bayesian approach typically begins with some assumed prior probability distribution for the parameter of interest and then determines how the observed data changes our understanding and provides a measure in the form of a posterior probability.

Given a hypothesis with parameter θ, such as the probability of turning up 1 in the random roll of a fair die, and data D, a sequence of ten rolls in which exactly three turn

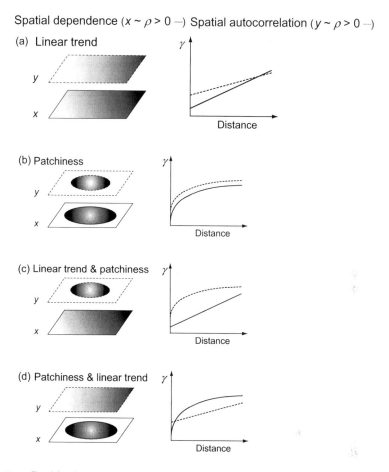

Figure 12.10 Combinations of different forms of spatial autocorrelation (ρ of the response variable y) and spatial dependence (ρ of the predictor variable x), whether continuing to decrease linearly with distance or reaching a plateau. The left panels show the spatial structure of the ecological and environmental data in a landscape. The right panels show the degree of spatial variance (semi-variance, γ) of the response variable. (a) Linear spatial autocorrelation and linear spatial dependence; (b) patchy spatial autocorrelation and patchy spatial dependence; (c) patchy spatial autocorrelation and linear spatial dependence; (d) linear spatial autocorrelation and patchy spatial dependence.

up 1, we are familiar with calculating the conditional probability of D given θ, $P(D|\theta)$. For this instance, $P(3\text{ 1's in 10 trials} \mid \theta = 1/6)$ is:

$$P(D \mid \theta) = \binom{10}{3}\left(\frac{1}{6}\right)^3\left(\frac{5}{6}\right)^7. \tag{12.2}$$

This conditional term also appears in the equation relating it to the joint probability of D and θ:

$$P(D, \theta) = P(\theta)P(D \mid \theta). \tag{12.3}$$

The Bayesian approach requires finding an expression for $P(\theta|D)$:

$$P(\theta \mid D) = \frac{P(\theta)P(D \mid \theta)}{P(D)}. \qquad (12.4)$$

This looks easy to solve, but a few wrinkles make it more difficult than it looks. The same equation is sometimes expressed in terms of likelihood, denoted L, which is an inverse of probability so that the likelihood of a hypothesis, given the data, is proportional to the probability of the data given the hypothesis:

$$\theta \xrightarrow{P(D|\theta)} D$$
$$D \xrightarrow{L(\theta|D)} \theta.$$

The relationship is simple:

$$P(D \mid \theta) \propto L(\theta \mid D). \qquad (12.5)$$

The posterior probability can be expressed by likelihoods or probabilities, but $P(D)$, the marginal probability, must be calculated over all possible values for θ, designated Θ:

$$P(\theta \mid D) = \frac{P(\theta)L(\theta \mid D)}{\int_{\Theta} P(\theta)L(\theta \mid D)d\theta}. \qquad (12.6)$$

The two wrinkles to be addressed are (a) determining an appropriate prior probability distribution for the parameter, $P(\theta)$, because it can affect that posterior probability to be calculated and (b) calculating the integral may not have an analytic solution. The first wrinkle may be solved by using the most 'neutral' prior assumption, such as equal probabilities for all values (a uniform distribution), or by using additional information about the system. The second can be solved by numerical methods, such as the commonly used Markov Chain Monte Carlo (MCMC) iterative procedure to determine the appropriate value (see chapter 9 of McElreath 2020). We will illustrate the basic approach with a simple example ...

Introductory Example *(paraphrased, with some liberties, from Eddy 2004)*
A cue ball is rolled from one end onto a billiard table. The location where it stops follows a uniform distribution along the table's long axis (Figure 12.11); its position is noted but is treated as an unknown. A second ball is repeatedly rolled in the same way, and each roll is scored as A if it stops to the left of the cue ball's marked position and as B if it is to the right. The probability of A depends on the cue ball's position and occurs with probability q; B occurs with probability $(1 - q)$. Eddy (2004) provided the following problem: After eight trials, $A = 5$ and $B = 3$, and although q is unknown, we

12.6 Looking Ahead

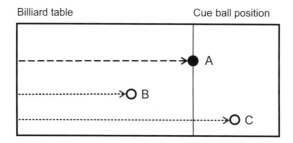

Figure 12.11 The physical arrangement of a billiard table and billiard balls as an illustration for Bayesian statistics.

want to calculate the probability that three more B's occur before a single additional A occurs; call that outcome W. Based on a uniform prior distribution for q, which can be justified by the trial's design, we have

$$P(q \mid A = 5, B = 3) = \frac{P(q)P(A = 5, B = 3 \mid q)}{\int_\Theta P(q)P(A = 5, B = 3 \mid q)dq}. \quad (12.7)$$

With some substitution and rearrangement, this gives

$$P(W) = \frac{\int_0^1 q^5(1-q)^6 dq}{\int_0^1 q^5(1-q)^3 dq}. \quad (12.8)$$

The convenient fact about this example is that the two integrals have analytical solutions: (5!6!/12!) and (5!3!/9!). This allows us to determine the expected $P(W) = 1/11$. Based on this approach, and the fact that the expected $P(W)$ is $(1-q)^3$, the estimate of q is $1 - (1/11)^{1/3} = 0.55031$; the maximum likelihood estimate of q is $5/8 = 0.625$, which is clearly different from the Bayesian result. In most ecological applications, analytical solutions for the integrals will not be available and will have to be solved by numerical methods. Markov Chain Monte Carlo is a common choice (see McElreath 2020, chapter 9), but some newer methods have advantages, as we mention in the examples below.

The billiard table example provides a conceptual introduction to how simple Bayesian analysis proceeds, with inverse probability and a posterior distribution based on a prior distribution. Much debate about the advantages and disadvantages of Bayesian methods can be found in the literature, but a brief and clear assessment is provided by Wasserman (2004) in the book with the alluring title *All of Statistics*. Recently, we have been encouraging our students to work through the book on *Statistical Rethinking* by McElreath (2020). We now turn to the hierarchical version and its application in spatial analysis.

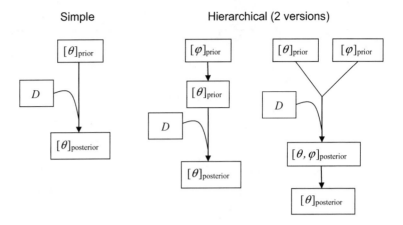

Figure 12.12 Simple Bayesian analysis with parameter θ (left): a prior distribution of a parameter is combined with the data to produce a refined posterior distribution for the same parameter. Hierarchical Bayesian analysis includes another parameter, φ, (two versions on the right), the parameter of immediate interest, θ, has a distribution that depends on parameter φ.

12.6.2.2 Hierarchical Bayesian Analysis

Starting with the simple form

$$P(\theta \mid D) \propto L(\theta \mid D)P(\theta), \tag{12.9}$$

the model becomes hierarchical in the sense of statistical levels when parameter θ is conditional upon another parameter, φ. The posterior distribution then depends on the prior distribution of θ given φ, and φ has its own prior too (Figure 12.12). The posterior probability becomes:

$$P(\theta, \varphi \mid D) \propto L(\theta \mid D)P(\theta \mid \varphi)P(\varphi). \tag{12.10}$$

The data might consist of counts of individuals in sampling units, with the counts following a Poisson distribution with parameter λ. That parameter could vary across time or space or could itself follow another distribution, such as a Gamma distribution with parameter β (see example 1 by Beckage *et al.* 2007).

The hierarchical form can also be based on an intervening process, rather than on parameters of parameters. Many ecological studies use that structure, with square brackets indicating the probability distribution (see Wikle *et al.* 2019), it is the following:

Data model: $[D|Y, \theta_D]$ or $[D|Y, \theta]$
Process model: $[Y|\theta_P]$ or $[Y|\theta]$
Parameter model: $[\theta_D, \theta_P]$ or $[\theta]$.

Then,

$$[Y, \theta \mid D] = \frac{[D \mid Y, \theta][Y \mid \theta][\theta]}{[D]}. \tag{12.11}$$

12.6 Looking Ahead

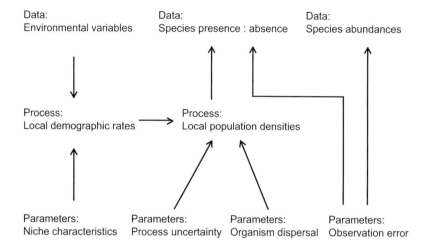

Figure 12.13 The schematic structure of a three-level Bayesian approach to spatial population dynamics (based on Pagel & Schurr 2012).

In summary, the three-level hierarchical scheme for ecological applications is:

$$\text{data} \quad [D \mid Y, \theta]$$
$$\updownarrow$$
$$\text{process} \quad [Y \mid \theta]$$
$$\updownarrow$$
$$\text{parameter} \quad [\theta]$$

This three-level model can be expanded by including several categories or compartments in each level. Pagel and Schuur (2012) used a process-based dynamic range model that combines the niche concept with spatial population dynamics, as shown in Figure 12.13 (Pagel & Schurr 2012).

A hierarchical model can have more layers or more levels of dependence, but two or three levels is a common format in ecological models, of which we will provide some examples.

12.6.2.3 Empirical Hierarchical Models

One alternative to specifying the parameter model, described for the Bayesian hierarchical model approach, is to estimate the parameters from the data and then use those estimates in the models for the layers above. This approach is called the empirical hierarchical model approach but it is sometimes referred to as the empirical Bayesian model, which can be confusing (see Wikle *et al.* 2019). The estimates are derived by one of several methods, but they are substituted into the data models as if they were *known*, which can give misleading results, especially using transformations of the estimates, which have variability associated with them (because they are estimates!).

12.6.2.4 Hierarchical Bayesian Analysis Examples

Here, we provide some examples of hierarchical Bayesian analysis in ecological studies; with the reminder that the hierarchies are not 'structural' like 'leaf, branch, tree, ...' but represent a series of statistical levels. We do not give all the details of the models or of the results (details available in the references cited). McCaslin et al. (2020) provided a list of such examples by category in various subdisciplines of ecology (fish and wildlife ecology, integrated population models, animal movement, ...), many of which have a spatial component (see their table 1).

Example 12.1

Change point analysis of seedling counts in transects of contiguous quadrats in gaps and under canopy (Beckage et al. 2007). This example provides a good link between Bayesian approaches and the more familiar methods for patch:gap analysis in quadrat-transect data.

> Level 1: Independent seedling counts, X_i with Poisson parameters λ_c (under canopy) and λ_g (in gaps). Two versions (canopy versus gap) of $P(X \mid \lambda)$.
>
> Level 2: Link λ_c and λ_g in several different transects with similar, but not identical, values; the λ_c and λ_g follow Gamma distributions, Γ_c (under canopy) and Γ_g (in gaps), 'borrowing strength' across different sample units. Two versions of $P(\lambda \mid a, b^{-1})$.
>
> Level 3: The parameter b^{-1} of the Gamma distribution itself follows an Inverse Gamma distribution with parameters γ and ς. Two versions of $P(b^{-1} \mid \gamma, \varsigma)$.

The parameters at successively higher levels of this kind of hierarchy are sometimes called *hyperparameters*, with the matching priors being *hyperpriors* (Gill 2002).

Correlations between seedling counts in adjacent quadrats (spatial autocorrelation) were permitted by including a Gaussian random field prior to random effects on the λs of quadrats in the same transect.

Example 12.2: Harbour Seal Haul-out Counts in Prince William Sound (Cressie et al. 2009)

The data are the counts of seals at specific sites around the Sound, observed at several dates over a number of years, and so the predictor variables include site, but also the year (for trend), date (for seasonal effects) and tide status (for phase).

The hierarchy in this example, and in the two that follow, is based on the following construction: the data, D, are dependent on ecological processes, E, and the parameters, P_D, for the relationship between the data and the processes, a data model. In turn, P_E are the parameters that characterize E, giving a process

(cont.)

model. The third level is the parameter models for the P_E, with parameters (or hyperparameters) of their own. Each of the levels can be further decomposed into components. Spatial (or temporal) structure can be included by relaxing the assumptions of independence.

> Level 1: Data: independent site counts, Y_{ij} for site i and time j, with Negative Binomial parameters λ_{ij} and κ. This gives, with square brackets indicating the distribution:
>
> $$[D \mid E, P_D] = [\{Y_{ij}\} \mid \{\lambda_{ij}\}, \kappa]. \quad (12.12)$$
>
> Level 2: Process: giving the log counts a Normal distribution $\log(\lambda_{ij}) = \mu_{ij} + \varepsilon_{ij}$, with the mean related to time-dependent site factors by the vector of weights, θ, and the error terms being normal i.i.d. with variance σ^2, we get:
>
> $$[E \mid P_E] = [\{\lambda_{ij}\} \mid \{\theta\}, \sigma^2]. \quad (12.13)$$
>
> Level 3: The parameter model for $\mu_{ij} = \theta_{0i} + \theta_{1i}\,year + \theta_{2i}\,date + \cdots$ The times and sites are initially treated as independent, but dependence can be introduced by making the outcome conditional on the same true underlying process (Cressie et al. 2009). Thus, if the data are decomposed into three sets, like three days of observations, we can compare having three potentially different parameter sets, $P_{D,1}$, $P_{D,2}$ and $P_{D,3}$, with the condition that they are all equal: $P_{D,1} = P_{D,2} = P_{D,3} = P_D$. Similarly, the ecological processes can be decomposed into different numbers of components, which might be site-specific.

A Level 4 is certainly possible, if the elements of θ, which are the parameters of the linear dependence model, have their own distributions.

Example 12.3

The spread of the House Finch in Eastern USA (Wikle 2003), with counts in geographically defined squares over several census years.

This follows the structure of (1) Data model; (2) Process model; and (3) Parameter model.

> Level 1: $X_t(s_i)$ is the count of birds at time t at location s_i. X has a Poisson distribution with parameter $\lambda_t(s_i)$.
>
> Level 2: $\log(\lambda_t(s_i)) = \mu_t + k'_{it} u_i \varepsilon_{ij}$, where u is a vector of a gridded latent spatio-temporal dynamic process, and k' maps u onto the spatial locations. Year-to-year autocorrelation is provided by $\mu_t = \mu_{t-1} + \eta_t$. The vector u

(cont.)

provides a kind of reaction-diffusion model, creating a Spatio-Temporal Autoregressive (STAR) model with diffusion parameter δ and error term γ. Level 3: a model for parameter δ that includes dependence on the correlation matrix of sites, with a spatial dependence parameter like the familiar autocorrelation parameter ρ. Space is included explicitly and by necessity in this analysis.

Example 12.4
Using non-homogeneous species abundance data to reconstruct interaction networks (Aderhold et al. 2012). These networks represent the full set of interactions among the species at a particular site, including competition, predation, parasitism, mutualism and facilitation. The nodes of the network (a directed graph) are the species and the edges between pairs of nodes represent particular interactions, such as species A is a parasite of species B. These interactions are detected in a spatially explicit context and can be analysed using the hierarchical approach.

Level 1: Abundances are observed for N species on a grid of locations, s_i, with coordinates (x_1, x_2), giving $\{Y_k(s_i)\}$ for $k =$ to N. The network of interactions is a directed graph with the species as nodes and a directed edge from each member of the 'parent set' of k, called π_k. Change points occur in each of the two directions dimensions, called $\{\xi_1\}$ and $\{\xi_2\}$.

Level 2: Latent effects in close proximity are usually similar, but several change points are possible along each of the two orthogonal directions. The abundance of any species depends on the abundance of its parent set, with a set of parameters determining a linear response. For each segment h, a segment-specific linear regression is:

$$Y_k(s_i) = b_{k0}(h) + \sum_{m \in \pi_k} b_{km}(h) Y_m(s_i) + \varepsilon(s_i). \qquad (12.14)$$

The latent variable, h, depends on the location and on the locations of the change points.

Level 3: Spatial autocorrelation in the response variable, the density of the focal species, is explicitly included, based on the values at the four nearest neighbours weighted by the inverse of the Euclidean neighbour distance.

$$A_k(s_i) = \sum_j d_{ij}^{-1} Y_k(s_i) \div \sum_j d_{ij}^{-1}, \qquad (12.15)$$

where s_j is in the neighbour set of s_i, currently the four nearest sample locations. Then,

> (cont.)
>
> $$Y_k(s_i) = b_{k0}(h) + \sum_{m \in \pi_k} b_{km}(h) Y_m(s_i) + b_{kA}(h) A_k(s_i) + \varepsilon(s_i). \quad (12.16)$$
>
> Notice that the effect of the autocorrelation can vary in strength depending on location, indexed by variable h. The details of this approach are provided in the paper cited. This example has a somewhat complex set-up, but it shows the flexibility of the hierarchical approach in including both change points and spatial autocorrelation.

12.6.2.5 Other Examples Cited

Chapter 8 mentioned two further examples of the Bayesian approach.

Aing et al. (2011) used a Bayesian hierarchical occupancy model to assess the occurrence of river otter from snow-track data. The model had three levels: occupancy by otters; availability of tracks for detection, conditional upon occupancy; and records of tracks' presence or absence, conditional upon their availability. Spatial dependence was included using an intrinsic conditional autoregressive (CAR) model followed by an MCMC procedure. They found that a spatial model gave better estimates for the detection parameters and better credibility intervals for the spatially autocorrelated data.

Beguin et al. (2012) suggested an alternative to MCMC procedure, integrated nested Laplace approximations (INLA), for fitting Bayesian hierarchical spatial models with general covariance structures. They used both the conditional autoregressive model (CAR) and the Matérn model (see Minasny & McBratney 2005) in a study investigating woodland caribou in Eastern Canada's boreal forest, and they found that the INLA method and the Matérn model provided several advantages over the MCMC approach. It was accurate and rapid, effectively removing spatial autocorrelation from the model residuals, and gave a good evaluation of the uncertainty of the parameter estimates.

Muff et al. (2019) also used INLA as the Bayesian comparator with the frequentist Template Model Builder (TMB) version of Generalized Linear Mixed Models (GLMM) for habitat selection studies using data from mountain goats (*Oreamnos americanus*) and Eurasian otters (*Lutra lutra*). They provided coded examples of applications in habitat selection studies with resource-selection functions (RSF) and step-selection functions (SSF) which were efficiently estimated at the population and individual levels.

McCaslin et al. (2020) described a recursive Bayesian (RB) computation for such models in a version that is transformation assisted (hence TARB), with an application to the migration trajectories of white storks (*Ciconia ciconia*) at both the population and individual levels. TARB was shown to have advantages, too, compared to the standard MCMC approach, and this provides context for the examples already-mentioned. For details on these alternative methods, INLA and TARB introduced here, see the cited works and their sources.

12.6.2.6 Comment on Hierarchical Bayesian Spatial Analysis

We have given a full treatment of the hierarchical Bayesian approach because of its applicability to spatial analysis in many contexts, employing several techniques to include spatial structure in the analysis. Harrison et al. (2011) presented a sophisticated example of butterfly metapopulation dynamics in which spatial structure is included implicitly in the dispersal between habitat patches, as affected by inter-patch distances. Gelfand (2012) has provided some further insights and discussion that focuses on the use of hierarchical models for examining explicitly spatial data, suggesting that the approach is both useful and very flexible and that it will prove its worth as more research includes spatially referenced data. The last decade has shown that to be true. The previous studies cited (Muff et al. 2019; McCaslin et al. 2020) indicate the rapid growth of both the technical sophistication and effectiveness of these methods and the range of ecological applications for which they are used.

12.6.3 Spatial Causal Inference

Let us return to the topic of causal inference in the spatial context (Section 7.1) and provide some reminders about this challenging aspect of spatial analysis. In the general case, for outcome Y and covariate X, if knowledge about X does not significantly improve predictions about Y, then X is not a cause of Y. If knowledge about X significantly improves predictions about Y, we may not conclude that X is a cause of Y, because both X and Y may be caused by Z; X may act through a third factor, M; or there may be an unobserved factor, U, at work (Figure 12.14). In Chapter 7, we discussed this last possibility with the concept of spatial confounding due to an influential unobserved or unmeasured factor with spatial structure of its own.

The problem with causal inferences based on spatial data is that they are often not consistent with the assumptions on which non-spatial causal analysis is founded (Akbari et al. 2023). Spatial autocorrelation, spatial heterogeneity and spatial scale effects like MAUP (Modifiable Areal Unit Problem) can all affect inferences based on such data.

Reich et al. (2021) have described models of the spatial structure of inference including spatial network inference which includes only the network nodes joined to

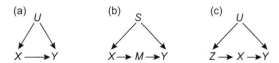

Figure 12.14 Hypothesized causal networks, with the nodes being processes or variables and the directional edges showing conditional dependence or conditional independence. Y is the outcome process or variable, with X as a possible cause. (a) U is an unobserved factor that may affect both X and Y; (b) M is a mediator for the effect of X on Y (indirect effects), but the spatial confounder is measured as S; (c) Z is a third observed variable (instrumental variable) that only affects X, so Y and Z are independent.

each other by network edges. Spatial heterogeneity can be addressed by evaluating local average treatment effects instead of global averages (Imbens & Angris 1994). For localized inference, Delgado and Florax (2015) suggested the 'difference of differences' approach to assess local autocorrelation and spatial interactions. For any of these approaches, care must be taken to ensure the context of the conditions under which these methods are valid.

For spatio-temporal analysis, the directionality of time may allow the use of the Granger causality test for the temporal component and Reich *et al.* (2021) suggested that extending this test to the spatio-temporal context is straightforward with the spatial dependence following a SAR or CAR distribution (see Chapter 8).

For the purely spatial situation, it seems that most solutions to the problems for causal inference still have assumptions that probably will not hold for all ecological spatial data. We suggest continuing with the familiar local statistics and restricted randomizations to understand the effects of heterogeneity and 'Model-and-Monte-Carlo' exercises to investigate interactions among factors.

12.6.4 Artificial Intelligence: From Machine Learning to Deep Learning to AI

Everyone sees enormous potential in artificial intelligence (AI), which is transforming not only all science but everything in our lives. Its impact on ecology will be profound and in ways we may not currently imagine (see Desjardins-Proulx *et al.* 2019, among many). Pichler and Hartig (2022) provided a very helpful guide and review for ecologists on machine learning (ML) and deep learning (DL) as important aspects of AI, which is a broad set of algorithms intended to match human performance in tasks like recognition or decision-making. Machine learning is a subset of general AI with algorithms making predictive models based on data. Its methods include regression models (e.g. LASSO), random forest and boosted regression trees (Pichler & Hartig 2022, table 1). In considering the spatial context for any of these ML methods, adding x–y coordinates as predictors improves the fit (i.e. percentage of variance explained) but not necessarily our understanding of the processes that produce the patterns detected.

Then, deep learning is a family of methods within machine learning, based on various forms of artificial neural networks (Pichler & Hartig 2022, table 1):

- a *deep neural network* consists of an input layer of 'neurons' and an output layer with several hidden layers (acting as latent variables) between them;
- a *convolutional neural network*, usually for image analysis, uses small kernels or filters as convolutional layers to pre-process the data, identifying features like edges before full processing;
- a *recurrent neural network*, usually for sequences, acts iteratively by reusing the hidden states of the network cells; and
- a *graph neural network*, for tasks like classification and regression, operates on the nodes and edges of a graph, allowing for non-Euclidean and non-raster data. These networks may also have convolutional layers, creating graph convolutional

networks (GCN). Ecologists will be especially interested in potential applications of *spatial* GCNs (Danel *et al.* 2019), *spatio-temporal* GCNs (Yan *et al.* 2018) and GCNs for dynamic graphs (see Zhang *et al.* 2019) and for multilayer graphs such as those described in Chapter 11 (Ma *et al.* 2019).

Because neural networks have been developed both to analyse and identify images (which are in essence spatial) and to analyse sequence data like languages (which are temporal), it is an obvious choice for both spatial (Fischer 2009) and spatio-temporal analysis and modelling (Wikle & Zammit-Mangion 2022). Deep learning has already been used for many applications in ecology (Borowiec *et al.* 2022) with good examples available as models to follow at the level of individuals, groups, populations, communities, landscapes and ecosystems and help and advice for future applications readily (Christin *et al.* 2019). For ecology, DL has generally been useful for tasks such as the identification and classification of organisms, counts and density estimation, as well as detecting and modelling disease distribution (Christin *et al.* 2019, figure 3). These can all contribute to spatial ecology and there are already good examples of deep learning for specifically spatial tasks; we will describe just two of the many.

The first is a study by Rammer and Seidl (2019) on predicting bark beetle outbreaks (*Ips typographus* on *Picea abies*) in the Bavarian Forest National Park. They used a 23-year time series from an area of 13,319 ha with 30 m resolution, giving 1.87 million data points to train a deep neural network (DNN). Their approach included five convolutional layers, making it a version of a convolutional neural network with a kernel to pre-process the spatial data. For a 19×19 moving window with local tree distribution and preceding outbreak activity and auxiliary variables, the output was the state in the current year. The DNNs were able to predict the short-term infestation risk at the local scale and the long-term outbreak dynamics at the landscape scale, providing convincing evidence of the predictive power of such deep learning approaches for spatial and spatio-temporal analysis. In the case of a similar research project, based on irregularly placed locations rather than a grid, the same sort of approach could be used but based on a graph neural network (GNN) that includes convolutional layers.

The second example used a convolutional neural network (CNN) for species distribution modelling (Deneu *et al.* 2021). CNN uses a kernel to pre-process spatial or image data for distinct features like shapes or edges (Pichler & Hartig 2022). They found that including local information on landscape heterogeneity and structure crucially improved the predictive performance of the CNN species distribution models, highlighting the importance of spatially structured environmental factors. Perhaps most importantly, the predictive gain was noticeable for rare species, suggesting a practical advantage for conservation efforts.

Clearly, deep learning methods have much to contribute to spatial and spatio-temporal analysis in ecological studies. It is also clear that, while these are still early days for these applications, much has already been accomplished. Without going into greater detail here, we refer to the extensive review by Wikle and Zammit-Mangion (2022) for the wealth of information there.

12.6 Looking Ahead

The most recent developments in AI (at time of writing) are based on models called *transformers* that use *self attention*, a form of internal-to-sequence semantic and contextual evaluation, instead of recursion, to weigh the importance of positional relationships among words in a sequence (Vaswani *et al.* 2017). Enormous amounts of textual material are required for training transformer-based programs like ChatGPT, which is by now a familiar name. The technical details are beyond the requirements here, but there are many helpful tutorials available (the series by Lucidate, for example, at www.youtube.com/watch?v=6XLJ7TZXSPg).

Reminder: interpretation requires some caution because predictions from ML and DL are usually accurate only for the near term (Dietze *et al.* 2018). These methods are not process-based but phenomenological, and we should watch for potential mismatches between the ML and DL predictions and actual responses to novel conditions (Cuddington *et al.* 2013; Boettiger 2022). This concern could create serious difficulties, or even dangers, depending on the application.

In our discussion of AI, we should also include *knowledge graphs* (KG) and their potential impact on all sciences. The original 'knowledge graphs' in information science (James 1992) were maps of words and meanings but now depict networks of entities and their relationships together with related concepts (Guan *et al.* 2019; Kejriwal *et al.* 2021). The entities in a knowledge graph can include almost anything and we can speculate about such a complex knowledge graph that is a network of all of ecology (Cousens & Dale 2023). There, we suggested that it might be best as a multilayer network (Chapter 11) with layers corresponding to different categories (scientists, topics, publications, methods, etc.) with interlayer edges showing intertype relationships (Cousens & Dale 2023). We know that knowledge graphs have been fundamental to organizing and representing scientific information in a structured manner (Kejriwal *et al.* 2021). Knowledge graphs do not currently contain the knowledge that flows in the collection they depict, but they could! Consider combining this approach with AI for analysis and synthesis. Desjardin-Proulx *et al.* (2019) suggested that recent AI work allows the creation of rich knowledge bases that combine the ability to represent complex ideas with modelling uncertainty. Pichler and Hartig (2022) noted that machine learning and deep learning can be used for analytical tasks usually handled by statistical models. They conclude that the trends of scientific and causal machine learning, together with advances in 'explainable' and 'responsible' artificial intelligence, have the potential for a significant impact on ecological predictive modelling and data analysis in the future. Can that then inform a true knowledge graph? The answer seems to be 'yes' and (at time of writing) many blogs and websites emphasize the facts that (1) AI applications can build and reinforce the effectiveness of Knowledge Graphs by finding and organizing the information they contain and (2) Knowledge Graphs can make enormous contributions to the development and application of AI by providing a rich library of relevant data (see for example https://neo4j.com/blog/future-ai-machine-learning-knowledge-graphs/). One version of the relationship is that Knowledge Graphs and Machine Learning have the directional interactions 'built from' and 'applied to', but they work together to support knowledge-enabled AI (Gartner Report,

September 2022). The future of this interaction will have large implications for all our endeavours.

The last element for this discussion of AI and ecological research is the development of *Literature-based Discovery* (LBD). This is the concept that the body of published knowledge contains more information and more useful information than has already been extracted or made explicit (Henry & McInnes 2017). Neural networks and other forms of AI can effectively extract and summarize what is in the literature to identify gaps in knowledge and generate new discoveries within it (Chrichton *et al.* 2020). Not only that, but this approach, when combined with appropriate natural language models, should be able to suggest or demonstrate new directions that scientists should explore, *contextual literature-based discovery* (Wang *et al.* 2023). The effectiveness of LBD has been demonstrated most obviously in biomedical fields but there is no reason to doubt its potential for ecological research. We believe that the combination of forms of AI, including the related phenomena of KG and LBD, will have a significant impact on the spatial aspects of ecological research.

12.6.5 Geometric Algebra

More than once in this book we have suggested that, if things (spatial structures, objects) become too complicated for easy analysis, one option is to convert the *things* into clouds of points and then analyse those point clouds. There is a powerful alternative to converting objects to point-clouds, known as *geometric algebra*, and its use is growing rapidly in popularity and sophistication among physicists, engineers, and computer scientists (Doran & Lasenby 2003; Vince 2008; Joot 2021). These are methods based on algebras developed by mathematicians in parallel or in competition with the vector calculus we learned in high school, and they are known generally as Clifford algebras (Taylor 2021, and the previous citations). The methodology deals with objects with size and orientation of any dimension: lines, areas, volumes, hypervolumes, etc., and has the flexibility to include objects of different dimensions (in *multivectors*). It is also ideal for object manipulations such as projections and rotations. Geometric algebra is now one of the most effective tools available in mathematics for a very wide range of applications from electromagnetism and spacetime models to robotics and neural computing (Wang *et al.* 2019, among many). The approach has many important advantages for physics and engineering but, for our purposes, the applications in computer graphics, image analysis, and geospatial information processing suggest promising contributions to spatial analysis, especially when combined with aspects of AI (Wang *et al.* 2019; Bhatti *et al.* 2020). We will not describe the theoretical background or detail the methods here, we suggest that there are several video series that provide the basics, and that there are many excellent technical texts available (Doran & Lasenby 2003; Vince 2008; Joot 2021; Taylor 2021). We note also that a range of these methods is available in many software packages, such as *clifford* for Python. We have introduced geometric algebra briefly as a separate and important topic because it has the power to change the way we do spatial analysis in the future.

p.s. If it is still available, you should watch the excellent interview with Joan Lasenby: www.youtube.com/watch?v=ikCIUzX9myY.

12.7 Other Future Directions

Throughout this book, we have ended many discussions with the comment that the area under discussion is one worthy of future research and further investigation for ecological applications. This was true of fibre analysis, multilayer networks and spatial causality to name just three. Clearly much is still to be done in almost every area. Let us emphasize a few:

- Spatial graphs and networks: a very promising area for more research and application in ecological studies. They can be used to test hypotheses and to study spatial dynamic behaviour of processes (Silk et al. 2017).
- Spatio-temporal graphs and networks: here too, many good questions remain to be understood and methodologies to be absorbed (Paley et al. 2007; Del Mondo et al. 2010; Bianconi 2018) and then used in ecological research; and, beyond these networks, more is to be done for spatio-temporal analysis (see Wikle et al. 2019), The study of polygon change analysis is just one example, but consider also network automata for evolving spatial networks (Anderson & Dragićević 2020).
- Fibre analysis and related methodologies go beyond zero-dimensional events to include the analysis of linear objects, whether straight or curving, simple or branched, to study a wide range of structures of ecological interest. The substantial literature on point patterns and point processes can serve as the basis for methods dealing with linear features. The range of ecological hypotheses that can be tested with the approach is wide and linear events can also be combined with point events for greater scope and flexibility.
- Spatial analysis of biological diversity deserves much more attention and development and urgently for conservation efforts. A multi-scale spatially explicit (and temporally indexed) approach should answer many questions about the structure of diversity and lead to a better understanding of causes and responses, especially if this can be related to the functional characteristics of spatio-temporal connectivity (see Cunillera-Montcusí et al. 2023). This approach to understanding spatial diversity at local and larger scales will contribute to our progress in clarifying the nature and role of complexity in ecological systems (Riva et al. 2022). Very specifically, we have mentioned the area of multispecies spatial point pattern analysis.
- Spatial structure is more than spatial autocorrelation and understanding it requires more than just exploratory analysis and the application of spatial statistics. It should be easy to incorporate spatial thinking into conceptual models of ecological systems but incorporating spatial structure into ecological studies can be difficult and complex, with the potential to be very expensive in resources. The detection of the important scales for ecological systems and the effects of scaling will be critical.

This is the real challenge for future work: effectively applying the knowledge, techniques and understanding of spatial analysis we currently have.
- All these areas offer the opportunity to exploit a hierarchical approach to testing a nested series of null and alternative hypotheses, but the details need to be worked out based on the greatest ecological interest and importance.
- There is also a great opportunity to exploit developing technology and techniques; the most important being AI, discussed in Section 12.6.4. We have already mentioned LiDAR (Light detection and ranging; see Jackson et al. 2020) and eDNA as examples of spatial analysis (Lamy et al. 2021; Cantera et al. 2022), which promise much for the future.

We have been very selective in the topics included in this book, but it is still far from short! The 'book' format may become outdated, as the rapidly growing field becomes more and more difficult to summarize and as it becomes easier and easier to fall behind as the relevant studies grow at an ever-increasing rate. The latter is especially daunting as our view of what is relevant becomes more and more inclusive, as it should. Obviously, some technology can be helpful in this endeavour, even if the ideal knowledge graph is never complete, and with limits, we are sure. Despite the challenges, spatial analysis and spatio-temporal analysis in ecological research continues to be a fascinating and exciting endeavour. There is lots more to do!

References

Abrams, J. F., Vashishtha, A., Wong, S. T., Nguyen, A., Mohamed, A., Wieser, S., Kuijper, A., Wilting, A. & Mukhopadhyay, A. (2019). Habitat-Net: Segmentation of habitat images using deep learning. *Ecological Informatics*, **51**, 121–128.

Aderhold, A, Husmeier, D., Lennon, J. J., Beale, C. M. & Smith, V. A. (2012). Hierarchical Bayesian models in ecology: Reconstructing species interaction networks from non-homogeneous species abundance data. *Ecological Informatics*, **11**, 55–64.

Aing, C., Halls, S., Oken, K., Dobrow, R. & Fieberg, J. (2011). A Bayesian hierarchical occupancy model for track surveys conducted in a series of linear, spatially correlated, sites. *Journal of Applied Ecology*, **48**, 1508–1517.

Akaike, H. (1974). A new look at the statistical model identification. *IEEE Transactions on Automatic Control*, **19**, 716–723.

Akbari, K., Winter, S. & Tomko, M. (2023). Spatial causality: A systematic review on spatial causal inference. *Geographical Analysis*, **55**, 56–89.

Albery, G. F., Kirkpatrick, L., Firth, J. A. & Bansal, S. (2021). Unifying spatial and social network analysis in disease ecology. *Journal of Animal Ecology*, **90**, 45–61.

Allesina, S. & Pascual, M. (2009). Food web models: A plea for groups. *Ecology Letters*, **12**, 652–662.

Almeida-Neto, M., Guimaraes, P., Guimaraes Jr, P. R., Loyola, R. D. & Ulrich, W. (2008). A consistent metric for nestedness analysis in ecological systems: Reconciling concept and measurement. *Oikos*, **117**, 1227–1239.

Alpargu, G. & Dutilleul, P. (2003). To be or not to be valid in testing the significance of the slope in simple quantitative linear models with autocorrelated errors. *Journal of Statistical Computation and Simulation*, **73**, 165–180.

Alpargu, G. & Dutilleul, P. (2006). Stepwise regression in mixed quantitative linear models with autocorrelated errors. *Communications in Statistics – Simulation and Computation*, **35**, 79–104.

Andersen, I. T. & Hahn, U. (2015). Matérn thinned Cox processes. *Spatial Statistics*, **15**, 1–21.

Anderson, M. J., Crist, T. O., Chase, J. M., Vellend, M., Inouye, B. D., Freestone, A. L., Sanders, N. J., Cornell, H. V., Comita, L. S., Davies, K. F., Harrison, S. P., Kraft, N. J. B., Stegen, J. C. & Swenson, N. G. (2011). Navigating the multiple meanings of beta diversity: A roadmap for the practicing ecologist. *Ecology Letters*, **14**, 19–28.

Anderson, T. & Dragićević, S. (2020). Representing complex evolving spatial networks: Geographic network automata. *International Journal of Geo-Information*, **9**, 270.

Anselin, L. (1995). Local indicators of spatial association: LISA. *Geographical Analysis*, **27**, 93–115.

Anselin, L. (2001). Spatial econometrics. In *A Companion to Theoretical Econometrics*, Baltagi, B. (ed.), pp. 310–330. Oxford: Basil Blackwell.

References

Anselin, L. (2009). Spatial regression. In *The SAGE Handbook of Spatial Analysis*, Fotheringham, A. S. & Rogerson, P. A. (eds.), pp. 255–275. London: SAGE Publications Ltd.

Anselin, L. & Li, X. (2019). Operational join count statistics for cluster detection. *Journal of Geographical Systems*, **21**, 189–210.

Ashenfelter, O. (1978). Estimating the effect of training programs on earnings. *Review of Economic Statistics*, **60**, 47–57.

Austin, M. P. (1987). Models for the analysis of species' response to environmental gradients. *Vegetatio*, **69**, 35–45.

Baddeley, A. J., Howard, C. V., Boyde, A. & Reid, S. (1987). Three-dimensional analysis of the spatial distribution of particles using the tandem-scanning reflected light microscope. *Acta Stereologica*, **6**, 87–100.

Baddeley, A., Nair, G., Raksit, S., McSwiggan, G. & Davies, T. M. (2021). Analysing point patterns on networks: A review. *Spatial Statistics*, 42, 100435.

Baddeley, A., Rubak, E. & Turner, R. (2015). *Spatial Point Patterns: Methodology and Applications with R*. Boca Raton: Chapman & Hall.

Bader, M. Y., Llambi, L. D., Case, B. S., Buckley, H. L., Toivonen, J. M., Camarero, J. J., Cairns, D. M., Brown, C. D., Wiegand, T. & Resler, L. M. (2021). A global framework for linking alpine-treeline ecotone patterns to underlying processes. *Ecography*, **44**, 265–292.

Ballani, F., Pommerening, A. & Stoyan, D. (2019). Mark–mark scatterplots improve pattern analysis in spatial plant ecology. *Ecological Informatics*, **49**, 13–21.

Barabási, A. (2009). Scale-free networks: A decade and beyond. *Science*, **325**, 412–413.

Baranyi, G., Saura, S., Podani, J. & Jordán, F. (2011). Contribution of habitat patches to network connectivity: Redundancy and uniqueness of topological indices. *Ecological Indicators*, **11**, 1301–1310.

Barbujani, G., Oden, N. N. & Sokal, R. R. (1989). Detecting regions of abrupt change in maps of biological variables. *Systematic Zoology*, **38**, 376–389.

Barbujani, G. & Sokal, R. R. (1991). Zones of sharp genetic change in Europe are also linguistic boundaries. *Proceedings of the National Academy of Sciences of the United States of America*, **87**, 1816–1819.

Barot, S., Gignoux, J. & Menaut, J.-C. (1999). Demography of a savanna palm tree: Predictions from comprehensive spatial pattern analysis. *Ecology*, **80**, 1987–2005.

Barrat, A., Barthelemy, M. & Vespignani, A. (2008). *Dynamical Processes on Complex Networks*. Cambridge: Cambridge University Press.

Barter, E. & Gross, T. (2017). Spatial effects in meta-foodwebs. *Scientific Reports*, **7**, 9980.

Barthelemy, M. (2011). Spatial networks. *Physics Reports*, **499**, 1–101.

Barthelemy, M. (2018). *Morphogenesis of Spatial Networks*. Cham: Springer.

Bartlett, M. S. (1935). Some aspects of the time correlation problem in regard to tests of significance. *Journal of the Royal Statistical Society*, **98**, 536–543.

Bartlett, M. S. (1948). Smoothing periodograms from time series with continuous spectra. *Nature*, 161, 686–687.

Bascompte, J. (2009). Disentangling the web of life. *Science*, **325**, 416–419.

Bascompte, J. & Solé, R. V. (1998). Spatiotemporal patterns in nature. *Trends in Ecology and Evolution*, **13**, 173–174.

Baselga, A. (2010). Partitioning the turnover and nestedness components of beta diversity. *Global Ecology and Biogeography*, **19**, 134–143.

References

Baskerville, E. B., Dobson, A. P., Bedford, T., Allesina, S., Anderson, T. M. & Pascual, M. (2011). Spatial guilds in the Serengeti food web revealed by a Bayesian group model. *PLoS Computational Biology*, **7**, e1002321.

Batschelet, E. (1981). *Circular Statistics in Biology*. London: Academic Press.

Bayles, B. R., Thomas, S. M., Simmons, G. S., Grafton-Cardwell, E. E. & Daugherty, M. P. (2017). Spatiotemporal dynamics of the Southern California Asian citrus psyllid (*Diaphorina citri*) invasion. *PLoS ONE*, **12**, e0173226.

Beale, C. M., Lennon, J. J., Yearsley, J. M, Brewer, M. J. & Elston, D. A. (2010). Regression analysis of spatial data. *Ecology Letters*, **13**, 246–264.

Beckage, B., Joseph, L., Belisle, P., Wolfson, D. B. & Platt, W. J. (2007). Bayesian change-point analyses in ecology. *New Phytologist*, **174**, 456–467.

Beguería, S. & Pueyo, Y. (2009). A comparison of simultaneous autoregressive and generalized least squares models for dealing with spatial autocorrelation. *Global Ecology and Biogeography*, **18**, 273–279.

Beguin, J., Martino, S., Rue, H. & Cumming, S. G. (2012). Hierarchical analysis of spatially autocorrelated data using integrated nested Laplace approximation. *Methods in Ecology and Evolution*, **3**, 921–929.

Benes, V. & Rataj, J. (2004). *Stochastic Geometry: Selected Topics*. Boston: Kluwer Academic Publisher.

Ben-Said, M. (2021). Spatial point-pattern analysis as a powerful tool in identifying pattern-process relationships in plant ecology: An updated review. *Ecological Processes*, **10**, 56.

Berman, M. (1986). Testing for spatial association between a point process and another stochastic process. *Applied Statistics*, **35**, 54–62.

Berthelot, G., Said, S. & Bansaye, V. (2020). How to use random walks for modeling the movement of wild animals. *bioRxiv*, https//doi.org/10.1101/2020.03.11.986885.

Bertness, M. D. & Calloway, R. (1994). Positive interactions in communities. *Trends in Ecology and Evolution*, **9**, 191–193.

Bézier, P. (1977). Essai de définition numérique des courbes et des surfaces expérimentales. Thèse de doctorat ès-sciences, Université Pierre et Marie Curie, Paris.

Bhatti, U. A., Yu, Z., Yuan, L., Zeeshan, Z., Nawaz, S. A., Bhatti, M., Mehmood, A., Ain, Q. U. & Wen, L. (2020). Geometric algebra applications in geospatial artificial intelligence and remote sensing image processing. *IEE Access*, **8**, 155783–155796.

Bianconi, G. (2018). *Multilayer Networks: Structure and Function*. Oxford: Oxford University Press.

Biggs, N. L., Lloyd, E. K. & Wilson, R. J. (1976). *Graph Theory 1736–1936*. Oxford: Oxford University Press.

Birre, D., Feuillet, T., Lagalis, R., Milian, J., Alexandre, F., Sheeren, D., Serrano-Notivoli, R., Vignal, M. & Bader, M.Y. (2023). A new method for quantifying treeline-ecotone change based on multiple spatial pattern dimensions. *Landscape Ecology*, **38**, 779–796.

Bivand, R. (1980). A Monte Carlo study of correlation coefficient estimation with spatially autocorrelated observations. *Quaestiones Geographicae*, 6, 5–10.

Bivand, R. S., Pebesma, E. J. & Gómez-Rubio, V. (2008). *Applied Spatial Data Analysis with R*. New York: Springer.

Bjørnstad, O. N. & Falck, W. (2001). Nonparametric spatial covariance functions: Estimation and testing. *Environmental and Ecological Statistics*, **8**, 53–70.

Bjørnstad, O. N., Ims, R. A. & Lambin, X. (1999). Spatial population dynamics: Analyzing patterns and processes of population synchrony. *Trends in Ecology and Evolution*, **14**, 427–432.

Blackman, R. C., Ho, H.-C., Walser, J.-C. & Altermatt, F. (2022). Spatio-temporal patterns of multi-trophic biodiversity and food-web characteristics uncovered across a river catchment using environmental DNA. *Communications Biology*, **5**, 259.

Blanchet, F. G., Legendre, P. & Borcard, D. (2008). Modelling directional spatial processes in ecological data. *Ecological Modelling*, **215**, 325–336.

Blanchet, F. G., Legendre, P., Maranger, R., Monti, D. & Pepin, P. (2011). Modelling the effect of directional spatial ecological processes at different scales. *Oecologia*, **166**, 357–368.

Boccaletti, S., Bianconi, G., Criado, R., del Genio, C., Gómez-Gardeñes, J., Romance, M., Sendiña-Nadal, I., Wang, Z. & Zanin, M. (2014). The structure and dynamics of multilayer networks. *Physics Reports*, **544**, 1–122.

Bode, M., Burrage, K. & Possingham, H. P. (2008). Using complex network metrics to predict the persistence of metapopulations with asymmetric connectivity patterns. *Ecological Modelling*, **214**, 201–209.

Boettiger, C. (2022). The forecast trap. *Ecology Letters*, **25**, 1655–1664.

Bolker, B. M., Brooks, M. E., Clark, C. J., Geange, S. W., Poulsen, J. R., Stevens, M. H. H. & White, J. S. (2009). Generalized linear mixed models: A practical guide for ecology and evolution. *Trends in Ecology and Evolution*, **24**, 127–135.

Boots, B. N. (2002). Local measures of spatial association. *Écoscience*, **9**, 168–176.

Borcard, D., Gillet, F. & Legendre, P. (2018). *Numerical Ecology with R*. 2nd ed. New York: Springer.

Borcard, D. & Legendre, P. (2002). All-scale spatial analysis of ecological data by means of principal coordinates of neighbour matrices. *Ecological Modelling*, **153**, 51–68.

Borcard, D. & Legendre, P. (2012). Is the Mantel correlogram powerful enough to be useful in ecological analysis? A simulation study. *Ecology*, **93**, 1473–1481.

Borcard, D., Legendre, P. & Drapeau, P. (1992). Partialling out the spatial component of ecological variation. *Ecology*, **73**, 1045–1055.

Borowiec, M. L., Dikow, R. B., Frandsen, P. B., McKeeken, A., Valentini, G. & White, A. E. (2022). Deep learning as a tool for ecology and evolution. *Methods in Ecology and Evolution*, **13**, 1640–1660.

Boulanger, E., Dalongeville, A., Andrello, M., Mouillot, D. & Manel, S. (2020). Spatial graphs highlight how multi-generational dispersal shapes landscape genetic patterns. *Ecography*, **43**, 1167–1179.

Brandtberg, T. (1999). Automatic individual tree based analysis of high spatial resolution aerial images on naturally regenerated boreal forests. *Canadian Journal of Forest Research*, 29, 1464–1478.

Brice, M. H., Cazelles, K., Legendre, P. & Fortin, M. J. (2019). Disturbances amplify tree community responses to climate change in the temperate–boreal ecotone. *Global Ecology and Biogeography*, **28**, 1668–1681.

Brown, J. H. (1995). *Macroecology*. Chicago: University of Chicago Press.

Brumback, B. A. (2022). *Fundamentals of Causal Inference: With R*. Boca Raton: CRC Press.

Buettel, J. C., Cole, A., Dickey, J. M. & Brook, B. W. (2018). Analyzing linear spatial features in ecology. *Ecology*, **99**, 1490–1497.

Burnham, K. P. & Anderson, D. R. (2002). *Model Selection and Multimodel Inference: A Practical Information-Theoretic Approach*. Heidelberg: Springer.

Canny, J. (1986). A computational approach to edge detection. *IEEE Transitional Pattern Analysis of Machine Intelligence*, **8**, 679–698.

Cantera, I., Decotte, J.-B., Dejean, T., Murienne, J., Vigouroux, R., Valentini, A. & Brosse, S. (2022). Characterizing the spatial signal of environmental DNA in river systems using a community ecology approach. *Molecular Ecology Resources*, **22**, 1274–1283.

Capblancq, T. & Forester, B. R. (2021). Redundancy analysis: A Swiss Army Knife for landscape genomics. *Methods in Ecology and Evolution*, **12**, 2298–2309.

Carl, G., Doktor, D., Schweger, O. & Kuhn, I. (2016). Assessing relative variable importance across different spatial scales: A two-dimensional wavelet analysis. *Journal of Biogeography*, **43**, 2502–2512.

Carl, G., Dormann, C. F. & Kühn, I. (2008). A wavelet-based method to remove spatial autocorrelation in the analysis of species distributional data. *Web Ecology*, **8**, 22–29.

Carl, G. & Kühn, I. (2008). Analyzing spatial ecological data using linear regression and wavelet analysis. *Stochastic Environmental Research and Risk Assessment*, 22, 315–324.

Carl, G. & Kühn, I. (2010). A wavelet-based extension of generalized linear models to remove the effect of spatial autocorrelation. *Geographical Analysis*, 42, 323–337.

Castillo-Paez, S., Fernandez-Casal, R. & Garcia-Soidan, P. (2019). A nonparametric bootstrap method for spatial data. *Computational Statistics and Data Analysis*, **137**, 1–15.

Cerioli, A. (1997). Modified tests of independence in 2×2 tables with spatial data. *Biometrics*, **53**, 619–628.

Cerioli, A. (2002). Testing mutual independence between two discrete-valued spatial processes: A correction to Pearson chi-squared. *Biometrics*, **58**, 888–897.

Ceron, K., Provete, D. B., Pires, M. M., Araújo, A. C., Blüthgen, N. & Santana, D. J. (2021). Differences in prey availability across space and time lead to interaction rewiring and reshape a predator-prey metaweb. *Ecology*, **103**, e3716.

Chao, A. N., Chiu, C. H. & Jost, L. (2014). Unifying species diversity, phylogenetic diversity, functional diversity, and related similarity and differentiation measures through Hill numbers. *Annual Review of Ecology, Evolution, and Systematics*, **45**, 297–324.

Chase, J. M. & Knight, T. M. (2013). Scale-dependent effect sizes of ecological drivers on biodiversity: Why standardised sampling is not enough. *Ecology Letters*, **16**, 17–26.

Chatfield, C. (1975). *The Analysis of Time Series: Theory and Practice*. London: Chapman & Hall.

Chilès, J.-P. & Delfiner, P. R. (2012). *Geostatistics: Modeling Spatial Uncertainty*. 2nd ed. New York: Wiley.

Chrichton, G., Baker, S., Guo, Y. & Korhonen, A. (2020). Neural networks for open and closed literature-based discovery. *PLoS ONE*, **15**, e0232891.

Christie, M. R. & Knowles, L. L. (2015). Habitat corridors facilitate genetic resilience irrespective of species dispersal abilities or population sizes. *Evolutionary Applications*, **8**, 454–463.

Christin, S., Hervet, E. & Lecompte, N. (2019). Applications for deep learning in ecology. *Methods in Ecology and Evolution*, **10**, 1632–1644.

Chuine, I. & Régnière, J. (2017). Process-based models of phenology for plants and animals. *Annual Review of Ecology, Evolution, and Systematics*, **48**, 159–182.

Chung, F. R. K. (1997). *Spectral Graph Theory*. Providence, RI: American Mathematical Society.

Claramunt, C. & Thériault, M. (1997). Towards semantics for modeling spatio-temporal processes within GIS. In *Advances in GIS Research II*, Kraak, M. J. & Molenaar, M. (eds.), pp. 47–63. London: Taylor & Francis.

Clark, A. T. (2020a). General scaling laws for stability in ecological systems. *Ecology Letters*, **24**, 1474–1486.

References

Clark, J. S. (2020b). *Models for Ecological Data, An Introduction*. Princeton: Princeton University Press.

Clark, J. S., Gelfand, A. E., Woodall, C. W. & Zhu, K. (2014). More than the sum of the parts: Forest climate response from joint species distribution models. *Ecological Applications*, **24**, 990–999.

Clements, F. E. (1916). *Plant Succession*. Washington DC: Carnegie Institute.

Cliff, A. D. & Ord, J. K. (1981). *Spatial Processes: Models and Applications*. London: Pion.

Clifford, P., Richardson S. & Hémon, D. (1989). Assessing the significance of correlation between two spatial processes. *Biometrics*, **45**, 123–134.

Cohn, R. D. (1999). Comparisons of multivariate relational structures in serially correlated data. *Journal of Agricultural, Biological & Environmental Statistics*, **4**, 238–257.

Condit, R., Ashton, P. S., Baker, P., Bunyavejchewin, S., Gunatilleke, S., Gunatilleke, N., Hubbell, S., Foster, R. B., Itoh, A., Lafrankie, J. V., Lee, H. S., Losos, E., Manokaran, N., Sukumar, R. & Yamakura, T. (2000). Spatial patterns in the distribution of tropical tree species. *Science*, **288**, 1414–1417.

Costa, A. Gonzalez, A. M. M., Giuzien, K., Doglioli, A. M., Gomez, J. M., Petrenko, A. A. & Allesina, S. (2019). Ecological networks: Pursuing the shortest path, however narrow and crooked. *Scientific Reports*, **9**, 17826.

Cousens, R. D. & Dale, M. R. T. (2023). The evolution of ecology. In *Effective Ecology: Seeking Success in a Hard Science*, Cousens, R. D. (ed.), pp. 13–31. Milton Park, UK: CRC Press.

Couteron, P. & Ollier, S. S. (2005). A generalized variogram-based framework for multiscale ordination. *Ecology*, **86**, 828–834.

Cozzo, E., Kivelä, M., De Domenico, M., Solé-Ribalta, A., Arenas, A., Gómez, S., Porter, M. A. & Moreno, Y. (2015). Structure of triadic relations in multiplex networks. *New Journal of Physics*, **17**, 073029.

Crabot, J., Clappe, S., Dray, S. & Datry, T. (2019). Testing the Mantel statistic with a spatially-constrained permutation procedure. *Methods in Ecology and Evolution*, **10**, 532–540.

Cressie, N. A. C. (1993). *Statistics for Spatial Data*, revised ed. New York: Wiley.

Cressie, N. A. C. (1996). Change of support and the modifiable areal unit problem. *Geographical Systems*, **3**, 159–180.

Cressie, N., Calder, C. A., Clark, J. S., Ver Hoef, J. M. & Wikle, C. K. (2009). Accounting for uncertainty in ecological analysis: The strengths and limitations of hierarchical statistical modeling. *Ecological Applications*, **19**, 553–570.

Cressie, N. & Wikle, C. K. (2011). *Statistics for Spatio-Temporal Data*. Hoboken, NJ: Wiley.

Crist, T. O., Veech, J. A., Gerring, J. C. & Summerville, K. S. (2003). Partitioning species diversity across landscapes and regions: A hierarchical analysis of alpha, beta, and gamma diversity. *American Naturalist*, **162**, 734–743.

Csillag, F., Boots, B., Fortin, M.-J., Lowell, K. & Potvin, F. (2001). Multiscale characterization of boundaries and landscape ecological patterns. *Geomatica*, 55, 291–307.

Csillag, F. & Kabos, S. (1996). Hierarchical decomposition of variance with applications in environmental mapping based on satellite images. *Mathematical Geology*, 28, 385–405.

Csillag, F. & Kabos, S. (2002). Wavelets, boundaries, and the spatial analysis of landscape pattern. *Écoscience*, **9**, 177–190.

Cuddington, K., Fortin, M.-J., Gerber, L. R., Hastings, A., Liebhold, A., O'Connor, M. & Ray, C. (2013). Process-based models are required to manage ecological systems in a changing world. *Ecosphere*, **4**, 1–12.

Cunillera-Montcusí, D., Fernández-Calero, J. M., Pölsterl, S., Argelich, R., Fortuño, P., Cid, N., Bonada, N. & Cañedo-Argülles, M. (2023). Navigating through space and time: A methodological approach to quantify spatiotemporal connectivity using stream flow data as a case study. *Methods in Ecology and Evolution*, **14**, 1780–1795.

Dale, M. R. T. (1985). Graph theoretical methods for comparing phytosociological structures. *Vegetatio*, **63**, 79–88.

Dale, M. R. T. (1986). Overlap and spacing of species' ranges on an environmental gradient. *Oikos*, **47**, 303–308.

Dale, M. R. T. (1988). The spacing and intermingling of species boundaries on an environmental gradient. *Oikos*, **53**, 351–356.

Dale, M. R. T. (1995). Spatial pattern in communities of crustose saxicolous lichens. *Lichenologist*, 27, 495–503.

Dale, M. R. T. (1999). *Spatial Pattern Analysis in Plant Ecology*. Cambridge: Cambridge University Press.

Dale, M. R. T. (2017). *Applying Graph Theory in Ecological Research*. Cambridge: Cambridge University Press.

Dale, M. R. T., Blundon, D. J., MacIsaac, D. A. & Thomas, A. G. (1991). Multiple species effects and spatial autocorrelation in detecting species associations. *Journal of Vegetation Science*, **2**, 635–642.

Dale, M. R. T., Dixon, P., Fortin, M.-J., Legendre, P., Myers, D. E. & Rosenberg, M. (2002). The conceptual and mathematical relationships among methods for spatial analysis. *Ecography*, **25**, 558–577.

Dale, M. R. T. & Fortin, M.-J. (2002). Spatial autocorrelation and statistical tests in ecology. *Écoscience*, **9**, 162–167.

Dale, M. R. T. & Fortin, M.-J. (2009). Spatial autocorrelation and statistical tests: Some solutions. *Journal of Agricultural, Biological, and Environmental Statistics*, **14**, 188–206.

Dale, M. R. T. & Fortin, M.-J. (2010). From graphs to spatial graphs. *Annual Review of Ecology, Evolution, and Systematics*, **41**, 21–38.

Dale, M. R. T. & Fortin, M.-J. (2014). *Spatial Analysis: A Guide for Ecologists*. 2nd ed. Cambridge: Cambridge University Press.

Dale, M. R. T. & Fortin, M.-J. (2021). *Quantitative Analysis of Ecological Networks*. Cambridge: Cambridge University Press.

Dale, M. R. T. & Mah, M. (1998). The use of wavelets for spatial pattern analysis in ecology. *Journal of Vegetation Science*, **9**, 805–814.

Dale, M. R. T. & Powell, R. D. (1994). Scales of segregation and aggregation of plants of different kinds. *Canadian Journal of Botany*, **72**, 448–453.

Dale, M. R. T. & Powell, R. D. (2001). A new method for characterizing point patterns in plant ecology. *Journal of Vegetation Science*, **12**, 597–608.

Dale, M. R. T. & Zbigniewicz, M. W. (1995). The evaluation of multi-species pattern. *Journal of Vegetation Science*, **6**, 391–398.

Damgaard, C. (2019). A critique of the space-for-time substitution practice in community ecology. *Trends in Ecology & Evolution*, **34**, 416–421.

Danel, T., Spurek, P., Tabor, J., Smieja, M., Lukasz, S., Slowik, A. & Maziarka, L. (2019). Spatial graph convolutional networks. *arXiv*: 1909.05310v2.

Darsanj, M., Basiri, R. & Moradi, M. (2021). Intraspecific interaction comparison in pure and mixed *Populus euphratica* stands using Mark correlation function in Behbahan Chaharasyab area. *Journal of Wood & Forest Science and Technology*, **27**, 81–95.

Dastour, H., Ghaderpour, E., Zaghloul, M. S., Farjad, B., Gupta, A., Eum, H., Achari, G. & Hassan, Q. K. (2022). Wavelet-based spatiotemporal analyses of climate and vegetation for the Athabasca river basin in Canada. *International Journal of Applied Earth Observation and Geoinformation*, **114**, 103044.

Daubechies, I. (1992). *Ten Lectures on Wavelets*. Philadelphia: Society for Industrial and Applied Mathematics.

De Domenico, M., Solé-Ribalta, A., Cozzo, E., Kivelä, M., Moreno, Y., Porter, M., Gomez, S. & Arenas, A. (2013). Mathematical formulation of multilayer networks. *Physical Review X*, **3**, 041022.

De Domenico, M., Solé-Ribalta, A., Ormodei, E., Gomez, S. & Arenas, A. (2015). Ranking in interconnected multilayer networks reveals versatile nodes. *Nature Communications*, **6**, 6868.

Dee, L. E., Ferraro, P. J., Severen, C. N., Kimmel, K. A., Borer, E. T., Byrnes, J. E., Clark, A. T., Hautier, Y., Hector, A., Raynaud, X., Reich, P. B., Wright, A. J., Arnillas, C. A., Davies, K. F., MacDougall, A., Mori, A. S., Smith, M. D., Adler, P. B., Bakker, J. D., Brauman, K. A., Cowles, J., Komatsu, K., Knops, J. M. H., McCulley, R. L., Moore, J. L., Morgan, J. W., Ohlert, T., Power, S. A., Sullivan, L. L., Stevens, C. & Loreau, M. (2023). Clarifying the effect of biodiversity on productivity in natural ecosystems with longitudinal data and methods for causal inference. *Nature Communications*, **14**, 2607.

DeFord, D. R. & Pauls, S. D. (2019). Spectral clustering methods for multiplex networks. *Physica A*, **533**, 121949.

Del Mondo, G., Stell, J. G., Claramunt, C. & Thibaud, R. (2010). A graph model for spatio-temporal evolution. *Journal of Universal Computer Science*, **16**, 1452–1477.

Delgado, M. S. & Florax, R. J. G. M. (2015). Difference-in-differences techniques for spatial data: Local autocorrelation and spatial interaction. *Economics Letters*, **137**, 123–126.

Deneu, B., Servajean, M., Bonnet, P., Botella, C., Munoz, F. & Joly, A. (2021). Convolutional neural networks improve species distribution modelling by capturing the spatial structure of the environment. *PLoS Computational Biology*, **17**, e1008856.

Desjardins-Proulx, P., Poisot, T. & Gravel, D. (2019). Artificial intelligence for ecological and evolutionary synthesis. *Frontiers in Ecology and Evolution*, **7**, a402.

Desrochers, A., Renaud, C., Hochachka, W. M. & Cadman, M. (2010). Area-sensitivity by forest songbirds: Theoretical and practical implications of scale dependency. *Ecography*, **33**, 921–931.

DeWitt, T. J., Fuentes, J. I., Ioerger, T. R. & Bishop, M. P. (2021). Rectifying *I*: Three point and continuous fit of the spatial autocorrelation metric, Moran's *I*, to ideal form. *Landscape Ecology*, **36**, 2897–2918.

Dietze, M. C., Fox, A., Beck-Johnson, L. M., Betancourt, J. L., Hooten, M. B., Jarnevich, C. S., Keitt, T. H., Kenney, M. A., Laney, C. M., Larsen, L. G. & Loescher, H. W. (2018). Iterative near-term ecological forecasting: Needs, opportunities, and challenges. *Proceedings of the National Academy of Sciences*, **115**, 1424–1432.

Diggle, P. J. (1979). Statistical methods for spatial point patterns in ecology. In *Spatial and Temporal Analysis in Ecology*, Cormack, R. M. & Ord, J. K. (eds.), pp. 95–150. Fairland, MD: International Cooperative Publishing House.

Diggle, P. J. (1983). *Statistical Analysis of Spatial Point Patterns*. London: Academic Press.

Diggle, P. J. (2013). *Statistical Analysis of Spatial and Spatio-temporal Point Patterns*. 3rd ed. Boca Raton: CRC Press.

Diggle, P. J. & Chetwynd, A. G. (1991). Second-order analysis of spatial clustering for inhomogeneous populations. *Biometrics*, **47**, 1155–1163.

Diggle, P. J., Chetwynd, A. G., Haggkvist, R. & Morris, S. (1995). Second-order analysis of space-time clustering. *Statististical Methods in Medical Research*, **4**, 124–136.

Ding, Z. & Ma, K. (2021). Identifying changing interspecific associations along gradients at multiple scales using wavelet correlation networks. *Ecology*, **102**, e03360.

Diniz-Filho, J. A. F. & Bini, L. M. (2005). Modelling geographical patterns in species richness using eigenvector-based spatial filters. *Global Ecology and Biogeography*, **14**, 177–185.

Dong, X., Frossard, P., Vandergheynst, P. & Nefedov, N. (2012). Clustering with multi-layer graphs: A special perspective. *IEEE Transactions in Signal Processing*, 60, 5820–5831.

Donges, J. F., Zou, Y., Marwan, N. & Kurtis, J. (2009). The backbone of the climate network. *Europhysics Letters*, **87**, 48007.

Doran, C. & Lasenby, A. (2003). *Geometric Algebra for Physicists*. Cambridge: Cambridge University Press.

Dormann, C. F., McPherson, J. M., Araújo, M. B., Bivand, R., Bolliger, J., Carl, G., Davies, R. G., Hirzel, A., Jetz, W., Kissling, W. D., Kühn, I., Ohlemüller, R., Peres-Neto, P. R., Reineking, B., Schröder, B., Schurr, F. M. & Wilson, R. (2007). Methods to account for spatial autocorrelation in the analysis of species distributional data: A review. *Ecography*, **30**, 609–628.

Dray, S. (2011). A new perspecive about Moran's *I* coefficient: Spatial autocorrelation as a linear regression problem. *Geographical Analysis*, **43**, 127–141.

Dray, S., Legendre P. & Peres-Neto, P. R. (2006). Spatial modeling: A comprehensive framework for principal coordinate analysis of neighbor matrices (PCNM). *Ecological Modelling*, **196**, 483–493.

Dray, S., Pélissier, R., Couteron, P., Fortin, M.-J., Legendre, P., Peres-Neto, P. R., Bellier, E., Bivand, R., Blanchet, F. G., De Cáceres, M., Dufour, A.-B., Heegaard, E., Jombart, T., Munoz, F., Oksanen, J., Thioulouse, J. & Wagner, H. H. (2012). Community ecology in the age of multivariate multiscale spatial analysis. *Ecological Monographs*, **82**, 257–275.

Duczmal, L., Kuldorff, M. & Huang, L. (2006). Evaluation of spatial scan statistics for irregularly shaped clusters. *Journal of Computational and Graphical Statistics*, **15**, 428–442.

Dungan, J. L., Perry, J. N., Dale, M. R. T., Legendre, P., Citron-Pousty, S., Fortin, M.-J., Jakomulska, A., Miriti, M. & Rosenberg, M. S. (2002). A balanced view of scale in spatial statistical analysis. *Ecography*, **25**, 626–640.

Dunne, J. (2006). The network structure of food webs. In *Ecological Networks: Linking Structure to Dynamics in Food Webs*. Pascual, M. & Dunne, J. A. (eds.), pp. 27–86. Oxford: Oxford University Press

Dupont, E., Wood, S. N. & Augustin, N. H. (2022). Spatial+: A novel approach to spatial confounding. *Biometrics*, 78, 1279–1290.

Dupont, Y. L., Trøjelsgaard, K., Hagen, M., Henriksen, M. V., Olesen, J. M., Pedersen, N. M. E. & Kissling, W. D. (2014). Spatial structure of an individual-based plant-pollinator network. *Oikos*, 123, 1301–1310.

Dutilleul, P. (1993). Modifying the *t* test for assessing the correlation between two spatial processes. *Biometrics*, **49**, 305–314.

Dutilleul, P. R. L. (2011). *Spatio-temporal Heterogeneity: Concepts and Analyses*. Cambridge: Cambridge University Press.

Eddy, S. R. (2004). What is Bayesian statistics? *Nature Biotechnology*, **22**, 1177–1178.

Efron, B. & Tibshirani, R. J. (1993). *An Introduction to the Bootstrap*. New York: Chapman & Hall.

Erdös, P. & Rényi, A. (1960). On the evolution of random graphs. *Publications of the Mathematical Institute of the Hungarian Academy of Science*, **5**, 17–61.

References

van Es, H. M. & van Es, C. L. (1993). The spatial nature of randomization and its effects on the outcome of field experiments. *Agronomy Journal*, **85**, 420–428.

Estes, L., Elsen, P. R., Treuer, T., Ahmed, L., Caylor, K., Chang, J., Choi, J. J. & Ellis, E. C. (2018). The spatial and temporal domains of modern ecology. *Nature Ecology & Evolution*, **2**, 819–826.

Evans, J. C., Fisher, D. N. & Silk, M. J. (2020). The performance of permutations and exponential random graph models when analyzing animal networks. *Behavioral Ecology*, **31**, 1266–1276.

Faghih, F. & Smith, M. (2002). Combining spatial and scale-space techniques for edge detection to provide a spatially adaptive wavelet-based noise filtering algorithm. *IEEE Transactions on Image Processing*, **11**, 1069–1071.

Fall, A., Fortin, M.-J., Manseau, M. & O'Brien, D. (2007). Spatial graphs: Principles and applications for habitat connectivity. *Ecosystems*, **10**, 448–461.

Fan, J., Meng, J., Ashkenazy, Y., Havlin, S. & Schnellnhuber, H. J. (2018). Climate network percolation reveals the expansion and weakening of the tropical component under global warming. *Proceedings of the National Academy of Sciences*, **115**, e12128–e12134.

Fan, J., Meng, J., Ludesscher, J., Chen, X., Ashkenazy, Y., Kurtis, J., Havlin, S. & Schnellnhuber, H. J. (2021). Statistical physics approaches to the complex Earth system. *Physics Reports*, **96**, 1–84.

Farage, C., Edler, D., Eklöf, A., Rosvall, M. & Pilosof, S. (2021). Identifying flow modules in ecological networks using Infomap. *Methods in Ecology and Evolution*, **12**, 778–786.

Farine, D. R. (2017). A guide to null models for animal social network analysis. *Methods in Ecology and Evolution*, **8**, 1309–1320.

Fattorini, S. (2007). Non-randomness in the species-area relationship: Testing the underlying mechanisms. *Oikos*, 116, 678–689.

Ferro, I. & Morrone, J. J. (2014). Biogeographical transition zones: A search for conceptual synthesis. *Biological Journal of the Linnean Society*, **113**, 1–12.

Finn, K. R., Silk, M. J., Porter, M. A. & Pinter-Wollman, N. (2017). Novel insights into animal sociality from multilayer networks. arXiv:1712.01790v1.

Fiorentino, D., Lecours, V. & Brey, T. (2018). On the art of classification in spatial ecology: Fuzziness as an alternative for mapping uncertainty. *Frontiers in Ecology and Evolution*, **6**, 231.

Fischer, M. M. (2009). Neural networks for spatial data analysis. In *Sage Handbook of Spatial Analysis*, Fotheringham, A. S. & Rogerson, P. A. (eds.), pp. 375–396. Los Angeles: Sage.

Fitzpatrick, M. C., Preisser, E. L., Porter, A., Elkinton, J., Waller, L. A., Carlin, B. P. & Ellison, A. M. (2010). Ecological boundary detection using Bayesian areal wombling. *Ecology*, **91**, 3448–3455.

Fletcher Jr, R. & Fortin, M.-J. (2018). *Spatial Ecology and Conservation Modelling*. Cham, Switzerland: Springer.

Fletcher Jr, R. J., Betts, M. G., Damschen, E. I., Hefley, T. J., Hightower, J., Smith, T. A., Fortin, M.-J. & Haddad, N. M. (2023). Addressing the problem of scale that emerges with habitat fragmentation. *Global Ecology and Biogeography*, **32**, 828–841.

Fletcher Jr, R. J., Hefley, T. J., Robertson, E. P., Zuckerberg, B., McCleery, R. A. & Dorazio, R. M. (2019). A practical guide for combining data to model species distributions. *Ecology*, **100**, e02710.

Floryan, D. & Graham, M. D. (2021). Discovering multiscale and self-similar structure with data-driven wavelets. *Proceedings of the National Academy of Sciences*, **118**, e2021299118.

Foltête, J.-C., Savary, P., Clauzel, C., Bourgois, M., Girardet, X., Sahroui, Y., Vuidel, G. & Garnier, S. (2020). Coupling landscape graph modeling and biological data: A review. *Landscape Ecology*, **35**, 1035–1052.

Fortin, M.-J. (1992). Detection of ecotones: Definition and scaling factors. PhD Thesis, Department of Ecology and Evolution, State University of New York at Stony Brook, pp. 258.

Fortin, M.-J. (1994). Edge detection algorithms for two-dimensional ecological data. *Ecology*, **75**, 956–965.

Fortin, M.-J. (1997). Effects of data types on vegetation boundary delineation. *Canadian Journal of Forest Research*, **27**, 1851–1858.

Fortin, M.-J. (1999a). Effects of sampling unit resolution on the estimation of the spatial autocorrelation. *Écoscience*, **6**, 636–641.

Fortin, M.-J. (1999b). The effects of quadrat size and data measurement on the detection of boundaries. *Journal of Vegetation Science*, **10**, 43–50.

Fortin, M.-J. & Dale, M. R. T. (2005). *Spatial Analysis: A Guide for Ecologists*. Cambridge: Cambridge University Press.

Fortin, M.-J. & Dale, M. R. T. (2009). Spatial autocorrelation in ecological studies: A legacy of solutions and myths. *Geographical Analysis*, **41**, 392–397.

Fortin, M.-J. & Drapeau, P. (1995). Delineation of ecological boundaries: Comparisons of approaches and significance tests. *Oikos*, **72**, 323–332.

Fortin, M.-J., Drapeau, P. & Jacquez, G. M. (1996). Quantification of the spatial co-occurrences of ecological boundaries. *Oikos*, **77**, 51–60.

Fortin, M.-J., Drapeau, P. & Legendre, P. (1989). Spatial autocorrelation and sampling design in plant ecology. *Vegetatio*, **83**, 209–222.

Fortin, M.-J. & Gurevitch, J. (2001). Mantel tests: Spatial structure in field experiments. In *Design and Analysis of Ecological Experiments*, 2nd ed., Scheiner, S. M. & Gurevitch, J. (eds.), pp. 308–326. Oxford: Oxford University Press.

Fortin, M.-J. & Jacquez, G. M. (2000). Randomization tests and spatially autocorrelated data. *Bulletin of the Ecological Society of America*, 81, 201–205.

Fortin, M.-J., Jacquez, G. M. & Shipley, B. (2012a). Computer-intense methods. In *Encyclopedia of Environmetrics*, 2nd ed., El-Shaarawi, A. H. and Piegorsch, W. W. (eds). pp. 489–493. Chichester: Wiley.

Fortin, M.-J., James, P. M., MacKenzie, A., Melles, S. J. & Rayfield, B. (2012b). Spatial statistics, spatial regression, and graph theory in ecology. *Spatial Statistics*, **1**, 100–109.

Fortin, M.-J., Keitt, T. H., Maurer, B. A., Taper, M. L., Kaufman, D. M. & Blackburn, T. M. (2005). Species ranges and distributional limits: pattern analysis and statistical issues. *Oikos*, **108**, 7–17.

Fortin, M.-J., Olson, R. J., Ferson, S., Iverson, L., Hunsaker, C., Edwards, G., Levine, D., Butera, K. & Klemas, V. (2000). Issues related to the detection of boundaries. *Landscape Ecology*, **15**, 453–466.

Fortin, M.-J. & Payette, S. (2002). How to test the significance of the relation between spatially autocorrelated data at the landscape scale: a case study using fire and forest maps. *Écoscience*, **9**, 213–218.

Fortuna, M. A. & Bascompte, J. (2008). The network approach in ecology. In *Unity in Diversity: Reflections on Ecology after the Legacy of Ramón Margalef*, Valladares, F., Camacho, A., Elosegi, A., Gracia, C., Estrada, M., Senar, J. C. & Gili, J. M. (eds.), pp. 371–392. Mexico: Fundacion BBVA.

References

Fortuna, M. A., Gómez-Rodriguez, C. & Bascompte, J. (2006). Spatial network structure and amphibian persistence in stochastic environments. *Proceedings of the Royal Society of London Series B, Biological Sciences*, **273**, 1429–1434.

Fotheringham, A. S. (2009). 'The problem of spatial autocorrelation' and local spatial statistics. *Geographical Analysis*, **41**, 398–403.

Fotheringham, A. S., Brunsdon, C. & Charlton, M. (2002). *Geographically Weighted Regression: The Analysis of Spatially Varying Relationships*. Chichester: Wiley.

Foxall, R. & Baddeley, A. (2002). Nonparametric measures of association between a spatial point process and a random set, with geological applications. *Applied Statistics*, **51**, 165–182.

Franckowiak, R. P., Panasci, M., Jarvis, K. J., Acuna-Rodriguez, I. S., Landguth, E. L., Fortin, M.-J. & Wagner, H. H. (2017). Model selection with multiple regression on distance matrices leads to incorrect inferences. *PLoS ONE*, **12**, e0175194.

Franco, M. & Harper, J. (1988). Competition and the formation of spatial pattern in spacing gradients: An example using *Kochia scoparia*. *Journal of Ecology*, **76**, 959–974.

Frazier, A. E. & Kedron, P. (2017). Landscape metrics: Past progress and future directions. *Current Landscape Ecology Reports*, **2**, 63–72.

Frydman, N., Freilikhman, S., Talpaz, I. & Pilosof, S. (2023). Practical guidelines and the EMLN R package for handling ecological multilayer networks. *Methods in Ecology and Evolution*, **14**, 2964–2973.

Gabriel, K. R. & Sokal, R. R. (1969). A new statistical approach to geographic variation analysis. *Systematic Ecology*, **18**, 259–270.

Gaggiotti, O., Chao, A., Peres-Neto, P., Chiu, C.-H., Edwards, C., Fortin, M.-J., Jost, L., Richards, C. & Selkoe, K. (2018). Diversity from genes to ecosystems: A unifying framework to study variation across biological metrics and scales. *Evolutionary Applications*, **11**, 1176–1193.

Galiana, N., Lurgi, M., Bastazini, V., Bosch, J., Cagnolo, L., Cazelles, K., Claramunt, B., Emer, C., Fortin. M.-J., Grass, I., Hernández-Castellano, C., Jauker, F., Leroux, S., McCann, K., McLeod, A., Montoya, D., Mulder, C., Osorio-Canadas, S., Reverté, S., Rodrigo, A., Steffan-Dewenter, I., Traveset, A., Valverde, S., Vázquez, D., Wood, S., Gravel, D., Roslin, T., Thuiller, W. & Montoya, J. M. (2022). Ecological network complexity scales with area. *Nature Ecology & Evolution*, **6**, s41559-021-01644-4.

Galiana, N., Lurgi, M., Claramunt-Lopez, B., Fortin, M.-J., Leroux, S., Cazelles, K., Gravel, D. & Montoya, J. M. (2018). The spatial scaling of species interaction networks. *Nature Ecology & Evolution*, **2**, 782–790.

Gao, B., Wang, J., Stein, A. & Chen, Z. (2022). Causal inference in spatial statistics. *Spatial Statistics*, **50**, 100621.

Gao, J., Li, D. & Havlin, S. (2014). From a single network to a network of networks. *National Science Review*, **1**, 346–356.

Garcia-Soidan, P., Menezes, R. & Rubiños, O. (2014). Bootstrap approaches for spatial data. *Stochastic Environment Research and Risk Assessment*, **28**, 1207–2019.

Gardner, M. (1970). The fantastic combinations of John Conway's new solitaire game 'life'. Mathematical games. *Scientific American*, **223**, 120–123.

Garroway, C. J., Bowman, J., Carr, D. & Wilson, P. J. (2008). Applications of graph theory to landscape genetics. *Evolutionary Applications*, **1**, 620–630.

Gaston, K. J. & Blackburn, T. M. (2000). *Patterns and Processes in Macroecology*. Malden: Blackwell Publishing Company.

Gauch, H. G. & Whittaker, R. H. (1972). Coenocline simulation. *Ecology*, **53**, 446–451.

Geary, R. C. (1954). The contiguity ratio and statistical mapping. *The Incorporated Statistician*, **5**, 115–145.

Gelfand, A. E. (2012). Hierarchical modeling for spatial data problems. *Spatial Statistics*, **1**, 30–39.

Gelfand, A. E. & Banerjee, S. (2015). Bayesian wombling: Finding rapid change in spatial maps. *WIRES Computational Statistics*, **7**, 307–325.

Gerstmann, H., Doktor, D., Glässer, C. & Möler, M. (2016). PHASE: A geostatistical model for the Kriging-based spatial prediction of crop phenology using public phenological and climatological observations. *Computers and Electronics in Agriculture*, **127**, 726–738.

Getis, A. (1991). Spatial interaction and spatial autocorrelation: a cross product approach. *Environment and Planning A*, **23**, 1269–1277.

Getis, A. & Franklin, J. (1987). Second-order neighborhood analysis of mapped point patterns. *Ecology*, **68**, 473–477.

Getis, A. & Ord, J. K. (1992). The analysis of spatial association by use of distance statistics. *Geographical Analysis*, **24**, 189–206.

Getis, A. & Ord, J. K. (1996). Local spatial statistics: an overview. In *Spatial Analysis: Modelling in a GIS Environment*, Longley, P. & Batty, M. (eds.), pp. 261–277. Cambridge: GeoInformation International.

Gilbert, B. & Bennett, J. R. (2010). Partitioning variation in ecological communities: Do the numbers add up? *Journal of Applied Ecology*, **47**, 1071–1082.

Gilbert, B., Datta, A., Casey, J. A. & Ogburn, E. L. (2022). Approaches to spatial confounding in geostatistics. *arXiv:*2112.14946v2 [stat.ME].

Gill, J. (2002). *Bayesian Methods: A Social and Behavioral Sciences Approach*. Boca Raton: Chapman & Hall/CRC.

Glaz, J., Naus, J. & Wallenstein, S. (2001). *Scan Statistics*. New York: Springer-Verlag.

Gleason, H. A. (1927). Further views of the succession concept. *Ecology*, **8**, 299–326.

Glenn-Lewin, D. C. & van der Maarel, E. (1992). Patterns and processes of vegetation dynamics. In *Plant Succession: Theory and Prediction*, Glenn-Lewin, D. C., Peet, R. K. & Veblen, T. T. (eds.), pp. 11–59. London: Chapman & Hall.

Godet, C. & Clauzel, C. (2021). Comparison of landscape graph modelling methods for analysing pond network connectivity. *Landscape Ecology*, **36**, 725–748.

Godfrey, S. S. (2013). Networks and the ecology of parasite transmission: A framework for wildlife parasitology. *International Journal for Parasitology: Parasites and Wildlife*, **2**, 235–245.

Good, P. (1993). *Permutation Tests: A Practical Guide to Resampling Methods for Hypothesis Testing*. New York: Springer-Verlag.

Goovaerts, P. & Jacquez, G. M. (2005). Detection of temporal changes in the spatial distribution of cancer rates using local Moran's I and geostatistically simulated spatial neutral models. *Journal of Geographical Systems*, **7**, 137–159.

Gordon, A. D. (1999). *Classification. Monographs on Statistics and Applied Probability 82*, 2nd ed. London: Chapman & Hall/CRC.

Goulard, M., Pagès, L. & Cabanettes, A. (1995). Marked point process: Using correlation functions to explore a spatial data set. *Biometrical Journal*, **37**, 837–853.

Gouveia, C., Moreh, A. & Jordán, F. (2021). Combining centrality measures: Maximizing the predictability of keystone species in food webs. *Ecological Indicators*, **126**, 107617.

References

Grace, J. B. (2020). Scientist's guide to developing explanatory statistical models using causal analysis principles. *Ecology*, 101, e02962.

Grace, J. B., Schoolmaster Jr, D. R., Guntenspergen, G. R., Little, A. M., Mitchell, B. R., Miller, K. M. & Schweiger, E. W. (2012). Guidelines for a graph-theoretic implementation of structural equation modeling. *Ecosphere*, **3**, 1–44.

Grant, E. H. C., Lowe, W. H. & Fagan, W. F. (2007). Living in the branches: Population dynamics and ecological processes in dendritic networks. *Ecology Letters*, **10**, 165–175.

Gravel, D., Massol, F. & Leibold, M. A. (2016). Stability and complexity in model meta-ecosystems. *Nature Communications*, **7**, 12457.

Greig-Smith, P. (1961). Data on pattern within plant communities. I. The analysis of pattern. *Journal of Ecology*, **49**, 695–702.

Griffith, D. A. (1981). Interdependence in space and time: Numerical and interpretative considerations. In *Dynamic Spatial Models*, Griffith, D. A. & MacKinnon, R. D. (eds.), pp. 258–287. Netherlands: Sijthoff & Noordhoff.

Griffith, D. A. (1996). Spatial autocorrelation and eigenfunctions of the geographic weights matrix accompanying geo-referenced data. *Canadian Geographer*, **40**, 351–367.

Griffith, D. A. (2008). Spatial-filtering-based contributions to a critique of geographically weighted regression (GWR). *Environment and Planning A*, **40**, 2751–2769.

Griffith, D. A. & Chun, Y. (2014). Spatial autocorrelation and spatial filtering. In *Handbook of Regional Science*, Fischer, M. M. & Nijkamp, P. (eds.), pp. 1477–1507. Berlin: Springer-Verlag.

Griffith, D. A. & Paelinck, J. H. P. (2009). Specifying a joint space- and time-lag using a bivariate Poisson distribution. *Journal of Geographical Systems*, **11**, 23–36.

Griffith, D. A. & Peres-Neto, P. R. (2006). Spatial modeling in ecology: The flexibility of eigenfunction spatial analyses. *Ecology*, **87**, 2603–2613.

Grillet, M.-E., Barrera, R., Martínez, J.-E., Berti, J. & Fortin, M.-J. (2010a). Disentangling the effect of local and global spatial variation on a mosquito-borne infection in a neotropical heterogeneous environment. *American Journal of Tropical Medicine and Hygiene*, **82**, 194–201.

Grillet, M.-E., Jordan, G. J. & Fortin, M.-J. (2010b). State transition detection in the spatio-temporal incidence of malaria. *Spatial and Spatio-Temporal Epidemiology Journal*, **1**, 251–259.

Guan, N., Song, D. & Liao, L. (2019). Knowledge graph embedding with concepts. *Knowledge Based Systems*, **64**, 38–44.

Guichard, F., Zhang, Y. & Lutscher, F. (2019). The emergence of phase asynchrony and frequency modulation in metacommunities. *Theoretical Ecology*, **12**, 329.

Guillot, G. & Rousset, F. (2013). Dismantling the Mantel tests. *Methods in Ecology and Evolution*, **4**, 336–344.

Guisan, A. & Rahbek, C. (2011). SESAM: A new framework integrating macroecological and species distribution models for predicting spatio-temporal patterns of species assemblages. *Journal of Biogeography*, **38**, 1433–1444.

Gurevitch, J., Koricheva, J., Nakagawa, S. & Stewart, G. (2018). Meta-analysis and the science of research synthesis. *Nature*, **555**, 175–182.

Haas, S. E., Hooten, M. B., Rizzo, D. M. & Meentemeyer, R. K. (2011). Forest species diversity reduces disease risk in a generalist plant pathogen invasion. *Ecology Letters*, 14, 1108–1116.

Haining, R. P. (2003). *Spatial Data Analysis: Theory and Practice*. Cambridge: Cambridge University Press.

References

Halder, A. (2020). Wombling for spatial and spatiotemporal processes: Applications to insurance data. PhD Dissertation, University of Connecticut.

Hall, K. R. (2008). Comparing geographic boundaries in songbird demography data with vegetation boundaries: A new approach to evaluating habitat quality. *Environmental and Ecological Statistics*, **15**, 491–521.

Hall, K. R. & Maruca, S. L. (2001). Mapping a forest mosaic: A comparison of vegetation and songbird distributions using geographic boundary analysis. *Plant Ecology*, **156**, 105–120.

Hammond, D. K., Vandergheynst, P. & Gribonval, R. (2011). Wavelets on graphs via spectral graph theory. *Applied and Computational Harmonic Analysis*, 30, 129–150.

Hansen, A. & di Castri, F. (1992). *Landscape Boundaries: Consequences for Biotic Diversity and Ecological Flows*. New York: Springer-Verlag.

Hanski, I. (2009). Metapopulations and spatial population processes. In *The Princeton Guide to Ecology*, Levin, S. A. (ed.), pp. 177–185. Princeton: Princeton University Press.

Hanski, I. & Woiwod, I. P. (1993). Spatial synchrony in the dynamics of moth and aphid populations. *Journal of Animal Ecology*, **62**, 656–668.

Harary, F. (1969). *Graph Theory*. Reading, MA: Addison-Wesley.

Harper, K. A., Macdonald, S. E., Burton, P. J., Chen, J., Brosofske, K. D., Saunders, S. C., Euskirchen, E. S., Roberts, D. A. R., Jaiteh, M. S. & Esseen, P. A. (2005). Edge influence on forest structure and composition in fragmented landscapes. *Conservation Biology*, **19**, 768–782.

Harrison, P. J., Hanski, I. & Ovaskainen, O. (2011). Bayesian state-space modeling of metapopulation dynamics in the Glanville fritillary butterfly. *Ecological Monographs*, **8**, 581–598.

Harrison, X. A., Donaldson, L., Correa-Cano, M. E., Evans, J., Fisher, D. N., Goodwin, C. E., Robinson, B. S., Hodgson, D. J. & Inger, R. (2018). A brief introduction to mixed effects modelling and multi-model inference in ecology. *PeerJ*, **6**, e4794.

Hayes, J. G. (1974). Numerical methods for curve and surface fitting. *Bulletin: Institute of Mathematics and Its Applications*, **10**, 144–162.

Heagerty, P. J. & Lumley, T. (2000). Window subsampling of estimating functions with application to regression models. *Journal of the American Statistical Society*, **95**, 197–211.

Hedley, S. L. & Buckland, S. T. (2004). Spatial models for line transect sampling. *Journal of Agricultural, Biological and Environmental Statistics*, **9**, 181–199.

Hefley, T. J., Broms, K. M., Brost, B. M., Buderman, F. E., Kay, S. L., Scharf, H. R., Tipton, J. R., Williams, P. J. & Hooten, M. B. (2017). The basis function approach for modeling autocorrelation in ecological data. *Ecology*, 98, 632–646.

Hendry, R. J. & McGlade, J. M. (1995). The role of memory in ecological systems. *Proceedings of the Royal Society of London Series B, Biological Sciences*, **259**, 153–159.

Henebry, G. M. (1995). Spatial model error analysis using autocorrelation indices. *Ecological Modelling*, **82**, 75–91.

Hengl, T., Heuvelink, G. B. M. & Rossiter, D. G. (2007). About regression-kriging: From equations to case studies. *Computers and Geosciences*, **33**, 1301–1315.

Hengl, T., Heuvelink, G. B. M. & Stein, A. (2004). A generic framework for spatial prediction of soil variables based on regression-kriging. *Geoderma*, **120**, 75–93.

Henry, S. & McInnes, B. T. (2017). Literature based discovery: Models, methods, and trends. *Journal of Biomedical Informatics*, **74**, 20–32.

References

Herrera, M., Mur, J. & Ruiz, M. (2016). Detecting causal relationships between spatial processes. *Papers in Regional Science*, 65, 577–594.

Hill, M. O. (1973). The intensity of spatial pattern in plant communities. *Journal of Ecology*, **61**, 225–235.

Hirt, M. R., Grimm, V., Li, Y., Rall, B. C., Rosenbaum, B. & Brose, U. (2018). Bridging scales: Allometric random walks link movement and biodiversity research. *Trends in Ecology & Evolution*, **33**, 701–712.

Holland, J. & Yang, S. (2016). Multi-scale studies and the ecological neighborhood. *Current Landscape Ecology Reports*, **1**, 135–145.

Huaylla, C. A., Nacif, M. E., Coulin, C., Kuperman, M. N. & Garibaldi, L. A. (2021). Decoding information in multilayer ecological networks: The keystone species case. *Ecological Modelling*, **460**, 109734.

Hufkens, K., Scheunders, P. & Ceulemans, R. (2009). Ecotones in vegetation ecology: Methodologies and definitions revisited. *Ecological Research*, **24**, 977–986.

Hui, C. & McGeoch, M. A. (2014). Zeta diversity as a concept and metric that unifies incidence-based biodiversity patterns. *The American Naturalist*, **184**, 684–694.

Hurlbert, S. H. (1984). Pseudoreplication and the design of ecological field experiments. *Ecological Monographs*, **54**, 187–211.

Huston, M. A. (1994). *Biological Diversity: The Coexistence of Species on Changing Landscapes*. New York: Cambridge University Press.

Hutchinson, M. C., Mora, B. B., Pilosof, S., Barner, A. K., Kéfi, S., Thébault, E., Jordano, P. & Stouffer, D. B. (2019). Seeing the forest for the trees: Putting multilayer networks to work for community ecology. *Functional Ecology*, **33**, 206–217.

Illian, J., Penttinen, A. & Stoyan, D. (2008). *Statistical Analysis and Modeling of Spatial Point Patterns*. Chichester: Wiley.

Imbens, G. & Angrist, J. (1994). Identification and estimation of local average treatment effects. *Econometrica*, **62**, 467–476.

Ives, A. R. & Zhu, J. (2006). Statistics for correlated data: Phylogenies, space, and time. *Ecological Applications*, **16**: 20–32.

Jackson, H. B. & Fahrig, L. (2015). Are ecologists conducting research at the optimal scale? *Global Ecology and Biogeography*, **24**, 52–63.

Jackson, T. D., Williams, G. J., Walker-Springett, G. & Davies, A. J. (2020). Three-dimensional digital mapping of ecosystems: A new era of spatial ecology. *Proceedings of the Royal Society, B*, **287**, 20192383.

Jacquemyn, H., Honnay, O. & Pailler, T. (2007). Range size variation, nestedness, and species turnover of orchid species along an altitudinal gradient on Reunion Island: Implications for conservation. *Biological Conservation*, **136**, 388–397.

Jacquez, G. M. (1995). The map comparison problem: Tests for the overlap of geographical boundaries. *Statistical Medicine*, **14**, 2343–2361.

Jacquez, G. M. (1996). A k nearest neighbour test for space–time interaction. *Statistical Medicine*, **15**, 1935–1949.

Jacquez, G. M., Kaufmann, A. & Goovaerts, P. (2008). Boundaries, links and clusters: A new paradigm in spatial analysis? *Environmental and Ecological Statistics*, **15**, 403–419.

Jacquez, G. M., Maruca, S. L. & Fortin, M.-J. (2000). From fields to objects: A review of geographic boundary analysis *Journal of Geographical Systems*, **2**, 221–241.

James, P. (1992). Knowledge graphs. In *Linguistic Instruments in Knowledge Engineering*, Van de Riet, R. P. & Meersman, R. A. (eds.), pp. 97–117. Amsterdam: Elsevier.

James, P. M. A., Fleming, R. A. & Fortin, M.-J. (2010). Identifying significant scale-specific spatial boundaries using wavelets and null models: Spruce budworm defoliation in Ontario, Canada as a case study. *Landscape Ecology*, **25**, 873–887.

James, P. M. A. & Fortin, M.-J. (2012). Ecosystems and spatial patterns. In *Encyclopedia of Sustainability Science and Technology*, Meyers, R. A. (ed.), pp. 3326–3342. New York: Springer Science+Business Media.

James, P. M. A., Sturtevant, B. R., Townsend, P., Wolter, P. & Fortin, M.-J. (2011). Two-dimensional wavelet analysis of spruce budworm host basal area in the Border Lakes landscape. *Ecological Applications*, **21**, 2197–2209.

Jennings, M. K., Zeller, K. A. & Lewison, R. (2020). Supporting adaptive connectivity in dynamic landscapes. *Land*, **9**, 295.

Jiménez, J., Augustine, B. C., Linden, D. W., Chandler, R. B. & Royle, J. A. (2020). Spatial capture–recapture with random thinning for unidentified encounters. *Ecology and Evolution*, **11**, 1187–1198.

Jiménez-Hernández, J. D. C., López-Cerino, M. & Aguirre-Salado, A. I. (2020). A Bayesian hierarchical model for the spatial analysis of carbon monoxide pollution extremes in Mexico City. *Hidawi*, a7135142.

Johnson, J. B. & Omland, K. S. (2004). Model selection in ecology and evolution. *Trends in Ecology & Evolution*, **19**, 101–108.

Johnston, C. A., Pastor, J. & Pinay, G. (1992). Quantitative methods for studying landscape boundaries. In *Landscape Boundaries*, Hansen, A. J. & di Castri, F. (eds.), pp. 107–125. New York: Springer-Verlag.

Johnstone, J. F., Allen, C. D., Franklin, J. F., Frelich, L. E., Harvey, B. J., Higuera, P. E., Mack, M. C., Meentemeyer, R. K., Metz, M. R., Perry, G. L. W., Schoennagel, T. & Turner, M. G. (2016). Changing disturbance regimes, ecological memory, and forest resilience. *Frontiers in Ecology and Environment*, **14**, 369–378.

Jombart, T., Dray, S. & Dufour, A. B. (2009). Finding essential scales of spatial variation in ecological data: A multivariate approach. *Ecography*, 32, 161–168.

Jones, E. L., Rendell, L., Pirotta, E. & Long, J. A. (2016). Novel application of a quantitative spatial comparison tool to species distribution data. *Ecological Indicators*, 70, 67–76.

Joo, R., Picardi, S., Boone, M. E., Clay, T. A., Patrick, S. C., Romero-Romero, V. S. & Baile, M. (2022). Recent trends in movement ecology of animals and human mobility. *Movement Ecology*, **10**, 26.

Joot, P. (2021). *Geometric Algebra for Electrical Engineers*. Toronto: Peter Joot.

Jost, L. (2006). Entropy and diversity. *Oikos*, **113**, 363–375.

Jost, L. (2007). Partitioning diversity into alpha and beta components. *Ecology*, **88**, 2427–2439.

Journel, A. G. & Huijbregts, C. (1978). *Mining Geostatistics*. London: Academia Press.

Juhasz-Nagy, P (1993). Notes on compositional diversity. *Hydrobiologia*, **249**, 173–182.

Kabos, S. & Csillag, F. (2002). The analysis of spatial association of nominal data on regular lattice by join-count-statistic without first-order homogeneity. *Computers and Geosciences*, **28**, 901–910.

Karami, J., Kavosi, M. R. & Babanzhad, M. (2015). Spatial pattern and disease severity of charcoal canker in Hyrcanian forests, north of Iran. *Journal of Forest Science*, **61**, 261–267.

Kareiva, P. & Shigesada, N. (1983). Analyzing insect movement as a correlated random walk. *Oecologia*, **56**, 234–238.

Karl, J. W. (2010). Spatial predictions of cover attributes of rangeland ecosystems using regression kriging and remote sensing. *Rangeland Ecology & Management*, **63**, 335–349.

Kéfi, S., Berlow, E. L., Wieters, E. A., Joppa, L. N., Wood, S. A., Brose, U. & Navarrete, S. A. (2015). Network structure beyond food webs: mapping non-trophic and trophic interactions on Chilean rocky shores. *Ecology*, **96**, 291–303.

Kéfi, S., Holmgren, M. & Scheffer, M. (2016). When can positive interactions cause alternative stable states in ecosystems? *Functional Ecology*, **30**, 88–97.

Kéfi, S., Thébault, E., Eklöf, A., Lurgi, M., Davis, A. J., Kondoh, M. & Krumins, J. A. (2018). Toward multiplex ecological networks: Accounting for multiple interaction types to understand community structure and dynamics. In *Adaptive Food Webs: Stability and Transitions of Real and Model Ecosystems*, Moore, J. C., de Ruiter, P. C., McCann, K. S. & Wolters, V. (eds.), pp. 74–87. Cambridge: Cambridge University Press.

Keil, P. & Chase, J. M. (2019). Global patterns and drivers of tree diversity integrated across a continuum of spatial grains. *Nature Ecology & Evolution*, **3**, 390–399.

Keil, P., Wiegand, T., Tóth, A. B., McGlinn, D. J. & Chase, J. M. (2021). Measurement and analysis of interspecific associations as a facet of biodiversity. *Ecological Monographs*, **91**, e10452.

Keitt, T. H. & Urban, D. L. (2005). Scale-specific inference using wavelet. *Ecology*, **86**, 2497–2504.

Keitt, T. H., Urban, D. L. & Milne, B. T. (1997). Detecting critical scales in fragmented landscapes. *Conservation Ecology* [online], 1, 4. www.consecol.org/vol1/iss1/art4/.

Kejriwal, M., Knoblock, C. A. & Szekely, P. (2021). *Knowledge Graphs: Fundamentals, Techniques, and Applications*. Cambridge, MA: MIT Press.

Khalighi, M., Sommeria-Klein, G., Gonze, D., Faust, K. & Lahti, L. (2022). Quantifying the impact of ecological memory on the dynamics of interacting communities. *PLoS Computational Biology*, **8**, e1009396.

Kim, D. (2021). Predicting the magnitude of residual spatial autocorrelation in geographical ecology. *Ecography*, **44**, 1121–1130.

Kim, J. & Lee, J.-G. (2015). Community detection in multi-layer graphs: A survey. *SIGMOD Record*, **44**, 37–48.

Kimmel, K., Dee, L. E., Avolio, M. L. & Ferraro, P. J. (2021). Causal assumptions and causal inference in ecological experiments. *Trends in Ecology & Evolution*, **36**, 1141–1152.

Kingman, J. F. C. (1993). *Poisson Processes*. Oxford: Oxford University Press.

Kinsey, A. C., Rossi, G, Silk, M. J. & VanderWaal, K. (2020). Multilayer and multiplex networks: An introduction to their use in veterinary epidemiology. *Frontiers in Veterinary Science*, **7**, 596.

Kivelä, M., Arenas, A., Barthelemy, M., Gleeson, J. P., Moreno, Y. & Porter, M. A. (2014). Multilayer networks. *Journal of Complex Networks*, **2**, 203–271.

Knox, E. G. (1964). The detection of space-time interactions. *Applied Statistics*, **13**, 25–29.

Koen, E. L., Garroway, C. J., Wilson, P. J. & Bowman, J. (2010). The effect of map boundary on estimates of landscape resistance to animal movement. *PLoS ONE*, **5**(7), e11785.

Koenig, W. D. & Knops, J. M. (1998). Testing for spatial autocorrelation in ecological time series. *Ecography*, **21**, 423–429.

Kolaczyk, E. D. & Csárdi, G. (2014). *Statistical Analysis of Network Data with R*. New York: Springer.

Kolasa, J. (2014). Ecological boundaries: A derivative of ecological entities. *Web Ecology*, **14**, 27–37.

Koleff, P., Gaston, K. J. & Lennon, J. J. (2003). Measuring beta diversity for presence absence data. *Journal of Animal Ecology*, 72, 367–382.

König, D., Carvajal-Gonzalez, S., Downs, A. M., Vassy, J. & Rigaut, J. P. (1991). Modeling and analysis of 3-D arrangements of particles by point processes with examples of application to biological data obtained by confocal scanning light microscopy. *Journal of Microscopy*, **161**, 405–433.

Krebs, C. J., Boutin, S. & Boonstra, R. (2001). *Ecosystem Dynamics of the Boreal Forest: The Kluane Project*. Oxford: Oxford University Press.

Krige, D. G. (1966). Two-dimensional weighted moving average trend surfaces for ore-evaluation. *Journal of the South Africa Institute of Mining and Metallurgy*, **66**, 13–38.

Ku, K., Bradley, J. R. & Niu, X. (2019). Boundary detection using a Bayesian hierarchical model for multiscale spatial data. *Technometrics*, **63**, 64–76.

Kükenbrink, D., Schneider, F. D., Leiterer, R., Schaepman, M. E. & Morsdorf, F. (2017). Quantification of hidden canopy volume of airborne laser scanning data using a voxel traversal algorithm. *Remote Sensing of Environment*, **194**, 424–436.

Kuldorff, M. (1997). A spatial scan statistic. *Communications in Statistics: Theory and Methods*, 26, 1481–1496.

Kuldorff, M., Heffernan, R., Hartman, J., Assunção, R. M. & Mostashari, F. (2005). A space-time permutation scan statistic for the early detection of disease outbreaks. *PLoS Medicine*, **2**, 216–224.

Kupfer, J. A. (2012). Landscape ecology and biogeography: Rethinking landscape metrics in a post-FRAGSTATS landscape. *Progress in Physical Geography*, **36**, 400–420.

Labonne, J., Ravigné, V., Parisi, B. & Gaucherel, C. (2008). Linking dendritic network structures to population demogenetics: The downside of connectivity. *Oikos*, **117**, 1479–1490.

Ladau, J. & Eloe-Fadrosh, E.A. (2019). Spatial, temporal, and phylogenetic scales of microbial ecology. *Trends in Microbiology*, **27**, 662–669.

Ladyman, J. & Wiesner, K. (2020). *What Is a Complex System?* New Haven: Yale University Press.

Lahiri, S. N. (1999). Theoretical comparisons of block bootstrap methods. *The Annals of Statistics*, **27**, 386–404.

Laliberté, E., Schweiger, A. K. & Legendre, P. (2020). Partitioning plant spectral diversity into alpha and beta components. *Ecology Letters*, **23**, 370–380.

Lambin, X., Elston, D. A., Petty, S. J. & MacKinnon, J. L. (1998). Spatial asynchrony and periodic traveling waves in cyclic populations of field voles. *Proceedings of the Royal Society of London Series B*, **265**, 1491–1496.

Lamy, T., Pitz, K. J., Chavez, F. P., Yorke, C. E. & Miller, R. J. (2021). Environmental DNA reveals the fine-grained and hierarchical spatial structure of kelp forest fish communities. *Nature Scientific Reports*, **11**, 14439.

Landi, P., Minoarivelo, H. O., Brännström, Å., Hui, C. & Dieckmann, U. (2018). Complexity and stability of ecological networks: A review of the theory. *Population Ecology*, **60**, 319–345.

Latombe, G., Hui, C. & Mcgeoch, M. A. (2017). Multi-site generalised dissimilarity modelling: Using zeta diversity to differentiate drivers of turnover in rare and widespread species. *Methods in Ecology and Evolution*, **8**, 431–442.

Ledo, A., Condés, S. & Montes, F. (2011). Intertype mark correlation function: A new tool for the analysis of species interactions. *Ecological Modelling*, **222**, 580–587.

Legendre, P. (1993). Spatial autocorrelation: Trouble or new paradigm? *Ecology*, **74**, 1659–1673.

Legendre, P. (2008). Studying beta diversity: Ecological variation partitioning by multiple regression and canonical analysis. *Journal of Plant Ecology*, **1**, 3–8.

Legendre, P. (2019). A temporal beta-diversity index to identify sites that have changed in exceptional ways in space-time surveys. *Ecology & Evolution*, **9**, 3500–3514.

Legendre, P., Borcard, D. & Peres-Neto, P. R. (2005). Analyzing beta diversity: Partitioning the spatial variation of community composition data. *Ecological Monographs*, **75**, 435–450.

Legendre, P. & Condit, R. (2019). Spatial and temporal analysis of beta diversity in the Barro Colorado Island forest dynamics plot, Panama. *Forest Ecosystems*, **6**, 7.

Legendre, P., Dale, M. R. T., Fortin, M.-J., Casgrain, P. & Gurevitch, J. (2004). Effects of spatial structures on the results of field experiments. *Ecology*, **85**, 3202–3214.

Legendre, P., Dale, M. R. T., Fortin, M.-J., Gurevitch, J., Hohn, M. & Myers, D. E. (2002). The consequences of spatial structure for the design and analysis of ecological field surveys. *Ecography*, **25**, 601–615.

Legendre, P. & De Cáceres, M. (2013). Beta diversity as the variance of community data: Dissimilarity coefficients and partitioning. *Ecology Letters*, **16**, 951–963.

Legendre, P. & Fortin, M.-J. (1989). Spatial pattern and ecological analysis. *Vegetatio*, **80**, 107–138.

Legendre, P. & Fortin, M.-J. (2010). Comparison of the Mantel test and alternative approaches for detecting complex multivariate relationships in the spatial analysis of genetic data. *Molecular Ecology Resources*, **10**, 831–844.

Legendre, P., Fortin, M.-J. & Borcard, D. (2015). Should the Mantel test be used in spatial analysis? *Methods in Ecology and Evolution*, **6**, 1239–1247.

Legendre, P., Lapointe, F.-J. & Casgrain, P. (1994). Modeling brain evolution from behavior: A permutational regression approach. *Evolution*, **48**, 1487–1499.

Legendre, P. & Legendre, L. (2012). *Numerical Ecology*, 3rd English edn. Amsterdam: Elsevier.

Leibold, M. A. & Chase, J. M. (2018). *Metacommunity Ecology*. Princeton: Princeton University Press.

Lele, S. (1991). Jackknifing linear estimating equations: Asymptotic theory and applications in stochastic processes. *Journal of the Royal Statistical Society B*, **53**, 253–267.

Levin, S. A. (1992). The problem of pattern and scale in ecology. *Ecology*, **73**, 1943–1967.

Levin, S. A. (2000). Multiple scales and the maintenance of biodiversity. *Ecosystems*, **3**, 498–506.

Lewis, M. A., Fagan, W. F., Auger-Methe, M., Frair, J., Fryxell, J. M., Gros, C., Gurarie, E., Healy, S. D. & Merkle, J. A. (2021). Learning and animal movement. *Frontiers in Ecology and Evolution*, **9**, 681704.

Lewis, P. & Shedler, G. (1979). Simulation of nonhomogeneous Poisson processes by thinning. *Naval Research Logistics Quarterly*, **26**, 403–413.

Liang, S. D., Banerjee, S. & Carlin, B. P. (2009). Bayesian Wombling for spatial point processes. *Biometrics*, **65**, 1243–1253.

Lichstein, J. W. (2007). Multiple regression on distance matrices: A multivariate spatial analysis tool. *Plant Ecology*, **188**, 117–131.

van Lieshout, M. N. M. & Baddeley, A. J. (1999). Indices of dependence between types in multivariate point patterns. *Scandinavian Journal of Statistics*, **26**, 511–532.

Lin, K.-P., Long, Z.-H. & Ou, B. (2011). The size and power of bootstrap tests for spatial dependence in a linear regression model. *Computational Economics*, **38**, 153–171.

Little, L. R. & Dale, M. R. T. (1999). A method for analysing spatio-temporal pattern in plant establishment, tested on a *Populus balsamifera* clone. *Journal of Ecology*, **87**, 620–627.

Liu, C. (2001). A comparison of five distance-based methods for pattern analysis. *Journal of Vegetation Science*, **12**, 411–416.

Long, J., Robertson, C. & Nelson, T. (2018). Stampr: Spatio-temporal analysis of moving polygons in R. *Journal of Statistical Software*, **84**, Code Snippet 1.

Lotwick, H. W. & Silverman, B. W. (1982). Methods for analysing spatial processes of several types of points. *Journal of the Royal Statistical Society B*, **44**, 406–413.

Lovejoy, T. E. & Hannah, L. (2019). *Biodiversity and Climate Change: Transforming the Biosphere*. New Haven, CT: Yale University Press.

Lowell, K. (1997). Effect(s) of the 'no-same-color-touching' constraint on the join-count statistic: A simulation study. *Geographical Analysis*, **29**, 339–353.

Lu, H. & Carlin, B. P. (2005). Bayesian areal wombling for geographical boundary analysis. *Geographical Analysis*, **37**, 265–285.

Ludwig, J. A. & Cornelius, J. M. (1987). Locating discontinuities along ecological gradients. *Ecology*, **68**, 448–450.

Lurgi, M., Galiana, N., Broitman, B. R., Kiel, S., Wieters, E. A. & Navarette, S. A. (2020). Geographical variation of multiplex ecological networks in marine intertidal communities. *Ecology*, **101**, e03165.

Lurgi, M., Galiana, N., Lopez, B. C., Joppa, L. N. & Montoya, J. M. (2014). Network complexity and species traits mediate the effects of biological invasions on dynamic food webs. *Frontiers in Ecology and Evolution*, **2**, 36.

Lv, L., Zhang, K., Zhang, T., Li, X., Sun, Q., Zhang, L. & Xue, W. (2021). Eigenvector-based centralities for multilayer temporal networks under the framework of tensor computation. *Expert Systems with Applications*, **184**, 115471.

Ma, Y., Wang, S., Aggrawal, C. C., Yin, D. & Tand, J. (2019). Multi-dimensional graph convolutional networks. Proceedings of the 2019 SIAM International Conference on Data Mining, pp. 657–665. Society for Industrial and Applied Mathematic. May 2019, Calgary, AB, Canada.

MacIsaac, D. A. (1989). Primary succession on proglacial deposits in the Canadian Rockies. MSc Thesis, University of Alberta.

MacKinnon, J. L., Petty, S. J., Elston, D. A., Thomas, C. J., Sherratt, T. N. & Lambin, X. (2001). Scale invariant spatio-temporal patterns of field vole density. *Journal of Animal Ecology*, **70**, 101–111.

Magurran, A. E. & McGill, B. J. (eds.) (2010). *Biological Diversity: Frontiers in Measurement and Assessment*. Oxford: Oxford University Press.

Manly, B. J. F. (2018). *Randomization, Bootstrap and Monte Carlo Methods in Biology*. 4th ed. Boca Raton: CRC Press.

Mantel, N. (1967). The detection of disease clustering and a generalized regression approach. *Cancer Research*, **27**, 209–220.

Marj, T., Tuia, D. & Ratle, F. (2006). Detection of clusters using space-time scan statistics. *International Journal of Wildland Fire*, **18**, 830–836.

Martensen, A. C., Saura, S. & Fortin, M.-J. (2017). Spatio-temporal connectivity: Assessing the amount of reachable habitat in dynamic landscapes. *Methods in Ecology and Evolution*, **8**, 1253–1264.

Mason, O. & Verwoerd, M. (2007). Graph theory and networks in biology. *IET Systems Biology*, **1**, 89–119.

Matheron, G. (1970). La théorie des variables réegionaliséees, et ses applications. *Les Cahiers du Centre de Morphologie Mathéematique de Fontainebleau*. Fontainebleau: Fascicule 5.

References

Matsuoka, S., Sugiyama, Y., Sato, H., Katano, I., Harada, K. & Doi, H. (2019). Spatial structure of fungal DNA assemblages revealed with eDNA metabarcoding in a forest river network in western Japan. *Metabarcoding & Metagenomics*, **3**, e36225.

Mayo, D. G. & Hand, D. (2022). Statistical significance and its critics: Practicing damaging science, or damaging scientific practice? *Synthese*, **200**, 220.

McBratney, A. B. & de Gruijter, J. J. (1992). A continuum approach to soil classification by modified fuzzy k-means with extra grades. *Journal of Soils Science*, **43**, 159–176.

McCaslin, H. M., Feuka, A. B. & Hooten, M. B. (2020). Hierarchical computing for hierarchical models in ecology. *Methods in Ecology and Evolution*, **12**, 245–254.

McCoy, E. D., Bell, S. S. & Walters, K. (1986). Identifying biotic boundaries along environmental gradients. *Ecology*, **67**, 749–759.

McElreath, R. (2020). *Statistical Rethinking*. 2nd ed. Boca Raton: CRC Press.

McGarigal, K., Wan, H. Y., Zeller, K. A., Timm, B. C. & Cushman, S. A. (2016). Multi-scale habitat selection modeling: A review and outlook. *Landscape Ecology*, **31**, 1161–1175.

McIntire, E. J. B. & Fajardo, A. (2009). Beyond description: The active and effective way to infer processes from spatial patterns. *Ecology*, **90**, 46–56.

McIntyre, N. & Wiens, J. (2000). A novel use of the lacunarity index to discern landscape function. *Landscape Ecology*, **15**, 313–321.

McLeish, M., Peláez, A., Pagán, I., Galiván, R., Fraile, A. & García-Arena, F. (2021). Structuring of plant communities across agricultural landscape mosaics: The importance of connectivity and the scale of effect. *BMC Ecology and Evolution*, **21**, 173.

McLeod, A. M. & Leroux, S. J. (2021). The multiple meanings of omnivory influence empirical modular theory and whole food web stability relationships. *Journal of Animal Ecology*, **90**, 447–459.

McRae, B. H., Dickson, B. G., Keitt, T. H. & Shah, V. B. (2008). Using circuit theory to model connectivity in ecology, evolution, and conservation. *Ecology*, **89**, 2712–2724.

Melián, C. J., Bascompte, J. & Jordano, P. (2005). Spatial structure and dynamics in a marine food web. In *Aquatic Food Webs: An Ecosystem Approach*, Belgrano, A., Scharler, U., Dunne, J. & Ulanowicz, R. E. (eds.), pp. 19–24. New York: Oxford University Press.

Melles, S. J., Badzinski, D., Fortin, M.-J., Csillag, F. & Lindsay, K. (2009). Disentangling habitat and social drivers of nesting patterns in songbirds. *Landscape Ecology*, **24**, 519–531.

Melles, S. J., Fortin, M.-J., Lindsay, K. & Badzinski, D. (2011). Expanding northward: Influence of climate change, forest connectivity, and population processes on a threatened species' range shift. *Global Change Biology*, **17**, 17–31.

Michelot, T., Langrock, R., Patterson, T. and Rexstad, E. (2015). *moveHMM: Animal Movement Modelling using Hidden Markov Models*. R package version 1.1.

Michelot, T., Langrock, R., Patterson, T. and Rexstad, E. (2016). moveHMM: Animal movement modelling using hidden Markov models. *Methods in Ecology and Evolution*, **7**, 1308–1315.

Milo, R., Shen-Orr, S., Itzkovitz, S., Kashtan, N., Chklovskii, D. & Alon, U. (2002). Network motifs: Simple building blocks of complex networks. *Science*, **298**, 824–827.

Mimet, A., Houet, T., Julliard, R. & Simon, L. (2013). Assessing functional connectivity: A landscape approach for handling multiple ecological requirements. *Methods in Ecology and Evolution*, **4**, 453–463.

Minasny, B. & McBratney, A. B. (2005). The Matérn function as a general model for soil variograms. *Geoderma*, **128**, 192–207.

Minchin, P. R. (1989). Montane vegetation of the Mt. Field Massif, Tasmania: A test of some hypotheses about properties of community patterns. *Vegetatio*, 83, 97–110.

Mizon, G. E. (1995). A simple message for autocorrelation correctors: Don't. *Journal of Econometrics*, **69**, 267–289.

Moran, P. A. P. (1948). The interpretation of statistical maps. *Journal of the Royal Statistical Society B*, **10**, 243–251.

Morrison, L.W. (2013). Nestedness in insular floras: Spatiotemporal variation and underlying mechanisms. *Journal of Plant Ecology*. doi: 10.1093/pe/rtt002.

Moss, R., Elston, D. A. & Watson, A. (2000). Spatial asynchrony and demographic traveling waves during red grouse population cycles. *Ecology*, **81**, 981–989.

Muff, S., Signer, J. & Fieberg, J. (2019). Accounting for individual-specific variation in habitat-selection studies: Efficient estimation of mixed-effects models using Bayesian or frequentist computation. *Journal of Animal Ecology*, **89**, 80–92.

Mugglestone, M. A. & Renshaw, E. (1996). A practical guide to the spectral analysis of spatial point processes. *Computational Statistics & Data Analysis*, **21**, 43–65.

Murakami, M. & Hirao, T. (2010). Lizard predation alters the effect of habitat area on the species richness of insect assemblages on Bahamian isles. *Diversity and Distributions*, 16, 952–958.

Myers, J. A., Chase, J. M., Crandall, R. M. & Jiménez, I. (2015). Disturbance alters beta-diversity but not the relative importance of community assembly mechanisms. *Journal of Ecology*, **103**, 1291–1299.

Nakagawa, S. & Schielzeth, H. (2013). A general and simple method for obtaining R2 from generalized linear mixed-effects models. *Methods in Ecology and Evolution*, **4**, 133–142.

Nathan, R., Getz, W. M., Revilla, E., Holyoak, M., Kadmon, R., Saltz, D. & Smouse, P. E. (2008). A movement ecology paradigm for unifying organismal movement research. *Proceedings of the National Academy of Sciences*, **105**, 19052–19059.

Nathan, R., Monk, C. T., Arlinghaus, R., Adam, T., Alós, J., Assaf, M., Baktoft, H., Beardsworth, C. E., Bertram, M. G., Bijleveld, A. I. & Brodin, T. (2022). Big-data approaches lead to an increased understanding of the ecology of animal movement. *Science*, **375**(6582), eabg1780.

Nekola, J. C. & White, P. S. (1999). The distance decay of similarity in biogeography and ecology. *Journal of Biogeography*, **26**, 867–878.

Newman, E. A., Kennedy, M. C., Falk, D. A. & McKenzie, D. (2019). Scaling and complexity in landscape ecology. *Frontiers in Ecology and Evolution*, **7**, 293.

Newman, M. E. J., Barabási, A.-L. & Watts, D. (eds.) (2006). *The Structure and Dynamics of Networks*. Princeton: Princeton University Press.

Niedballa, J., Axtner, J., Döbert, T. F., Tilker, A., Nguyen, A., Wong, S. T., Fiderer, C., Heurich, M. & Wilting, A. (2022). imageseg: An R package for deep learning-based image segmentation. *Methods in Ecology and Evolution*, **13**, 2363–2371.

Noy-Meir, I. & Anderson, D. (1971). Multiple pattern analysis or multiscale ordination: towards a vegetation hologram. In *Statistical Ecology, vol. 3: Populations, Ecosystems, and Systems Analysis*, Patil, G. P., Pielou, E. C. & Waters, W. E. (eds.), pp. 207–232. University Park: Pennsylvania State University Press.

Oden, N. L. (1984). Assessing the significance of a spatial correlogram. *Geographical Analysis*, **16**, 1–16.

Oden, N. L. & Sokal, R. R. (1986). Directional autocorrelation: An extension of spatial correlograms to two dimensions. *Systematic Zoology*, **35**, 608–617.

Oden, N. L. & Sokal, R. R. (1992). An investigation of 3-matrix permutation tests. *Journal of Classification*, **9**, 275–290.

Oden, N. L., Sokal, R. R., Fortin, M.-J. & Goebl, H. (1993). Categorical wombling: Detecting regions of significant change in spatially located categorical variables. *Geographical Analysis*, **25**, 315–336.

Ohlsson, M. & Eklöf, A. (2020). Spatial resolution and location impact group structure in a marine food web. *Ecology Letters*, **23**, 1451–1459.

Okabe, A., Boots, B., Sugihara, K. & Chiu, S. N. (2000). *Spatial Tessellations: Concepts and Applications of Voronoi Diagrams*, 2nd ed. Chichester: Wiley.

Okabe, A. & Yamada, I. (2001). The K-function method on a network and its computational implementation. *Geographical Analysis*, **33**, 271–290.

O'Neill, R. V., Hunsaker, C. T., Timmins, S. P., Jackson, B. L., Jones, K. B., Riitters, K. H. & Wickham, J. D. (1996). Scale problems in reporting landscape pattern at the regional scale. *Landscape Ecology*, **11**, 169–180.

Openshaw, S. (1984). *The Modifiable Areal Unit Problem*. Norwich: Geo Books.

Osborne, P. E., Foody, G. M. & Suárez-Seoane, S. (2007). Non-stationarity and local approaches to modelling the distributions of wildlife, *Diversity and Distributions*, **13**, 313–323.

Ostendorf, B. & Reynolds, J. F. (1998). A model of arctic tundra vegetation derived from topographic gradients. *Landscape Ecology*, **13**, 187–201.

Otto, S. P. & Day, T. (2011). Mathematical modeling in biology. In *A Biologist's Guide to Mathematical Modeling in Ecology and Evolution*, Otto, S. & Day, T. (eds.), pp. 1–16. Princeton: Princeton University Press.

Ovaskainen, O., Roy, D. B., Fox, R. & Anderson, B. J. (2016). Uncovering hidden spatial structure in species communities with spatially explicit joint species distribution models. *Methods in Ecology and Evolution*, **7**, 428–436.

Oyana, T. J. (2021). *Spatial Analysis with R*. 2nd ed. Boca Raton: CRC Press.

Pagel, J. & Schurr, F. M. (2012). Forecasting species ranges by statistical estimation of ecological niches and spatial population dynamics. *Global Ecology and Biogeography*, **21**, 293–304.

Paley, C. J., Taraskin, S. & Elliott, S. (2007). Temporal and dimensional effects in evolutionary graph theory. *Physical Review Letters*, **98**, 98–103.

Pang, S. E. H., Slik, J. W. F., Zurell, D. & Webb, E. L. (2023). The clustering of spatially associated species unravels patterns in tropical tree species distributions. *Ecosphere*, **14**, e4589.

Parrott, L., Proulx, R. & Thibert-Plante, X. (2008). Three-dimensional metrics for the analysis of spatiotemporal data in ecology. *Ecological Informatics*, **3**, 343–353.

Pascual, M. & Dunne, J. A. (eds.) (2006). *Ecological Networks: Linking Structure to Dynamics in Food Webs*. New York: Oxford University Press.

Pascual-Hortal, L. & Saura, S. (2006). Comparison and development of new graph-based landscape connectivity indices: Towards the prioritization of habitat patches and corridors for conservation. *Landscape Ecology*, **21**, 959–967.

Pasqueretta, C., Jeanson, R., Pansanel, J., Raine, N. E., Chittka, L. & Lihoreau, M. (2019). A spatial network analysis of resource partitioning between bumblebees foraging on artificial flowers in a flight cage. *Movement Ecology*, **7**, 4.

Pearl, J. & MacKenzie, D. (2018). *The Book of Why*. New York, Basic Books.

Pearl, J. (2009). *Causality: Models, Reasoning and Inference*. 2nd ed. Cambridge. Cambridge University Press.

Pearse, W. D., Davis, C. C., Inouye, D. W., Primack, R. B. & Davies, T. J. (2017). A statistical estimator for determining the limits of contemporary and historic phenology. *Nature Ecology & Evolution*, **1**, 1876–1882.

Pelletier, B., Fyles, J. W. & Dutilleul, P. (1999). Tree species control and spatial structure of forest floor properties in a mixed-species stand. *Écoscience*, **6**, 79–91.

Pelletier, D., Clark, M., Anderson, M. G., Rayfield, B., Walder, M. A. & Cardille, J. A. (2014). Applying circuit theory for corridor expansion and management at regional scales: Tiling, pinch points, and omnidirectional connectivity. *PLoS ONE*, **9**, e84135.

Penttinen, A. K., Stoyan, D. & Henttonen, H. M. (1992). Marked point processes in forest statistics. *Forest Science*, **38**, 806–824.

Perea, A. J., Wiegand, T., Garrido, J. L., Rey, P. J. & Alcántara, J. M. (2021). Legacy effects of seed dispersal mechanisms shape the spatial interaction network of plant species in Mediterranean forests. *Journal of Ecology*, **109**, 3670–3684.

Pereira, J., Saura, S. & Jordán, F. (2017). Single-node vs. multi-node centrality in landscape graph analysis: Key habitat patches and their protection for 20 bird species in NE Spain. *Methods in Ecology and Evolution*, 8, 1458–1467.

Peres-Neto, P. R. (2006). A unified strategy for estimating and controlling spatial, temporal and phylogenetic autocorrelation in ecological models. *Oecologia Brasiliensis*, **10**, 105–119.

Peres-Neto, P. R. & Legendre, P. (2010). Estimating and controlling for spatial structure in the study of ecological communities. *Global Ecology and Biogeography*, **19**, 174–184.

Perna, A. & Latty, T. (2014). Animal transportation networks. *Journal of The Royal Society Interface*, **11**, 20140334.

Perry, J. N., Liebhold, A. M., Rosenberg, M. S., Dungan, J., Miriti, M., Jakomulska, A. & Citron-Pousty, S. (2002). Illustrations and guidelines for selecting statistical methods for quantifying spatial pattern in ecological data. *Ecography*, **25**, 578–600.

Peters, V. S. (2003). Keystone processes affect succession in boreal mixed woods: The relationship between masting in white spruce and fire history. PhD Thesis, University of Alberta.

Peterson, E. E., Ver Hoef, J. M., Isaak, D. J., Falke, J. A., Fortin, M.-J., Jordan, C., McNyset, K., Monestiez, P., Ruesch, A. S., Sengupta, A., Som, N., Steel, A., Theobald, D. M., Torgersen, C. E. & Wenger, S. J. (2013). Modeling dendritic ecological networks in space: An integrated network perspective. *Ecology Letters*, **16**, 707–719.

Peterson, G. D. (2002). Contagious disturbance, ecological memory, and the emergence of landscape pattern. *Ecosystems*, **5**, 329–338.

Philibert, M. D., Fortin, M.-J. & Csillag, F. (2008). Spatial structure effects on the detection of patches boundaries using local operators. *Environmental and Ecological Statistics*, **15**, 447–467.

Pichler, M. & Hartig, F. (2022). Machine learning and deep learning: A review for ecologists. *Methods in Ecology and Evolution*, **14**, 994–1016.

Pielou, E. C. (1977). *Mathematical Ecology*. New York: Wiley.

Pilosof, S., Porter, M. A., Pascual, M. & Kéfi, S. (2017). The multilayer nature of ecological networks. *Nature Ecology & Evolution*, **1**, 0101.

Pinto, N. & Keitt, T. H. (2009). Beyond the least-cost path: Evaluating corridor redundancy using a graph-theoretic approach. *Landscape Ecology*, **24**, 253–266.

Pinto-Ledezma, J. N., Larkin, D. J. & Cavender-Bares, J. (2018). Patterns of beta diversity of vascular plants and their correspondence with biome boundaries across North America. *Frontiers in Ecology and Evolution*, **6**, 194.

Plant, R. E. (2019). *Spatial Data Analysis in Ecology and Agriculture using R*. 2nd ed. Boca Raton: CRC Press.

Plante, M., Lowell, L., Potvin, F., Boots, B. & Fortin, M.-J. (2004). Studying deer habitat on Anticosti Island, Québec: Relating animal occurrences and forest map information. *Ecological Modelling*, 174, 387–399.

Poisot, T., Canard, E., Mouillot, D., Mouquet, N. & Gravel, D. (2012). The dissimilarity of species interaction networks. *Ecology Letters*, **15**, 1353–1361.

Polakowska, A. E., Fortin, M.-J. & Couturier, A. (2012). Quantifying the spatial relationship between bird species' distributions and landscape feature boundaries in southern Ontario, Canada. *Landscape Ecology*, **27**, 1481–1493.

Proulx, S. R., Promislow, D. E. L. & Phillip, P. C. (2005). Network thinking in ecology and evolution. *Trends in Ecology and Evolution*, **20**, 345–352.

Radicchi, F. & Bianconi, G. (2017). Redundant interdependencies boost the robustness of multiplex networks. *Physical Review X*, **7**, 011013.

Rajala. T., Olhede, S. C. & Murrell, D. J. (2019). When do we have the power to detect biological interactions in spatial point patterns? *Journal of Ecology*, **107**, 711–721.

Rammer, W. & Seidl, R. R. (2019). Harnessing deep learning in ecology: An example predicting bark beetle outbreaks. *Frontiers in Plant Science*, **10**, a1327.

Ranta, E., Kaitala, V., Lindström, J. & Helle, E. (1997). The Moran effect and synchrony in population dynamics. *Oikos*, **78**, 136–142.

Rayfield, B., Fortin, M.-J. & Fall, A. (2011). Connectivity for conservation: A framework to classify network measures. *Ecology*, **92**, 847–858.

Read, S., Bath, P., Willett, P. & Maheswaran, R. (2010). Measuring the spatial accuracy of the spatial scan statistic. *Spatial and Spatio-temporal Epidemiology*, **2**, 69–78.

Redding, T. E., Hope, G. D., Schmidt, M. G. & Fortin, M.-J. (2004). Spatial patterns of N availability across forest-clearcut edges in the southern interior of British Columbia. *Canadian Journal of Forest Research*, **34**, 1018–1024.

Reich, B. J., Yang, S., Guan, Y., Giffin, A. B., Miller, M. J. & Rappold, A. (2021). A review of spatial causal inference methods for environmental and epidemiological applications. *International Statistical Review*, **89**, 605–634.

Remmel, T. K. & Fortin, M.-J. (2013). Categorical, class-focused map patterns: Characterization and comparison. *Landscape Ecology*, **28**, 1587–1599.

Richardson, M. C., Fortin, M.-J. & Branfireun, B. A. (2009). Hydrogeomorphic edge detection and delineation of landscape functional units from LiDAR digital elevation models. *Water Resources Research*, **45**, W10441, doi:10.1029/2008WR007518.

Ricotta, C. (2006). Towards a complex, plural and dynamic approach to diversity: Rejoinder to Myers and Patil, Podani, and Sarkar. *Acta Biotheoretica*, 54, 141–146.

Rietkerk, M., Dekker, S. C., de Ruiter, P. C. & van de Koppel, J. (2004). Self-organised patchiness and catastrophic shifts in ecosystems. *Science*, **305**, 1926–1929.

Ripley, B. D. (1976). The second-order analysis of stationary point processes. *Journal of Applied Probability*, **13**, 255–266.

Ripley, B. D. (1978). Spectral analysis and the analysis of pattern in plant communities. *Journal of Ecology*, **66**, 965–981.

Ripley, B. D. (1988). *Statistical Inference for Spatial Processes*. Cambridge: Cambridge University Press.

Riva, F., Graco-Roza, C., Daskalova, G. N., Hudgins, E. J., Lewthwaite, J. M. M., Newman, E. A., Ryo, M. & Mammola, S. (2022). Towards a cohesive understanding of ecological complexity. *Science Advances*, **9**, eabq4207.

Roberts, D. R., Bahn, V., Ciuti, S., Boyce, M. S., Elith, J., Guillera-Arroita, G., Hauenstein, S., Lahoz-Monfort, J. J., Schröder, B., Thuiller, W., Warton, D. I., Hartig, F. & Dormann, C. F. (2017). Cross-validation strategies for data with temporal, spatial, hierarchical, or phylogenetic structure. *Ecography*, **40**, 913–929.

Robertson, C., Nelson, T. A., Boots, B. & Wulder, M. A. (2007). STAMP: Spatial-temporal analysis of moving polygons. *Journal of Geographical Systems*, **9**, 207–227.

Robitaille, A. L., Webber, Q. M. R., Turner, J. W. & VanderWal, E. (2021). The problem and promise of scale in multilayer animal social networks. *Current Zoology*, **67**, 113–123.

Rogers, H. S., Beckman, N. G., Hartig, F., Johnson, J. S., Pufal, G., Shea, K., Zurell, D., Bullock, J. M., Cantrell, R. S., Loiselle, B., Pejchar, L., Razafindratsima, O. H., Sandor, M. E., Schupp, E. W., Strickland, W. C. & Zambrano, J. (2019). The total dispersal kernel: A review and future directions. *AoB Plants*, **11**, plz042.

Rosenzweig, L. (1995). *Species Diversity in Space and Time*. Gateshead: Cambridge University Press.

Rota, C. T., Ferreira, M. A., Kays, R. W., Forrester, T. D., Kalies, E. L., McShea, W. J., Parsona, A. W. & Millspaugh, J. J. (2016). A multispecies occupancy model for two or more interacting species. *Methods in Ecology and Evolution*, **7**, 1164–1173.

Rouquette, J. R., Dallimer, M., Armsworth, P. R., Gaston, K. J., Maltby, L. & Warren, P. H. (2013). Species turnover and geographic distance in an urban river network. *Diversity and Distributions*, **19**, 1429–1439.

Roushangar, K., Moghaddas, M., Ghasempour, R. & Alizadeh, F. (2021). Evaluation of spatiotemporal characteristics of precipitation using discrete maximal overlap wavelet transform and spatial clustering tools. *Hydrology Research*, **52**(2), 414–430.

Royle, J. A., Fuller, A. K. & Sutherland, C. (2017). Unifying population and landscape ecology with spatial capture-recapture. *Ecography*, **41**, 444–456.

Ruiz, E. J. L., Mozo, H. G., Vilches, E. D. & Galán, C. (2012). The use of geostatistics in the study of floral phenology of *Vulpia geniculata* (L.) Link. *The Scientific World Journal*, 624247.

Sadahiro, Y. & Umemura, M. (2002). A computational approach for the analysis of changes in polygon distributions. *Journal of Geographical Systems*, **3**, 137–154.

Sahneh, F. D. & Scoglio, C. (2014). Competitive epidemic spreading over arbitrary multilayer networks. *Physics Revue E*, **89**, 068217.

Sanuy, D. & Bovet, P. (1997). A comparative study on the paths of five anura species. *Behavioural Processes*, **41**, 193–199.

Saunders, S. C., Brosofske, K. D., Chen, J., Drummer, T. D. & Gustafson E. J. (2005). Identifying scales of pattern in ecological data: A comparison of lacunarity, spectral and wavelet analyses. *Ecological Complexity*, **2**, 87–105.

Saura, S., Bodin, Ö. & Fortin, M. J. (2014). Stepping stones are crucial for species' long-distance dispersal and range expansion through habitat networks. *Journal of Applied Ecology*, 51, 171–182.

Scheiner, S. M., Cox, S. B., Willig, M., Mittelbach, G. G., Osenberg, C. & Kaspari, M. (2000). Species richness, species-area curves and Simpson's paradox. *Evolutionary Ecology Research*, **2**, 791–802.

Schmera, D., Legendre, P., Erös, T., Tóth, M., Magyari, E. K., Baur, B. & Podani, J. (2022). New measures for quantifying directional changes in presence–absence community data. *Ecological Indicators*, **136**, 108618.

Schooler, S. L. & Zald, H. S. J. (2019). Lidar prediction of small mammal diversity in Wisconsin USA. *Remote Sensing*, **11**, 2222.

Schröder, W., Schmidt, G. & Schönrock, S. (2014). Modelling and mapping of plant phenological stages as bio-meterological indicators for climate change. *Environmental Sciences Europe*, **26**, 5.

Setzer, R. W. (1985). Spatio-temporal patterns of mortality in Pemphigus populicaulis and P. populitransversus on cottonwoods. *Oecologia*, 67, 310–321.

Shade, A., Dunn, R. R., Blowes, S. A., Keil, P., Bohannan, B. J., Herrmann, M., Küsel, K., Lennon, J. T., Sanders, N. J., Storch, D. & Chase, J. (2018). Macroecology to unite all life, large and small. *Trends in Ecology & Evolution*, **33**, 731–744.

Shih, F. Y. (2009). *Image Processing and Mathematical Morphology: Fundamentals and Applications.* Boca Raton: CRC Press.

Shimatani, K. (2001). Multivariate point processes and spatial variation of species diversity. *Forest Ecology and Management*, **142**, 215–229.

Shoemaker, L. G., Sullivan, L. L., Donohue, I., Cabral, J. S., Williams, R. J., Mayfield, M. M., Chase, J. M., Chu, C., Harpole, W. S., Huth, A. & HilleRisLambers, J. (2020). Integrating the underlying structure of stochasticity into community ecology. *Ecology*, **101**, e02922.

Silk, M. J., Croft, D. P., Delahay, R. J., Hodgson, D. J., Boots, M., Weber, N. & McDonald, R. A. (2017). Using social network measures in wildlife disease ecology, epidemiology, and management. *BioScience*, **67**, 245–257.

Smouse, P. E., Long, J. C. & Sokal, R. R. (1986). Multiple regression and correlation extensions of the Mantel test of matrix correspondence. *Systematic Zoology*, **35**, 627–632.

Soininen, J., McDonald, R. & Hillibrand, H. (2007). The distance decay of similarity in ecological communities. *Ecography*, 30, 3–12.

Sokal, R. R., Oden, N. L., Thomson, B. A. & Kim, J. (1993). Testing for regional differences in means: Distinguishing inherent from spurious spatial autocorrelation by restricted randomization. *Geographical Analysis*, 25, 199–210.

Sokal, R. R. & Rohlf, F. J. (1995). *Biometry*. 3rd ed. San Fransisco: Freeman.

Sokal, R. R. & Wartenberg, D. E. (1983). A test of spatial autocorrelation using an isolation-by-distance model. *Genetics*, **105**, 219–237.

Solé, R. (2007). Scaling laws in the drier. *Nature*, **449**, 151–153.

Solé, R. & Bascompte, J. (2006). *Self-Organization in Complex Ecosystems.* Princeton: Princeton University Press.

Somers, K. M. & Jackson, D. A. (2022). Putting the Mantel test back together again. *Ecology*, **103**, e3780.

Spear, S., Balkenhol, N., Fortin, M.-J., McRae, B. & Scribner, K. (2010). Use of resistance surfaces for landscape genetic studies: Considerations for parameterization and analysis. *Molecular Ecology*, **19**, 3576–3591.

Spooner, P. G., Lunt, I. D., Okabe, A. & Shiode, S. (2004). Spatial analysis of roadside Acacia populations on a road network using the network K-function. *Landscape Ecology*, **19**, 491–499.

Sprugel, D. G. (1976). Dynamic structure of wave-regenerated *Abies balsamea* forests in the northeastern United States. *Journal of Ecology*, **64**, 889–911.

St-Louis, V., Fortin, M.-J. & Desrochers, A. (2004). Association between microhabitat and territory boundaries of two forest songbirds. *Landscape Ecology*, **19**, 591–601.

Stein, A., Gerstner, K. & Kreft, H. (2014). Environmental heterogeneity as a universal driver of species richness across taxa, biomes and spatial scales. *Ecology Letters*, **17**, 866–880.

Steinhaeuser, K., Chawla, N. V. & Ganguly, A. R. (2010). Complex networks in climate science: progress, opportunities and challenges. *Proceedings of the 2010 Conference on Intelligent Data Understanding*, 16–26. 5–6 October 2010, Mountain View, CA.

Stoyan, D., Kendall, W. S. & Mecke, J. (1995). *Stochastic Geometry and Its Applications*. 2nd ed. Chichester: Wiley.

Stoyan, D. & Penttinen, A. (2000). Recent applications of point process methods in forestry statistics. *Statistical Science*, **15**, 61–78.

Strampelli, P., Anderson, L., Everatt, K. T., Somers, M. J. & Rowcliffe, J. M. (2018). Leopard *Panthera pardus* density in southern Mozambique: Evidence from spatially explicit capture-recapture in Xonghile Game Reserve. *Oryx*, **54**, 405–411.

Strydom, T. & Poisot, T. (2023). SpatialBoundaries.jl: Edge detection using spatial wombling. *Ecography*, **2023**, e06609.

Sturtevant, B. R. & Fortin M.-J. (2021). Understanding and modeling forest disturbance interactions at the landscape level. Special issue: Using landscape simulation models to help balance conflicting goals in changing forests. *Frontiers in Ecology and Evolution*, **9**, 653647.

Swetnam, T. L., Lynch, A. M., Falk, D. A., Yool, S. R. & Guertin, D. P. (2015). Discriminating natural variation from legacies of disturbance in semi-arid forests, Southwestern U.S.A. *Ecosphere*, **6**, 1–22.

Takahashi, K., Kulldorff, M., Tango, T. & Yih, K. (2008). A flexibly shaped space-time scan statistic for disease outbreak detection and monitoring. *International Journal of Health Geographics*, **7**, 14.

Tango, T. (2021). Spatial scan statistics can be dangerous. *Statistical Methods in Medical Research*, **30**, 75–86.

Tavaré, S. (1983). Serial dependence in contingency tables. *Journal of the Royal Statistical Society B*, **45**, 100–106.

Tavaré, S. & Altham, P. M. E. (1983). Serial dependence of observations leading to contingency tables, and corrections to chi-squared statistics. *Biometrika*, **70**, 139–144.

Taylor, M. D. (2021). *An Introduction to Geometric Algebra and Geometric Calculus*. Orlando, FL: M. D. Taylor.

Teng, S. N., Xu, C., Sandel, B. & Svenning, J. C. (2018). Effects of intrinsic sources of spatial autocorrelation on spatial regression modelling. *Methods in Ecology and Evolution*, **9**, 363–372.

Thaden, H. & Kneib, T. (2018). Structural equation models for dealing with spatial confounding. *The American Statistician*, **72**, 239–252.

Thiery, J. M., Dherbes, J. M. & Valentin, C. (1995). A model simulationg the genesis of banded vegetation patterns in Niger. *Journal of Ecology*, **83**, 497–507.

Tiefelsdorf, M. & Griffith, D. A. (2007). Semi-parametric filtering of spatial autocorrelation: The eigenvector approach. *Environmental Planning A*, **39**, 1193–1221

Timóteo, S., Correia, M., Rodriguez-Echeverria, S., Freitas, H. & Heleno, R. (2018). Multilayer networks reveal the spatial structure of seed-dispersal interactions across the Great Rift landscapes. *Nature Communications*, **9**, 140.

Tobin, P. C. & Bjørnstad, O. N. (2003). Spatial dynamics and cross-correlation in a transient predator–prey system. *Journal of Animal Ecology*, **72**, 460–467.

Tobler, W. F. (1970). A computer movie simulating urban growth in the Detroit region. *Economic Geography*, **46**, 234–240.

Torrence, C. & Compo, G. P. (1998). A practical guide to wavelet analysis. *Bulletin of the American Meteorological Society*, **79**, 61–78.

Toussaint, G. T. (1980). The relative neighbourhood graph of a finite planar set. *Pattern Recognition*, **12**, 261–268.

Tuia, D., Ratle, F., Lasaponara, R., Tleesca, L. & Kanevski, M. (2008). Scan statistics analysis of forest fire clusters. *Communications in Nonlinear Science and Numerical Simulation*, **13**, 1689–1694.

Turchin, P. (1998). *Quantitative Analysis of Movement*. Sunderland: Sinauer.

Turgeon, K., Turpin, C. & Gregory-Eaves, I. (2017). Boreal river impoundments caused nearshore fish community assemblage shifts but little change in diversity: A multiscale analysis. *Canadian Journal of Fisheries and Aquatic Science*, **76**, 740–752.

Turner, M. G. & Gardner, R. H. (2015). *Landscape Ecology in Theory and Practice: Pattern and Process*. 2nd ed. New York: Springer-Verlag.

Upton, G. J. G. & Fingleton, B. (1985). *Spatial Data Analysis by Example, vol. I: Point Pattern and Quantitative Data*. New York: Wiley.

Urban, D. L. (2023). *Agents and Implications of Landscape Pattern: Working Models for Landscape Ecology*. Springer Nature.

Urban, D. L. & Keitt, T. (2001). Landscape connectivity: A graph-theoretic perspective. *Ecology*, **82**, 1205–1218.

Urban, D. L., Minor, E. S., Treml, E. A. & Schick, R. S. (2009). Graph models of habitat mosaics. *Ecology Letters*, **12**, 260–273.

Urban, M. C., Strauss, S. Y., Pelletier, F., Palkovacs, E. P., Leibold, M. A., Hendry, A. P., De Meester, L., Carlson, S. M., Angert, A. L. & Giery, S. T. (2020). Evolutionary origins for ecological patterns in space. *Proceedings of the National Academy of Sciences*, **117**, 17482–17490.

Urdangarin, A., Goicoa, T. & Ugarte, M. D. (2023). Evaluating recent methods to overcome spatial confounding. *Revista Mathmática Complutense*, **36**, 333–360.

Van Strien, M. J., Keller, D. & Holderegger, R. (2012). A new analytical approach to landscape genetic modelling: Least-cost transect analysis and linear mixed models. *Molecular Ecology*, **21**, 4010–4023.

Vaswani, A., Shazeer, N., Parmar, N., Uszkoreit, J., Jones, L., Gomez, A. N., Kaiser, L. & Polosukhin, I. (2017). Attention is all you need. *31st Conference on Neural Information Processing Systems (NIPS 2017)*. 4–9 December 2017, Long Beach, CA, USA.

Velázquez, E., Martínez, I., Getzin, S., Moloney, K. A. & Wiegand, T. (2015). An evaluation of the state of point pattern analysis in ecology. *Ecography*, **39**, 1042–1055.

Vellend, M. (2016). *The Theory of Ecological Communities*. Princeton: Princeton University Press.

Vepakomma, U., Kneeshaw, D. & Fortin, M.-J. (2012). Spatial contiguity and continuity of disturbance in boreal forests: Gap persistence, expansion, shrinkage and displacement. *Journal of Ecology*, **100**, 1257–1268.

Ver Hoef, J. M. & Glenn-Lewin, D. C. (1989). Multiscale ordination: A method for detecting pattern at several scales. *Vegetatio*, **82**, 59–67.

Ver Hoef, J. M., Peterson, E. E. & Theobald, D. (2006). Spatial statistical models that use flow and stream distance. *Environmental and Ecological Statistics*, **13**, 449–464.

Ver Hoef, J. M., Peterson, E. E., Hooten, M. B., Hanks, E. M. & Fortin, M.-J. (2018). Spatial autoregressive models for statistical inference from ecological data. *Ecological Monographs*, **88**, 36–59.

Viana, D. S. & Chase, J. M. (2019). Spatial scale modulates the inference of metacommunity assembly processes. *Ecology*, **100**, e02576.

Vince, J. (2008). *Geometric Algebra for Computer Graphics*. London: Springer-Verlag.
Wackernagel, H. (2013). *Multivariate Geostatistics: An Introduction with Applications*. 2nd ed. Berlin: Springer.
Wagner, H. H. (2003). Spatial covariance in plant communities: Integrating ordination, geostatistics, and variance testing. *Ecology*, **84**, 1045–1057.
Wagner, H. H. (2004). Direct multi-scale ordination with canonical correspondence analysis. *Ecology*, **85**, 342–351.
Wagner, H. H. (2013). Rethinking the linear regression model for spatial ecological data. *Ecology*, **94**, 2381–2391.
Wagner, H. H. & Dray, S. (2015). Generating spatially constrained null models for irregularly spaced data using Moran spectral randomization methods. *Methods in Ecology and Evolution*, **6**, 1169–1178.
Wagner, H. H. & Fortin, M.-J. (2005). Spatial analysis of landscapes: Concepts and statistics. *Ecology*, **86**, 1975–1987.
Wakefield, J. C., Kelsall, J. E. & Morris, S. E. (2000). Clustering, cluster detection, and spatial variation in risk. In *Spatial Epidemiology: Methods and Applications*, Elliott, P., Wakefield, J. C., Best, N. G. & Briggs, D. J. (eds.), pp. 128–152. Oxford: Oxford University Press.
Wallerman, J., Joyce, S., Vencatasawmy, C. P. & Olsson, H. (2002). Prediction of forest stem volume using kriging adapted to detected edges. *Canadian Journal of Forest Research*, **32**, 509–518.
Wang, Q., Downey, D., Ji, H. & Hope, T. (2023). Learning to generate novel scientific directions with contextualized literature-based discovery. *arXiv*: 2305.14259v1.
Wang, U., Wang, K., Cao, W. & Wang, X. (2019). Geometric algebra in signal and image processing: A survey. *IEEE Access*, **7**, 156315–156325.
Wang, Y., Weinacker, H. & Koch, B. (2008). A LiDAR point cloud based procedure for vertical canopy strcture analysis and 3D single tree modelling in forest. *Sensors*, **8**, 3938–3951.
Warton, D. I., Blanchet, F. G., O'Hara, R. B., Ovaskainen, O., Taskinen, S., Walker, S. C. & Hui, F. K. (2015). So many variables: Joint modeling in community ecology. *Trends in Ecology & Evolution*, **30**, 766–779.
Wasserman, L. (2004). *All of Statistics: A Concise Course in Statistical Inference*. New York: Springer.
Wasserstein, R. L., Schirm, A. L. & Lazar, N. A. (2019). Moving to a world beyond 'p < 0.05'. *The American Statistician*, **73**, 1–19.
Watt, A. S. (1947). Pattern and process in the plant community. *Journal of Ecology*, **35**, 1–22.
Weaver, J. E., Conway, T. M. & Fortin, M.-J. (2012). An invasive species' relationship with environmental variables changes across multiple spatial scales. *Landscape Ecology*, **27**, 1351–1362.
Webster, R. (1973). Automatic soil-boundary location from transect data. *Mathematical Geology*, **5**, 27–37.
Webster, R. & Oliver, M. A. (2007). *Geostatistics for Environmental Scientists*. Chichester: Wiley.
White, E. P., Ernest, S. K. M., Adler, P. B., Hurlberts, A. H. & Lyons, S. K. (2010). Integrating spatial and temporal approaches to understanding species richness. *Philosophical Transactions of the Royal Society B*, **365**, 3633–3643.
Whittaker, R. H. (1977). Evolution of species diversity in land communities. In *Evolutionary Biology*, Hecht, M. K., Steere, W. C. & Wallace, B. (eds.), 1–67. New York: Plenum.
Whittle, P. (1954). On stationary processes in the plane. *Biometrika*, **41**, 434–449.

References

Wiegand, T., Kissling, W. D., Cipriotti, P. A. & Aguiar, M. R. (2006). Extending point pattern analysis for objects of finite size and irregular shape. *Journal of Ecology*, **94**, 825–837.

Wiegand, T. & Moloney, K. A. (2013). *Handbook of Spatial Point-Pattern Analysis in Ecology.* Boca Raton: CRC Press.

Wiens, J. A. (1989). Spatial scaling in ecology. *Functional Ecology*, **3**, 385–397.

Wiens, J. A., Crist, T. O. & Milne, B. T. (1993). On quantifying insect movements. *Environmental Entomology*, **22**, 709–715.

Wikle, C. K. (2003). Hierarchical Bayesian models for predicting the spread of ecological processes. *Ecology*, **84**, 1382–1394.

Wikle, C. K. & Zammit-Mangion, A. (2022). Statistical deep learning for spatial and spatio-temporal data. *arXiv*:2206.02218v1.

Wikle, C. K., Zammit-Mangion, A. & Cressie, N. (2019). *Spatio-Temporal Statistics with R.* Boca Raton: CRC Press.

Williams, B. K. & Brown, E. D. (2019). Sampling and analysis frameworks for inference in ecology. *Methods in Ecology and Evolution*, **10**, 1832–1842.

Williams, M. J. & Musolesi, M. (2016). Spatio-temporal networks: Reachability, centrality and robustness. *Royal Society Open Science*, **3**, 160196.

Wilson, J. B. & Agnew, A. D. Q. (1992). Positive-feedback switches in plant communities. *Advances in Ecological Research*, **23**, 263–336.

Windle, M. J. S., Rose, G. A., Devillers, R. & Fortin, M.-J. (2010). Exploring spatial non-stationarity of fisheries survey data using geographically weighted regression (GWR): An example from the Northwest Atlantic. *ICES Journal of Marine Science*, 67, 145–154.

With, K. A. (2019). *Essentials of Landscape Ecology.* Oxford: Oxford University Press.

Wolkovich, E. M. & Donahue, M. J. (2021). How phenological tracking shapes species and communities in non-stationary environments. *Biological Reviews*, **96**, 2810–2827.

Womble, W. H. (1951). Differential systematics. *Science*, **114**, 315–322.

Wu, J. (2004). Effects of changing scale on landscape pattern analysis: Scaling relations. *Landscape Ecology*, **19**, 125–138.

Wu, J. & Qi, Y. (2000). Dealing with scale in landscape analysis: an overview. *Geographic Information Sciences*, **6**, 1–5.

Wu, J. G., Shen, W. J. & Sun, W. Z. (2002). Empirical patterns of the effects of changing scale on landscape metrics. *Landscape Ecology*, **117**, 761–782.

Xuan, L. & Hong, Z. (2018). An improved canny edge detection algorithm. *8th IEEE International Conference on Software Engineering and Service Science (ICSESS).* 23–25 November 2018, Beijing.

Yan, S., Xiong, Y. & Lin, D. (2018). Spatial temporal graph convolutional networks for skeleton-based action recognition. *arXiv*: 1801.07455v2.

Yang, F. & Zhou, Y. (2017). Quantifying spatial scale of positive and negative terrains pattern at watershed scale: Case in soil and water conservation region on Loess Plateau. *Journal of Mountain Science*, **14**, 1642–1654.

Yarranton, G. A. & Morrison, R. G. (1974). Spatial dynamics of primary succession: Nucleation. *Journal of Ecology*, **62**, 417–428.

Young, C. G., Dale, M. R. T. & Henry, G. H. R. (1999). Spatial pattern of vegetation in high Arctic sedge meadows. *Écoscience*, **6**, 556–564.

Yu, D., Liu, Y., Xun, B. & Shao, H. (2013). Measuring landscape connectivity in an urban area for biological conservation. *Clean: Soil Air Water*, 41, 1–9.

Zeller, K. A., Jennings, M. K., Vickers, T. W., Ernest, H. B., Cushman, S. A. & Boyce, W. M. (2018). Are all data types and connectivity models created equal? Validating common connectivity approaches with dispersal data. *Diversity & Distributions*, **24**, 868–879.

Zeller, K. A., Lewison, R., Fletcher, R. J. Jr., Tulbure, M. G. & Jennings, M. K. (2020). Understanding the importance of dynamic landscape connectivity. *Land*, **9**, 303.

Zelnik, Y. R., Barbier, M., Shamafelt, D. W., Loreau, M. & Germain, R. M. (2024). Linking intrinsic scales of ecological processes to characteristic scales of biodiversity and functioning patterns. *Oikos*, **2024**, e10514.

Zhan, Q., Liang, Y. & Xiao, Y. (2011). Pattern detection in airborne LiDAR data using Laplacian of Gaussian filter. *Geo-spatial Information Science*, **14**, 184–189.

Zhang, S., Tong, H., Xu, J. & Maciejewski, R. (2019). Graph convolutional networks: A comprehensive review. *Computational Social Networks*, **6**, 11.

Zhang, Z., Liu, Y., Yuan, L., Weber, E. & van Kleunen, M. (2021). Effect of allelopathy on plant performance: A meta-analysis. *Ecology Letters*, **24**, 348–362.

Zinoviev, D. (2018). *Complex Network Analysis in Python*. Raleigh, NC: Pragmatic Bookshelf.

Zuur, A. F., Ieno, E. N. & Elphick, C. S. (2010). A protocol for data exploration to avoid common statistical problems. *Methods in Ecology and Evolution*, **1**, 3–14.

Zuur, A. F., Ieno, E. N., Walker, N., Saveliev, A. A. & Smith, G. M. (2009). *Mixed Effects Models and Extensions in Ecology with R*. New York: Springer.

Index

adjacency matrix, 45, 297, 321
aggregation
  indices, 32
Akaike information criterion (AIC), 116, 199, 208
angle
  classes, 107
  distribution, 243
  turning, 243
angular
  autocorrelation, 243, 341
  concordance, 241
anisotropy, 4, 28, 33, 48, 106–107, 123, 256, 333
  autocovariance, 209
  geometric, 116, 133
  zonal, 116, 133
artificial intelligence (AI), 355
assumption(s), 13, 17, 97, 309, 332, 334, 346, 351, 354–355
  independence, 248
  quasi-stationarity, 109
  stationarity, 2, 102, 109, 119, 122, 134, 136, 214
  weak stationarity, 109
autocorrelation
  cycles, 9–11, 18–19, 173, 180, 190, 252, 261, 316, 332
  cyclic, 180
  directional, 243
  induced, 168, 176
  inherent, 163, 168, 176
  matrix, 174, 177
  negative, 12
  serial, 76, 171
  spatial, 83, 138, 163, 183, 228, 270, 281, 332, 334–335, 339, 342, 352–353
  spatio-temporal, 18
  statistical nuisance, 163
  temporal, 12, 228, 247, 254, 335
autocovariance, 102, 206
autoregressive, 170, 176, 183
  conditional autoregressive (CAR), 353
  Durbin, 107
  spatial, 124

Bayesian analysis, 344
  hierarchical, 348
Bayesian hierarchical analysis, 208–209
Bayesian hierarchical model, 349
block size, 82–86
blocked quadrat methods, 82
boundary, 144
  angle, 151
  closed, 144
  clumping, 76
  cohesive, 153, 156, 162
  ecotones, 142, 161
  edge effect, 144, 158
  elements, 152, 156
  first-order derivatives, 149
  gradual, 144
  intermingling, 147
  kernel filters, 149, 160
  linear, 142
  nonlinear, 142
  open, 144
  patch, 138
  persistence, 153, 160
  properties, 138, 149, 152
  rate of change, 150, 154
  second-order derivatives, 159
  sharp, 144
  statistics, 153
  transition zones, 142, 144
  width, 144, 152
boundary detection
  Bayesian areal wombling, 156
  Bayesian point wombling, 155
  Canny adaptive filter, 160
  categorical-wombling, 155, 161
  convolution, 149, 356
  kernel filters, 159
  Laplacian filter, 159–160
  lattice-wombling, 149, 152–154, 158, 161
  moving window, 149, 356
  multivariate, 148
  scale-space approach, 160
  triangular-window, 154

Index

triangulation-wombling, 154, 161, 301, 341
boundary elements, 156
boundary statistics
　direct overlap, 156
　mean minimum nearest distance statistics, 156
　overall mean minimum nearest distance statistics, 157
　singletons, 140, 335
brousse tigrée, 12

causal inference, 189, 354
causality, 199, 359
　Granger, 192, 355
　inference, 202
chaos, 19
circumcircle method, 42, 80, 90, 123, 339
classes
　Sturge's rule, 46
clonal plant, 241, 244, 284, 286
clustering
　algorithms, 138, 140
　fuzzy k-means, 140, 142
　hierarchical agglomerative, 138, 140
　k-means, 139–140, 142
　spatial, 18, 21–22, 139–140, 152, 155–156, 161, 277
　spatial constraints, 139
coefficient(s)
　correlation, 268, 276
　determination, 195, 215
　dissimilarity, 266, 269
　Pearson correlation, 276
　scaling, 221
community, 1, 5–6, 9–10, 12–14, 16–18, 38–39, 44, 57, 60, 85, 133, 194, 229
competition, 8, 10, 13, 16, 344, 358
complementary combinations, 272
complete census, 62–63, 124, 261, 263
complete graph, 302, 306
complete randomness, 121
complete spatial randomness, 3, 52–53, 55, 61, 64–65, 69, 74, 79–80, 96, 99, 121, 287, 293, 296
concept
　cyclic change, 10
　mosaic cycle, 19, 59
　self-organization, 12–13, 19, 256, 282
Condit's Ω functions, 339
conditional randomness, 121
connectance, 297
connectedness, 297, 321, 335
connectivity, 18
　algorithms, 139, 155
　bishop, 99
　functional, 322, 328, 332, 335
　matrix, 46, 98, 103, 193, 214–215

contiguous quadrats, 33, 40, 45, 60, 86, 340
contingency tables, 178–179
convex hull, 18, 22, 247, 308
correlation, 26, 176, 178, 241, 311
　linear, 101, 196
　matrix, 170, 352
　partial, 174, 178, 197
　Pearson, 101, 103, 193–194, 317
　serial, 164–165, 179, 182
　Spearman, 196, 317
　spurious, 51, 197
correlogram
　cross-correlation, 118
　directional, 109
　Geary's c, 107
　modified, 255
　Moran's I, 107
　omnidirectional, 107, 109, 120, 340
　partial spatial, 107
　spatial, 107, 192
covariance matrix, 85, 130, 133, 177, 200, 206
covariate variables, 163, 204
cross-correlation, 254, 256
cycle, 254

data
　continuous, 24, 88–89, 97, 117, 214, 227–229, 241, 339
　discrete, 24, 89, 204, 227, 229, 241, 261
　lattice, 98–99
　missing, 33, 311
　multivariate, 61, 81, 161, 193, 196, 210–211, 288
　population, 29, 214
　qualitative, 25, 97, 116, 122, 133, 139, 148, 209, 214, 284, 288
　quantitative, 25, 29–30, 73, 97, 101, 122, 136, 140, 145, 148, 158, 218, 271, 288, 339
degree of freedom, 55, 163, 189, 208
　Clifford's correction, 177
　Dutilleul's correction, 177
　effective sample size, 192
Directed Acyclic Graph, 190
disease ecology, 324
dispersal, 4–5, 11, 13–14, 17–18, 20, 26, 30, 35, 39, 52, 54, 106, 163, 209, 251, 253, 261, 278, 280, 284, 286, 295, 297–298, 302–303, 306, 308–310, 318–319, 325, 354
distance, 45
　classes, 45–46, 71, 99, 101–103, 106–107, 118, 168, 177, 179, 193, 196, 228, 256, 339
　coefficient, 193
　distribution, 243
distance decay, 26
distance to independence, 169, 184, 188
distance-based MEM (dbMEM, PCNM), 214–215

Index

diversity
  alpha, 259, 262, 269
  beta, 17, 214, 226, 259, 261, 265–269, 280–281, 325
  biological, 282, 359
  combination, 271, 273
  combinatorial, 272, 274, 281
  compositional, 270, 274
  effective number, 264
  effective number of species, 273
  entropy, 87, 273–274
  evenness, 81, 85–86, 263, 273–274
  first-order, 109, 279
  florula, 281
  gamma, 259, 266, 269
  Gleason, 9
  Hill's index, 264, 273
  Jaccard, 268–269, 273, 279
  local contributions to beta-diversity, 269
  neighbours, 280
  nestedness, 275
  node overlap, decrease filling, 276
  second–order, 279
  Shannon-Weaver index, 263, 273
  Shimatani's K, 280–281
  Simpson's index, 258–259, 263, 265
  site-pair beta-diversity, 269
  Sørensen index, 268
  spatial, 8, 17, 258, 268, 280, 313, 359
  spatio-temporal, 283
  spectral, 270
  turnover, 151, 155, 266–267, 269–270
  zeta, 269

ecological fallacy, 39
ecological hypotheses, 52–53, 311, 342, 359
ecological memory, 12, 329
ecological questions, 19, 57, 139, 156, 162, 237, 338
ecotones, 1
edge, 143
  types, 143
edge effect, 32, 36, 53, 63, 235
  correction, 36, 64, 77, 287
effective number of combinations, 273–274
effective sample size, 169, 177–178
eigenanalysis, 85, 215, 226
  asymmetric Moran's Eigenvector Maps (AEM), 215
  Moran's Eigenvector Maps (MEM), 18, 214–215, 217–218, 226
empty space function, 293
environmental gradient(s), 1, 5, 14, 76, 147–148, 211, 260, 265, 270, 278
epidemiology, 20, 39, 68, 230, 237, 241, 285, 323
equidistance
  classes, 46

equifrequency
  classes, 46
error
  measurement, 111, 156
Euclidean distance, 40, 45–46, 99, 101, 128, 145, 156, 193, 196, 198, 209, 213–214, 216, 221, 310–311, 339
experimental
  design, 20, 55, 102, 163
  units, 81, 184, 186, 247
exploratory spatial data analysis (ESDA), 189, 207
exponential decay, 261
exponential decline, 175, 183
extreme values, 97

feedback loops, 11, 282
fibre pattern analysis, 286–287, 293
  angles, 286
  bivariate, 288
  marked, 291
  neighbours, 288
  parallelism, 286, 309
fibre patttern analysis
  Bézier curves, 291
Fourier analysis, 339
  transform, 88, 218
fuzzy
  boundary, 142
  k-means, 142
  set theory, 140, 142

gap size, 11, 147
Geary's c, 32
geostatistics, 109
Getis' statistics, 237
global spatial statistics, 38, 135, 226, 335
graph, 62, 285
  aspatial, 284
  bipartite, 300
  complete, 280, 302, 307, 311, 319
  connected, 302
  Delaunay triangulation, 40, 42, 80, 99, 138, 140, 154, 215, 235, 280–281, 296, 303
  digraph, 297, 300, 309–310
  directed, 352
  Dirichlet polygons, 43, 125
  dyadic, 324
  empty, 280
  Gabriel, 42, 215, 312
  graph of graphs, 300
  Minimum Spanning Tree, 40, 214, 216, 235, 280, 303, 313
  modularity, 282
  nearest neighbours, 280, 296
  neighbour networks, 40, 299
  networks-of-networks, 326
  planar, 235, 296

relative neighbourhood, 41
signed, 299
spatial, 62, 95, 280, 286, 295, 309–310
spatio-temporal, 229, 244, 249, 359
subgraph, 43, 91, 301–302
tessellation, 43
Thiessen polygons, 125
Ulam tree, 299
Voronoi polygons, 43, 125
graph metrics
bridge, 302
betweenness centrality, 298
centrality, 295, 308
closeness centrality, 298
cut-edges, 297, 302
cycles, 40
diameter, 298, 302, 307
eccentricity, 298, 302
edge matching, 323
link overlap, 323
minimum path length, 302
nestedness, 311
node degree, 297
path length, 298
graph theory, 60, 91, 143, 190, 284, 302
components, 297, 302–303
connectivity, 295, 297
edge connectivity, 302
edges, 284, 295
graph, 295
interaction networks, 295
modularity, 295, 299, 301
nearest point-to-line-neighbour distances, 293
network, 22, 295
node connectivity, 302
nodes, 284, 295
points and lines, 292, 294
random graphs, 285
scale-free, 285
small world, 285
spatial network, 295
spectral gap, 298
Griffith's index, 229

hierarchical
analysis, 299
hierarchical Bayesian analysis, 354
home range, 17, 241
homogeneity, 297
homogeneous, 2–3, 29
hot spots, 281
hypotheses
hierarchy, 248, 293

independent and identically distributed (i.i.d.), 38, 351
independent variables, 125, 174, 176, 204, 260, 311
index of dispersion, 81

inferential tests, 34, 163, 199
information theory, 21, 263
inhomogeneous, 4
integrated nested Laplace approximations (INLA), 209
interaction networks, 326
isotropy, 4, 28, 48, 103, 106, 136, 215

join count statistics, 98, 286
second-order neighbour, 99
spatio-temporal, 229

kernel
adaptive, 207
fixed, 207
kernel filter, 335
kernel function, 254
Kriging, 114, 341
blocked, 131
co-Kriging, 133, 253
conditional annealing, 124, 133
error, 129–130
indicator, 133
indicator function, 116
Lagrangian multiplier, 129
multivariate, 133
nonlinear, 132
ordinary, 132
punctual, 131
stratified, 133
three-dimensional, 128, 130
universal, 132
variance, 129, 131

lack of independence, 26, 44, 55, 83, 102, 106–107, 163, 179, 188, 199, 228, 273, 334, 342
lacunarity, 86
landscape
circuit theory, 309
circuitscape, 309
connectivity, 247, 251, 301, 308–309, 322, 328
corridor, 302, 309
dynamics, 144
fragmentation, 26, 39, 86, 301, 309
functional network, 301
genetics, 308
heterogeneous, 13
metrics, 30
node removal, 306–307
linear
collinear, 211
interpolation, 126
nonlinear, 126, 259
spatial dependence, 355
local
cold spots, 121
Geary's c, 122

local (cont.)
　Getis' statistics, 121, 339
　hot spots, 121
　Moran's *I*, 122
local analysis
　Getis' statistics, 336
local indicator of spatial association (LISA), 119
local Moran's *I*, 120
local spatial analysis, 134
local spatial statistics, 135, 335

machine learning (ML), 355
macroecology, 21, 270
Mantel test, 32, 193, 234, 266, 269
　correlogram, 196, 254
　Hadamard product, 194
　partial, 197, 213
mark correlation, 288, 339
Markov chain
　first-order, 246
Markov Chain Monte Carlo (MCMC), 208–209, 346–347, 353
Markov model, 179–180, 246
mark-recapture, 228
Matérn model, 209, 353
metacommunity, 13–14, 16–18, 209
　neutral dynamics, 14
　paradigms, 14
　patch dynamics, 14
　species sorting, 14
metapopulation, 12–13, 16–18, 209, 286, 310, 323, 327, 354
metaweb, 16, 325
modifiable area unit problem (MAUP), 39, 118, 191
Monte Carlo, 67
　simulation, 58, 182
Moran's *I*, 32
movement, 17–18, 35, 209, 229, 243, 261, 295, 297–298
　analysis, 241
　animal, 4, 227, 229, 241, 246, 249, 294, 296, 308–309, 311
　biased correlated random walk, 246
　complete avoidance, 248
　correlated random walk, 249
　complete randomness, 248
　crossing, 248
　ecology, 249
　path, 228, 243, 293, 300
　random walk, 244, 308, 318
　rapid, 248
　rapid crossing, 249
　segment, 246
moving window, 171, 231, 237
　isotropic, 109
multilayer, 313
　spatial graph, 296
　versatility, 318, 328
multilayer networks, 301
multiple tests
　Bonferroni correction, 106, 120, 169, 336
　Bonferroni progressive, 107
multiscale, 359
multiscale analysis, 17, 32, 62, 189, 268
　quadtree decomposition, 158
　wavelet decomposition, 158
multispecies
　analysis, 71, 85–86
　interactions, 8, 301, 313
multivariate analysis, 60, 79, 90, 225, 260, 299

nearest neighbours, 40, 75, 226, 276, 316, 352
　hierarchy, 40
　mutually, 40
neighbour
　hierarchy, 44, 296
　refined nearest, 63, 65
　search, 120
neighbour rules
　algorithms, 98
　bishop, 40, 45, 99
　queen, 99
　rook, 99, 180, 205
neighbourhood
　effects, 119
　matrix, 213
　penalties, 191
　search, 32, 129–130
network, 285
　connected, 317
　cycle, 317
　dynamic, 298
　dynamics, 17, 327
　edge matching, 317
　exponential random graph models, 324
　hierarchical, 43
　hierarchy of neighbours, 296, 301
　meta, 16
　multilayer, 253, 313
　multilayer temporal, 319
　multilink, 323
　multiplex, 315, 329
　network-of-networks, 16
　path, 317
　social, 324
　spatio-temporal, 234, 249–250, 313, 316, 318, 325
　spatio-temporal multilayer, 313
　supra-Laplacian, 321
　topology, 54, 100, 135, 235, 321–322
　triad completion, 316
network theory, 295
network-of-networks. *See* graph of graphs
nonrandom, 4, 16–17, 26, 249, 274
nonrandomness, 6, 67, 81, 338

non-stationarity, 2, 37–38, 75, 152, 162, 186, 189, 210, 237, 243, 333, 335–337, 339, 341
null model, 3–4, 10, 29, 65, 147, 230, 244, 322–323

ordination
  canonical correspondence analysis (CCA), 145, 210
  discriminant functions, 145
  horseshoe effect, 267
  Mahalanobis distances, 145
  multiscale (MSO), 85, 213, 339
  multivariate analysis, 148
  partial CCA, 211, 217
  partial RDA, 211, 214, 217
  principal components analysis (PCA), 85, 145, 211, 215, 223
  principal coordinate analysis (PCoA), 214
  redundancy analysis (RDA), 145, 197, 210

partitioning
  temporal, 282
patch, 138
patches, 138, 337
path, 246, 297, 302
  complexity, 248
  crossing, 248
  cycles, 246
  directional, 246
  least-cost, 309–310, 318
  length, 247, 281
  shortest, 234
  tortuosity, 246–247, 341
path analysis, 190, 198, 311
pattern, 296
  aggregated, 26, 67, 76, 339
  association, 98, 178
  brousse tigrée, 28, 253
  checkerboard, 149, 272
  clumping, 17, 52, 61–64, 68, 71, 75, 77, 81, 288, 296, 338
  Fibonacci spiral, 184
  heterogeneous, 14
  homogeneous, 30, 136, 151, 281, 331
  inhomogeneity, 37
  inhomogeneous, 75
  multispecies, 71
  nonrandomness, 145
  overdispersed, 26
  overdispersion, 52, 61–64, 70, 77, 81, 288, 291
  patchiness, 1–3, 5, 13, 17, 26–27, 29, 35, 50, 52, 99, 107, 163, 173, 180, 183–184, 212, 216, 261, 267, 342
  patchy, 9, 99, 332, 339–340
  random, 37, 99
  scale, 83, 215
  segregation, 66
  spatial, 221, 256, 276
  spatio-temporal, 229
  temporal, 256
  trend, 212
  underdispersion, 52, 61, 64
  uniform, 1, 14, 34, 99
  wave, 178, 220, 256
pilot study, 37, 184, 186, 336
point pattern analysis, 17, 61, 95, 230, 237, 287, 296
  bivariate, 8, 66
  Diggle's function, 63, 66
  empty space function, 63, 65
  event–event distance, 62
  event-to-nearest-event, 70
  marked, 17, 19, 75
  multivariate, 43, 69–70, 280
  nearest neighbour function, 293
  point events, 52
  point-event, 288
  point-to-nearest-event, 70
  refined nearest neighbour, 63
  thinning, 68
  univariate, 95
Poisson distribution, 2, 64–65, 81, 96, 348, 351
Poisson space-time joint index, 229
polygon change
  events, 240
polygon change analysis, 18, 22, 229, 240
population data, 339
Principal Coordinates of Neighbour Matrices (PCNM, dbMEM), 214
processes
  allelopathy, 8
  biological, 5
  competition, 6, 8, 19, 52, 57, 65, 71, 147–148, 286, 300, 352
  ecological, 4, 6, 14
  endogenous, 26
  exogenous, 26
  facilitation, 6–7, 9, 57, 352
  heterogeneous, 37
  homogeneous, 3, 61, 73
  induced, 26, 344
  inherent, 26, 344
  inhibition, 9
  inhomogeneous, 3–4, 73
  isolation by distance, 14
  marked, 60
  mass effects, 14
  mutualism, 57, 352
  non-stationary, 4
  parasitism, 352
  phenology, 251
  pollination, 286
  predation, 6, 286, 352
  spatial point, 1
  spatial stochastic, 4

processes (cont.)
  stochastic, 1
  succession, 9, 19
  tolerance, 10
proximity, 16
  matrix, 204
  sampling locations, 126
  spatial, 251, 271, 319, 324
  spatio-temporal, 234
pseudo-replication, 254, 256

quadrat variance analysis, 95
quadrat variance methods
  paired quadrat variance (PQV), 82
  random paired quadrat frequency, 230
  three term local quadrat variance (3TLQV), 82
  triplet quadrat variance (tQV), 82
  two term quadrat variance (TTLQV), 82
  two-dimensional, 86

random effect, 111, 209, 350
randomization, 67, 296, 323
  bootstrap, 54, 117, 182, 192
  caterpillar, 182
  complete, 120, 152, 162, 180, 199
  hierarchy, 52
  jackknife, 182
  permutation, 55
  reference distribution, 54–55
  restricted, 55, 57–58, 68, 84, 152, 162–163, 180–181, 192, 199, 275, 286, 290
  toroidal shift, 56–57, 68, 181
randomization tests
  torus distances, 36
randomness, 81
reference distribution, 179–180, 182, 195, 215, 230
regression
  autoregressive, 22, 162, 167, 204, 344
  autoregressive moving average (ARMA), 204
  complete independence, 164
  conditional autoregressive (CAR), 58, 124, 204–205, 208–209, 353
  double autoregression, 165
  double dependence, 175
  first-order autoregressive, 204
  generalized least-squares (GLS), 203, 208
  generalized linear mixed model (GLMM), 203, 208, 353
  geographically weighted regression (GWR), 22, 203, 206
  induced autoregression, 165
  induced dependence, 174–175
  inherent autoregression, 165
  inherent dependence, 169, 175
  Kriging, 203
  linear, 195, 197, 215
  Markov model, 204

moving average (MA), 107, 171, 205
moving average models, 124, 176
multiple, 50, 126, 210, 213–215, 217
multiple regression on distance matrices (MRDM), 199
ordinary least squares (OLS), 200
polynomial, 126, 213
random effect, 203
residual analysis, 201
resource-selection functions (RSF), 353
simultaneous autoregressive (SAR), 58, 124, 177, 204
spatial, 22, 191, 203, 269
spatial dependence, 353, 355
spatial error models (SEM), 22, 203
spatial filtering, 107, 204, 217
spatial independence, 164
spatial lag, 204
spatio-temporal autoregressive (STAR), 352
residual sums of squares (RSS), 199
response
  linear, 352
  nonlinear, 11, 282
restricted randomization, 180, 196, 215, 249, 342, 355
Ripley's K, 17, 32, 52, 63, 66, 72, 96, 123, 226, 235, 265, 287, 337–339
  bivariate, 66
  indicator function, 64
  three dimensions, 79
Ripley's L, 66, 71
robustness, 298

sample data, 133, 339
sampling, 184
sampling design, 20, 31, 34, 39, 97, 117, 139, 142, 216, 344
  spatial, 142, 145
sampling units
  aggregated, 33
  change of support, 118
  coarse graining, 39
  contiguous quadrats, 81, 97, 147, 178, 218, 230, 350
  grain, 31, 259
  modifiable areal unit problem (MAUP), 354
  resolution, 31
  voxel, 86
scale, 85, 89
  scaling, 38
  spatial, 136
  temporal, 39, 136
scales of pattern, 81–82, 340
scan statistics, 122
search
  neighbour, 120
search window, 35, 50, 60, 226

Index 401

segregation, 288, 339
semi-variance, 32, 111
similarity matrix, 215
spatial
  aggregation, 58
  autocovariance, 102, 111
  confounding, 201
  contingency, 212
  correlogram, 106
  dependence, 26, 51, 58
  edge, 252
  heterogeneity, 261
  lag, 33, 102, 331
  legacy, 212
  predictor, 206, 217, 225
  range, 15, 106–107, 117, 196
  resolution, 212
  scale, 259, 281, 335
  spillover, 191
  variance, 111
spatial autocorrelation, 17–18, 111, 163, 170, 176–177, 279
  confounding, 59
  correction, 177
  degree, 118, 192–193
  diagnostic indicator, 58
  estimation, 58
  Geary's c, 103, 111
  H Moran statistic, 122
  Moran's I, 102, 111, 196, 211, 215
  negative, 27, 59, 75–76, 103, 106, 173, 178, 215
  nuisance, 188
  parametric tests, 58
  positive, 83, 215
  predictor, 59
  proxy, 59
  range, 186
  sampling, 59
  second-order, 101
  shape, 28
  significance, 34
  Type I error, 186
spatial dynamics, 5, 13
spatial extent, 270, 336
spatial interpolation
  inverse distance weighting, 124, 126
  Kriging, 124, 128
  proximity polygons, 124
  trend surface analysis, 124, 126
spatial partitioning, 136, 301
spatial pattern, 6
spatial pattern analysis, 221
spatial sampling
  optimal, 131
spatial statistics
  first-order, 49
  global, 119
  local, 119
  second-order, 50, 65, 70
  spatial structure, 359
spatially constrained clustering, 299
spatio-temporal
  analysis, 5, 17, 21–22, 192, 229, 237, 249, 251, 286, 323, 330, 360
  association, 230
  autocorrelation, 228
  clustering, 234
  clusters, 237
  data, 156, 189, 256, 308, 330
  digraph, 237
  dimensions, 20
  domains, 23, 31, 336
  Griffith's index, 228
  scale, 270, 330
  Spatial-Temporal Analysis of Moving Polygons (STAMP), 240, 326, 328, 355–356, 359
spatio-temporal graph
  disease spread, 249
spatio-temporal network
  disease spread, 329
  spatial efficiency, 318
  temporal efficiency, 318
species
  composition, 5, 44, 189, 214, 229, 260, 265–266, 271, 279–280, 282, 299, 311
  diversity, 258
  interactions, 16
  multiple, 84
  response, 148
  unimodal curve response, 6, 211
species interaction, 18
  motifs, 91
species interaction network
  trophic network, 300
species-area curve, 259
spectral analysis, 90, 321
spectral decomposition, 214, 218, 226
stationarity
  intrinsic hypothesis, 38
  l, 243
  lack, 120
  second-order, 109, 132
  strong, 2
  weak, 2, 132
stationary, 37
stochastic spatial modelling, 342
study area
  extent, 259
survey
  design, 185, 342
synchrony, 252–253, 256

Tavaré's method, 179
temporal

temporal (cont.)
　analysis, 227
　cycles, 10, 253
　explicit, 12, 17, 20, 31
　lag, 331
　network, 318
　sampling interval, 229–230, 234–235, 310, 318
　scale, 13, 20, 189, 335
tessellation, 21, 40
three-dimensional, 1
threshold distance, 44, 216, 276, 302–303
time lag, 228, 254
time series, 107, 117, 122, 163, 204, 235, 252, 254, 256, 336, 356
Tobler's Law, 97, 261, 332
topology, 22, 285
transition probabilities, 179
travelling waves, 253, 256

variance partitioning, 216
variance–covariance matrix, 85, 129, 204–205, 210, 213
variance-to-mean ratio, 50, 81, 86
variogram
　bounded, 113
　cross-covariance, 118
　cross-validation, 130
　directional, 116
　empirical, 213
　experimental, 111, 128, 133
　head locations, 116
　linear, 208
　multivariate, 213
　nugget effect, 111
　omnidirectional, 86
　sample, 111
　semi-variance, 111
　sill, 111, 113, 128, 133
　spatial range, 111, 216
　spherical, 208, 216
　tail locations, 116
　theoretical, 111, 129, 133
　unbounded, 113
variography, 31
　semi-variance, 103, 111
Voronoi polygons, 43

wavelet, 32, 95
　boater, 80, 90
　continuous decomposition transform (CWT), 220
　correlation, 90
　covariance, 90, 340
　discrete decomposition transform (DWT), 220
　French top hat (FTH), 89
　Haar, 89, 146, 210
　hierarchical transform, 214
　maximum overlap discrete transform (MODWT), 221
　Mexican hat, 89, 340
　Morlet, 89–90
　multiscale analysis, 158
　scalogram, 221
　transform, 89, 158, 214, 220–221
　variance, 89–90, 158, 221
　Walsh transform, 88, 90
weight matrix, 46, 119, 205

zone of influence, 106, 333
zone of overlap, 6

Printed in the United States
by Baker & Taylor Publisher Services